I0823037

WEBB'S UNIVERSE

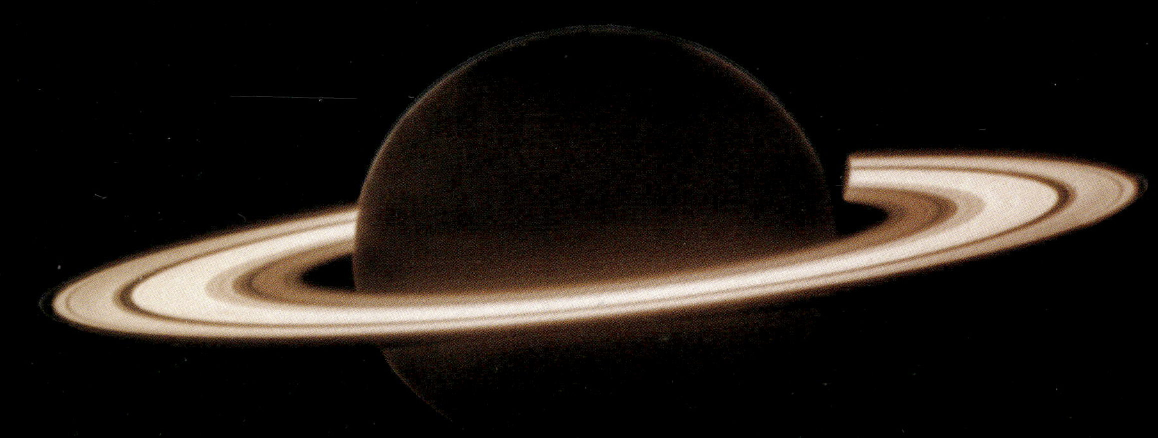

WEBB'S UNIVERSE

The Space Telescope Images That Reveal Our Cosmic History

MAGGIE ADERIN-POCOCK

Abrams, New York

To Lori, as you grow older you continue to fill my heart with joy, knowing that the world is a better place with you in it.

Acknowledgements

As a dyslexic, writing books is hard for me, so I would like to thank all the people who have made this one possible. First, I would like thank Sarah Wild for her keen insight and advice, taking my ramblings and bringing them into sharp focus. A picture is worth a thousand words, so thanks to illustrator Peter Liddiard and stellar designers Ana Bjezancevic and Barbara Ward, who enabled some of the more complex notions to be brought to life through their beautiful work. And a special thanks to my editors Jo Stansall and Lucy Stewardson for their hard work in putting this book together and the stress of my missed deadlines. Finally, thanks to our amazing universe, which never ceases to inspire and fill me with awe, due to its complexity and magnificent beauty, which we have tried to do justice in the pages that follow.

Editor: Michael Sand
Design Manager: Darilyn Lowe Carnes
Managing Editor: Annalea Manalili
Production Manager: Larry Pekarek

Library of Congress Control Number: 2024936151

ISBN: 978-1-4197-7469-0
eISBN: 979-8-88707-526-6

Illustrations by Peter Liddiard
Designed and typeset by Ana Bjezancevic and Barbara Ward

First published in Great Britain in 2024 by Michael O'Mara Books Limited
9 Lion Yard, Tremadoc Road
London SW4 7NQ

Printed and bound in China
10 9 8 7 6 5 4 3 2 1

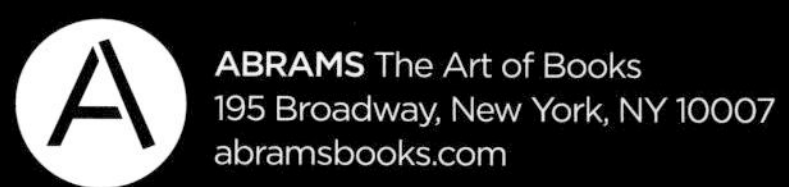

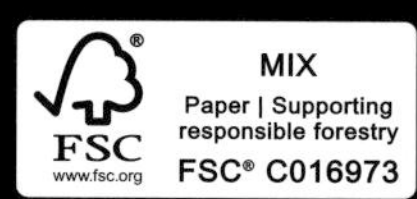

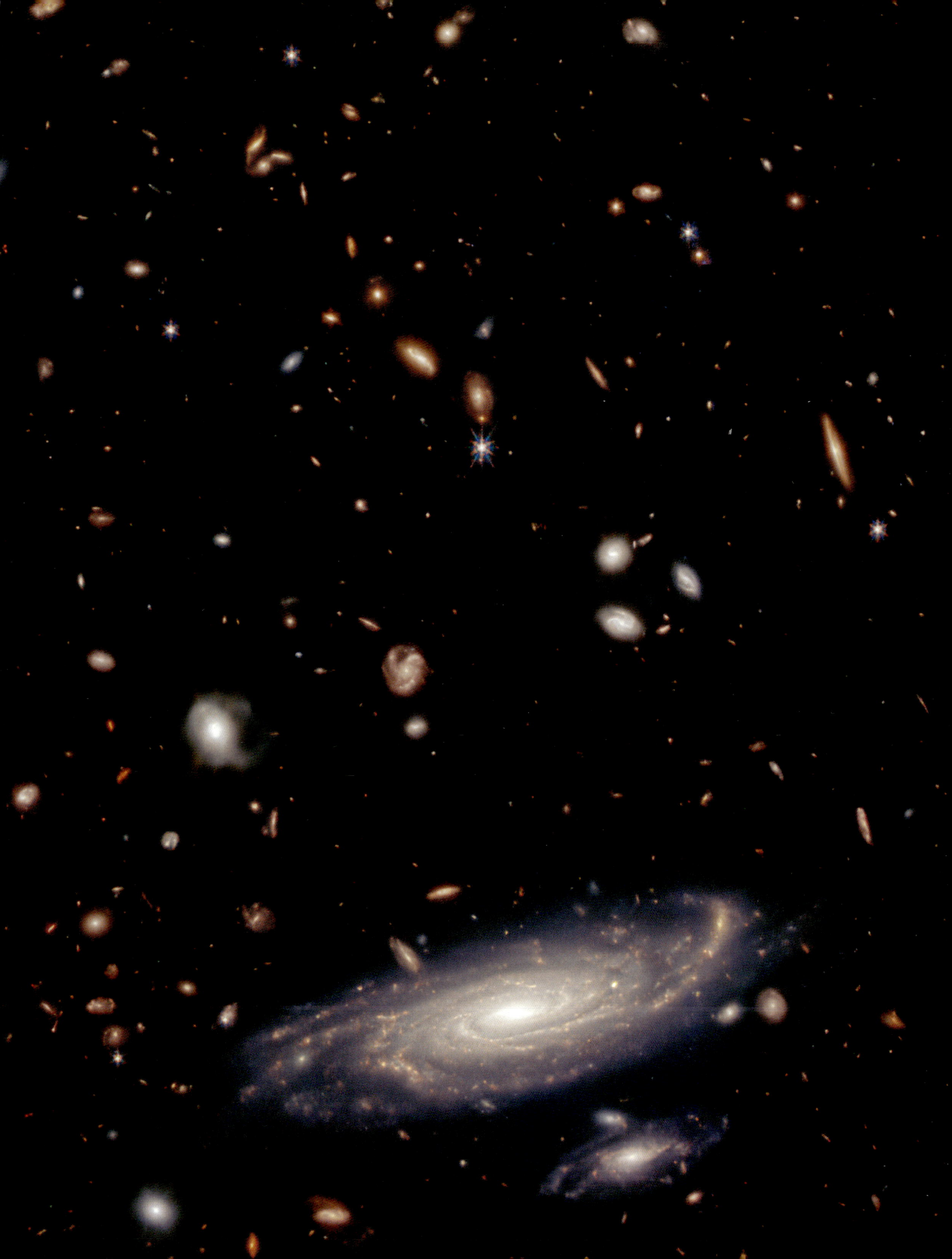

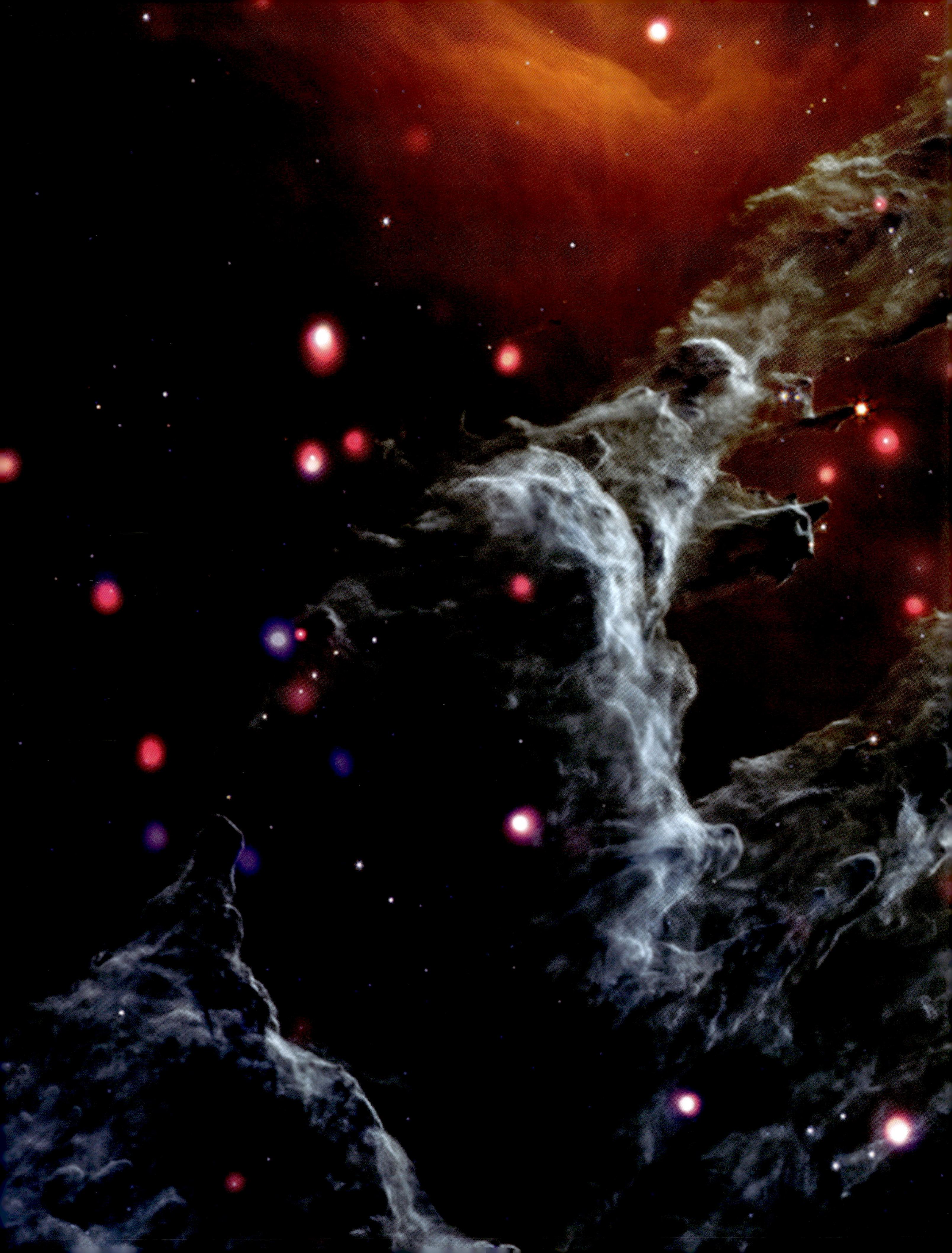

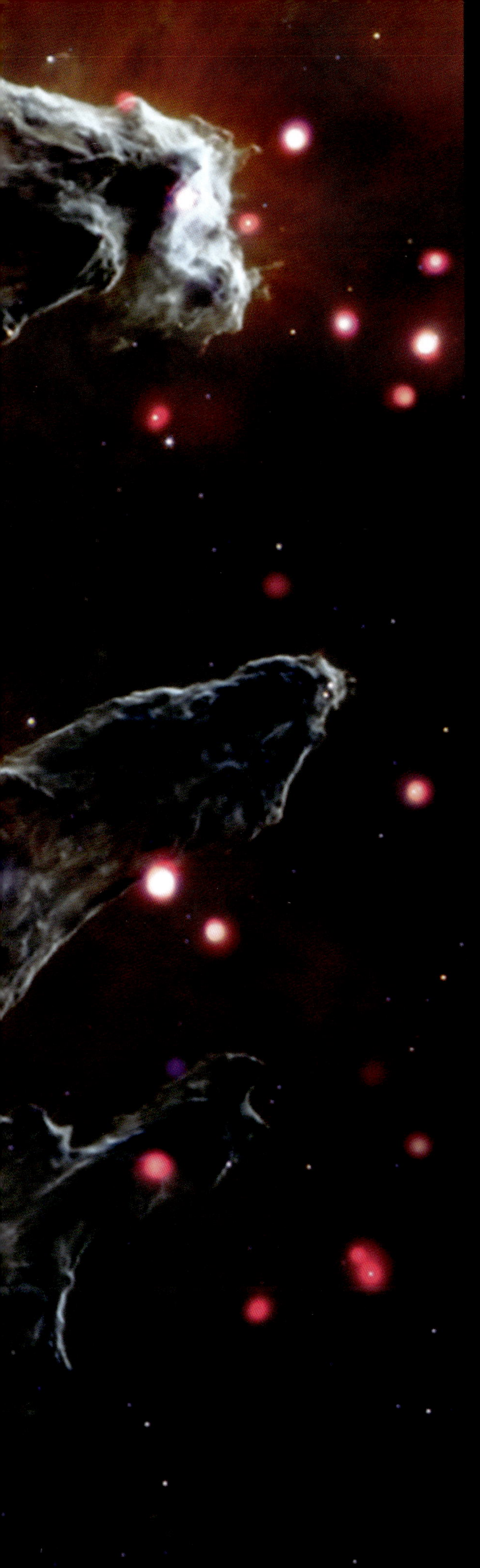

CONTENTS

Introduction 9

PART 1: THE JAMES WEBB SPACE TELESCOPE 15

Space telescopes 17

On the outer edges of the rainbow 18

Webb's predecessors 21

Enter the James Webb Space Telescope 29

Tricky infrared telescopes 32

Blossoming in space 35

Unstable location 36

An infrared space observatory 36

The anatomy of the observatory 38

PART 2: NEW FRONTIERS OF SPACE 41

New technologies and approaches 43

Webb's science goals 46

A cosmos of galaxies 49

The birth of stars 53

The origins of life 54

Webb's instruments 57

Painting the sky 60

PART 3: OUT OF THE DARK 63

Close to home: our solar system and beyond 67

Exoplanets 87

Nebulae 97

Stars 125

Galaxies 163

Picture credits 220

Index 222

LEFT: There is much more to the Pillars of Creation in the Eagle Nebula than meets the naked eye. This image combines data from Webb, Chandra, Hubble, Spitzer, XMM-Newton and European Southern Observatory telescopes.

PREVIOUS PAGE: Scientists collected the data for this image in Webb's early days to test its instruments.

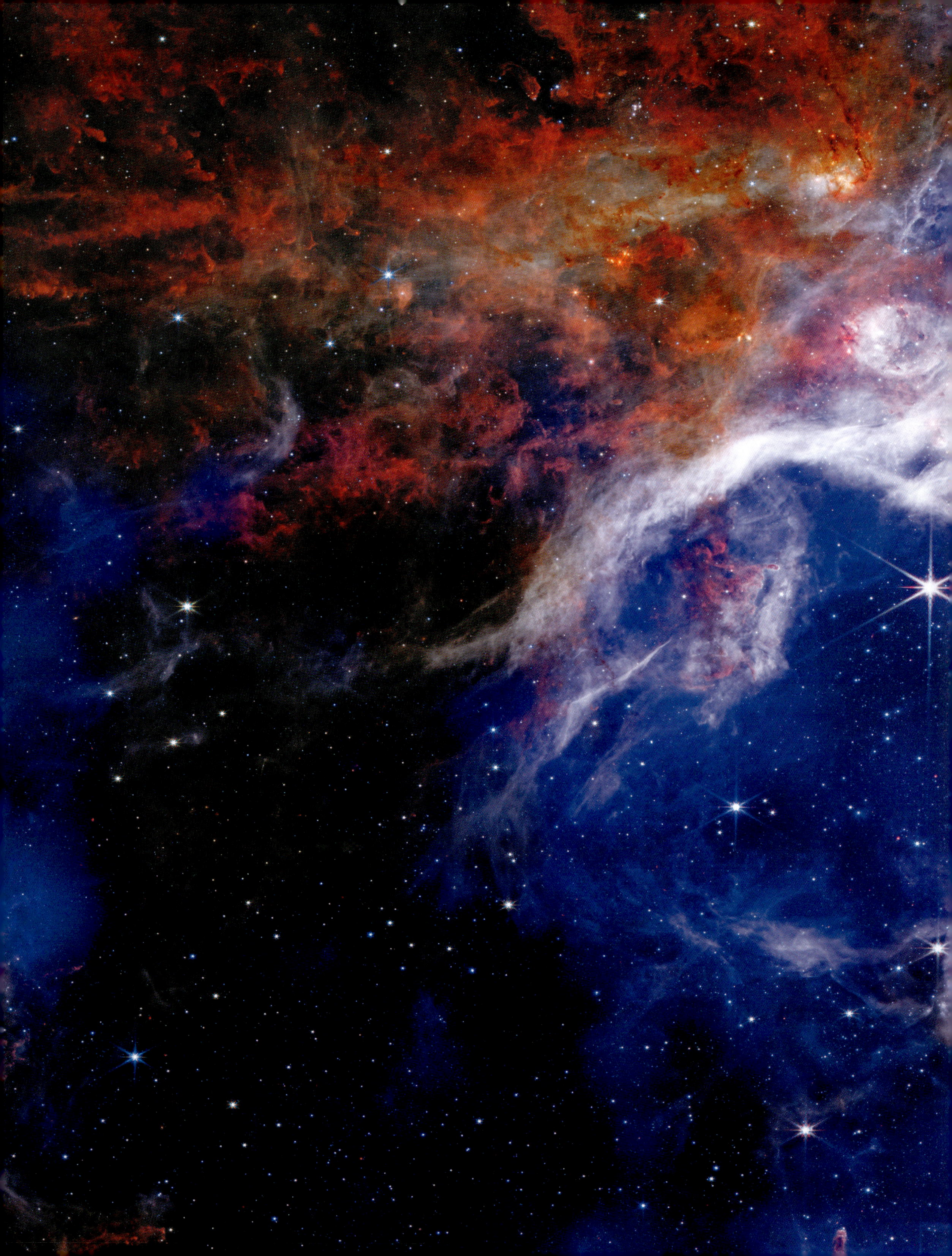

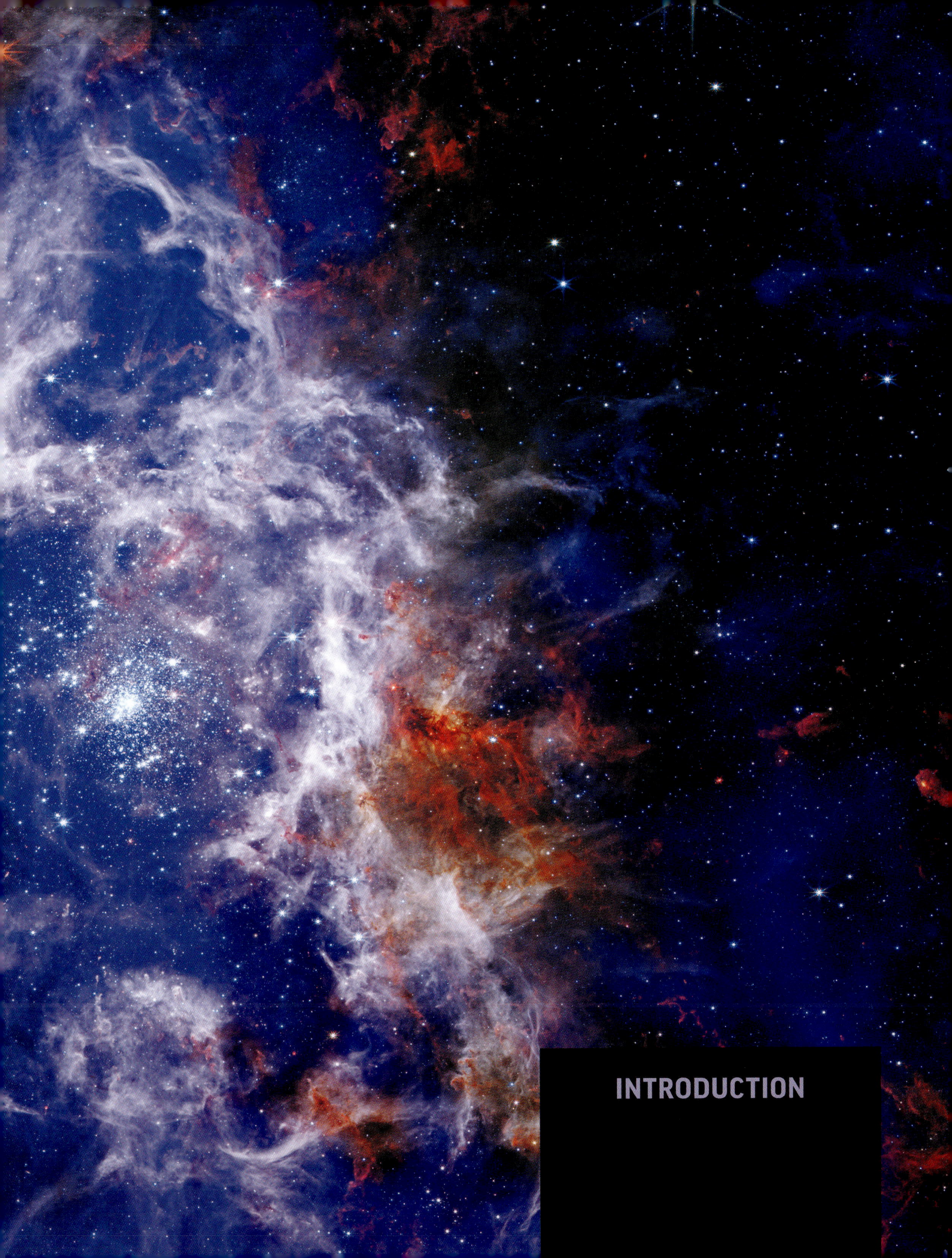

INTRODUCTION

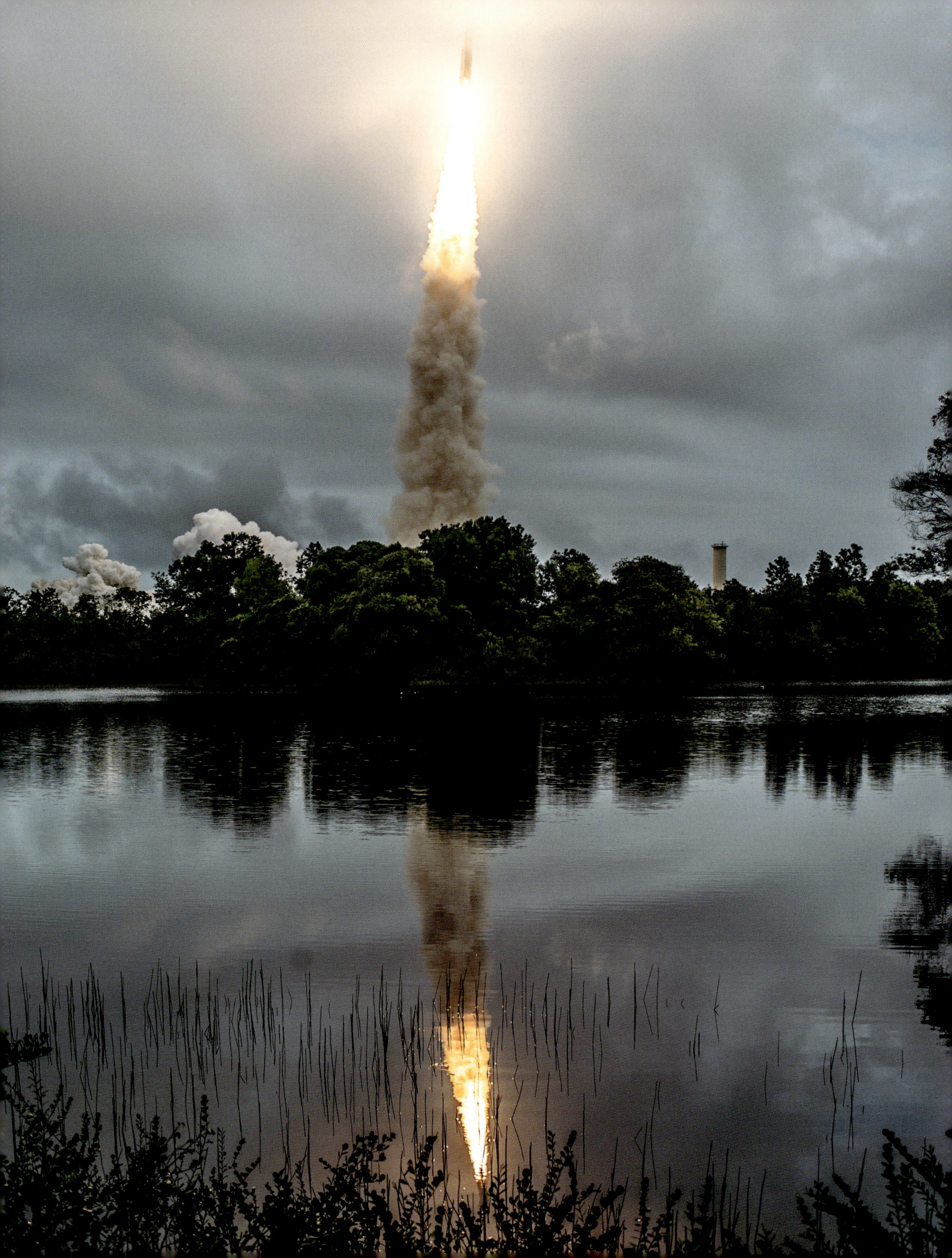

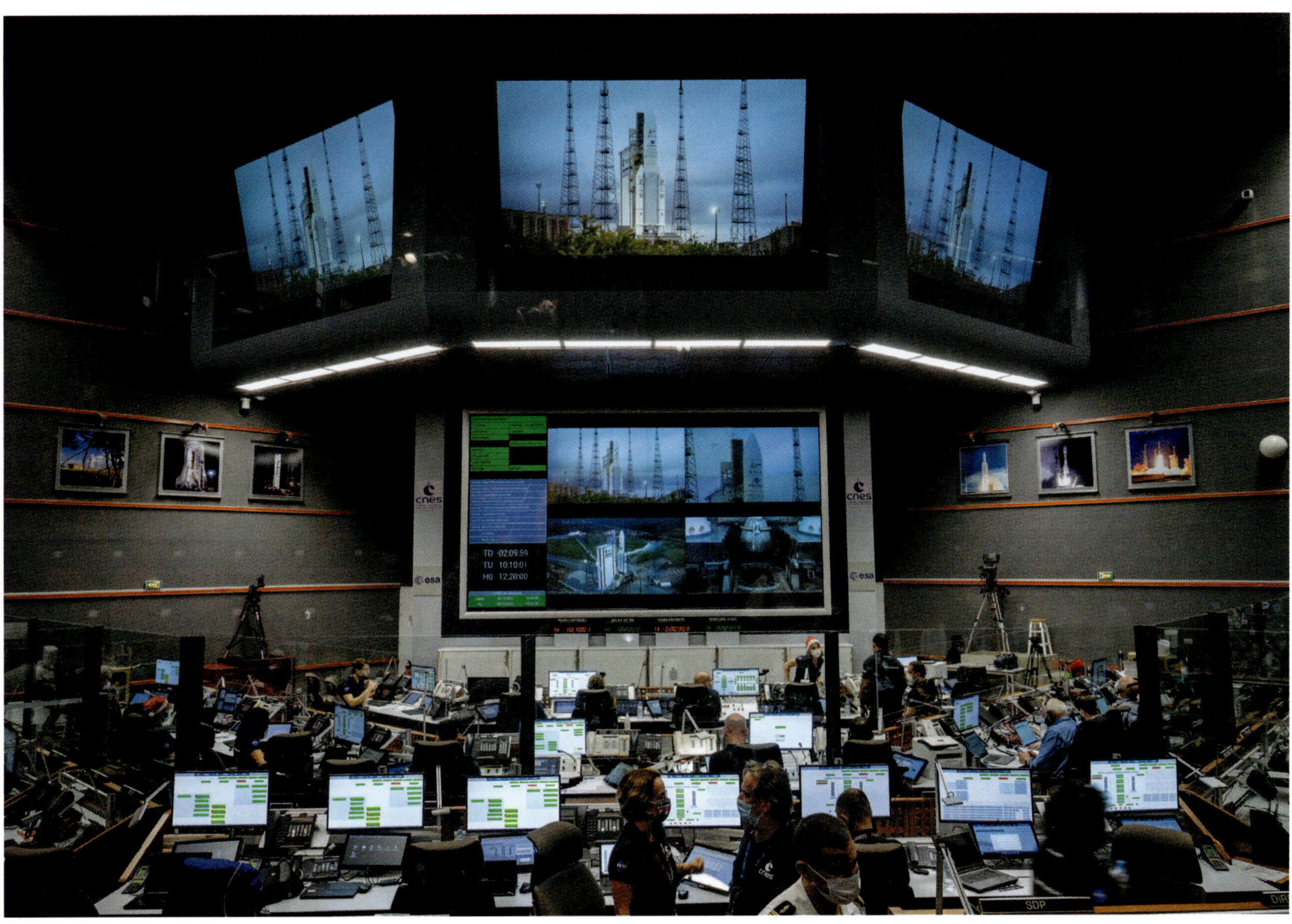

LEFT: Webb was launched on an ArianeSpace Ariane 5 Launch Vehicle on 25 December 2021.

ABOVE: Control room in the countdown to the launch. This was the culmination of more than thirty years of work on JWST.

CHRISTMAS DAY 2021 REMAINS VERY memorable for me, but not because of seeing my daughter enjoying presents from Santa or receiving wonderful gifts myself. In fact, this was a very stressful day for me and thousands of my fellow scientists, engineers and technocrats around the world. After years of delays, the ambitious Webb Observatory finally left Earth on board an Ariane 5 rocket. Decades in the making, the telescope was folded inside this launch vehicle and propelled through space at more than 35,400 kilometres (21,996 miles) per hour. After about thirty minutes, Webb separated from the rocket and began the careful process of unwrapping itself, just as many of us were doing with our Christmas presents that day.

The moment the James Webb Space Telescope left the Earth's surface, there were more than 300 single steps at which the multibillion-dollar observatory could fail. So I, and many others who had worked on the project, sat in our respective homes watching the launch and holding our breath, hoping that nothing would go wrong. Even after it was successfully launched into space, the nail-biting continued for another few weeks as we waited to see whether Webb would successfully complete its complex unfolding sequence. But there was nothing any of us could do at this point except hope that our brainchild, which had now travelled millions of kilometres away from Earth, would deploy and give us that new insight into the universe we so craved.

With only a few minor hitches, the telescope was finally ready for science on 8 January 2022. A long, tense wait, but the best delayed present many of us could have hoped for. Now, far beyond the Earth and our moon, Webb is ushering in a new era of astronomy.

Although the time I spent working on Webb was relatively short compared to the project's full duration, I had the feeling throughout my involvement of being part of something amazing. As a space

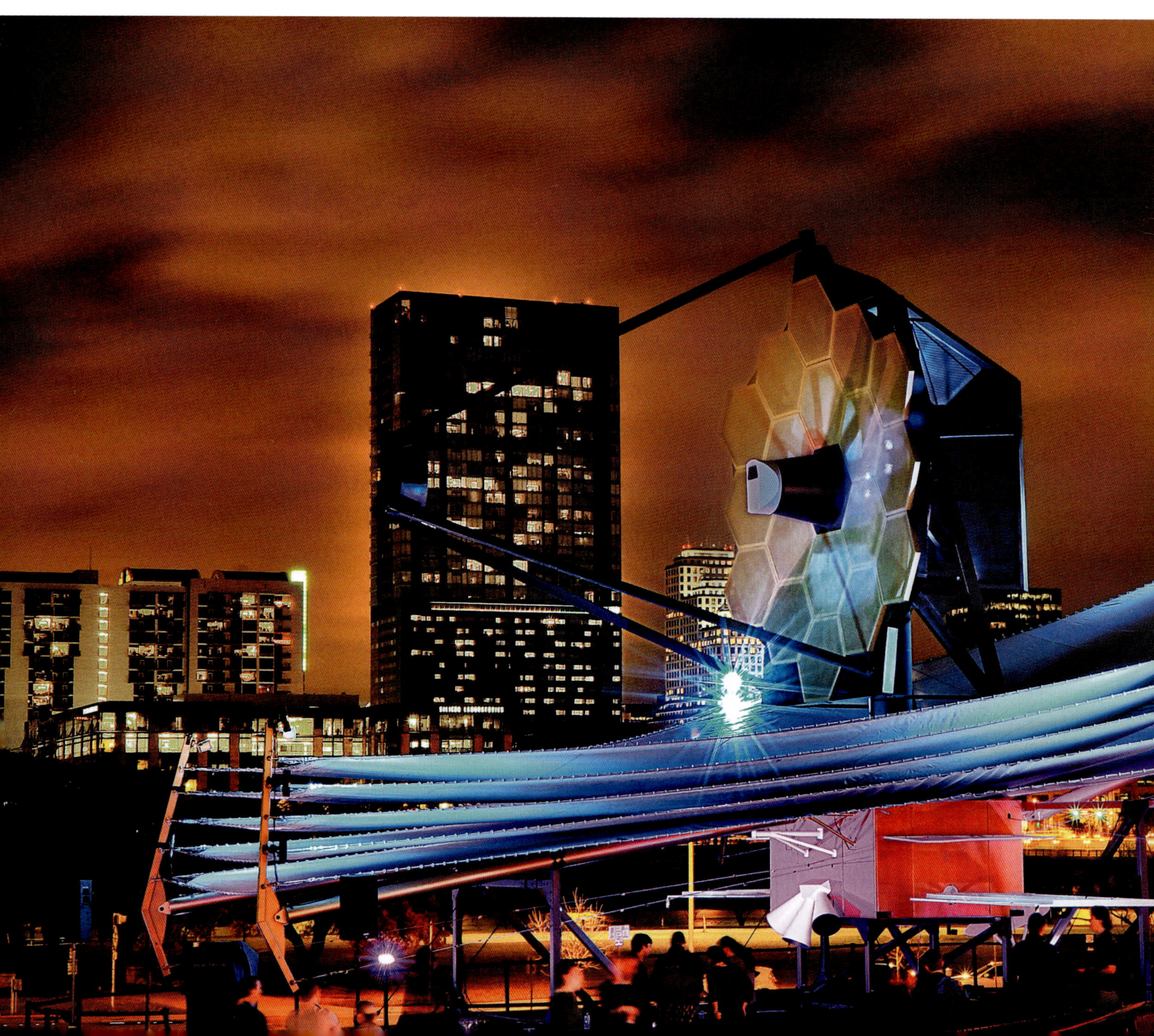

A full-scale model of the James Webb Space Telescope, on display in Texas.

scientist, I have had the privilege of working on a number of space missions – both looking down at Earth from above to understand the many changes happening to our planet, and looking out into space to understand our place in the cosmos. But Webb felt particularly special due to its sheer size, scope and potential.

Webb is the largest space telescope ever built, and all of us working on it – some 10,000 scientists and engineers around the world – knew that it had the potential to literally shed new light on the universe and help unravel some of the mysteries that we have been pondering for decades. My team and I worked on the NIRSpec instrument, a complex beast made up of many subsystems which analyses near-infrared light – which we'll look at in more detail later on.

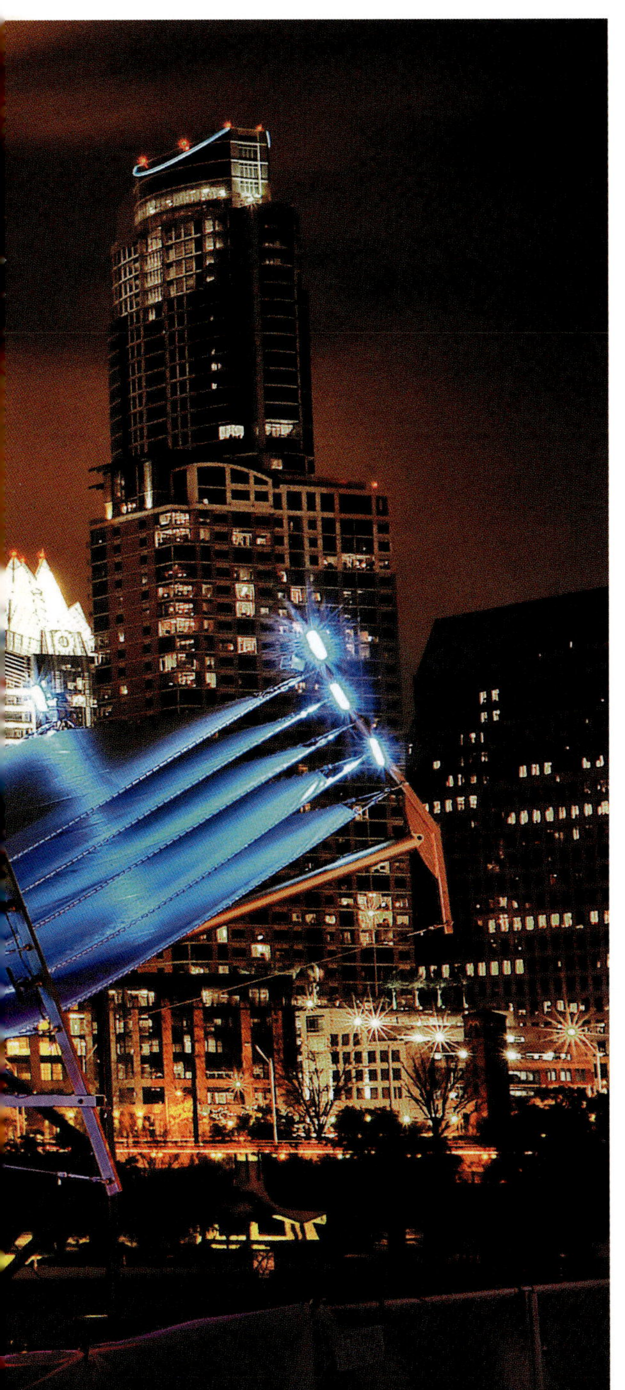

Since its launch, Webb has been snapping phenomenal pictures of our breathtaking universe. Designed to help us answer the many questions that we still have about the origins of our solar system and how the universe all hangs together, our newest space telescope is already living up to its potential. But how did the largest space telescope ever built come into being? After the success of the Hubble Space Telescope, why build an infrared telescope rather than another that detected visible light? How does Webb work, and what are its incredible images revealing to us about the universe?

Webb is our most ambitious space telescope to date. But it is a continuation of a journey in which virtually every culture throughout history and around the world has participated. People have always gazed at the stars and tried to unravel their secrets. Initially, we created myths to explain what we saw; but, over time, we began to want to understand the universe based on what we'd observed. In this sense, stargazing has developed alongside humanity. As our technology has improved, so has our stargazing – and often our desire to observe and understand more of the universe has pushed us to develop better and more sophisticated instruments.

Right now, Webb has an unprecedented view of the cosmos, observing it in wavelengths that are invisible to our eyes. Some of the most interesting places in the universe, such as dust clouds where new stars are being born, have been dark to us up until now.

My hope with this book is to explain the true wonder of Webb – its incredible design and potential, and the game-changing data and science that it is sending back to Earth through its images. From the drawing board to the universe beyond, and with an insider's insight, I'm excited to share Webb's journey.

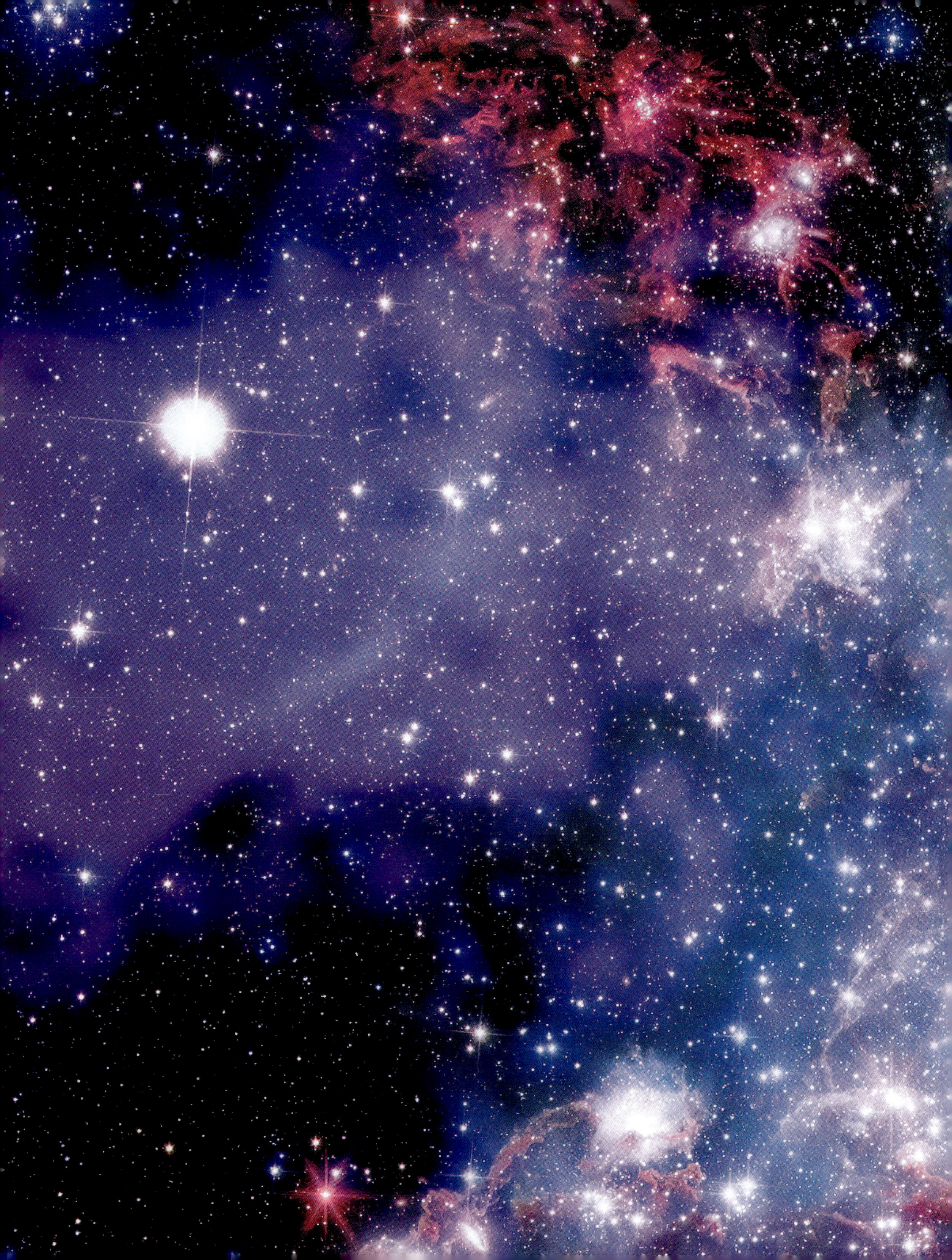

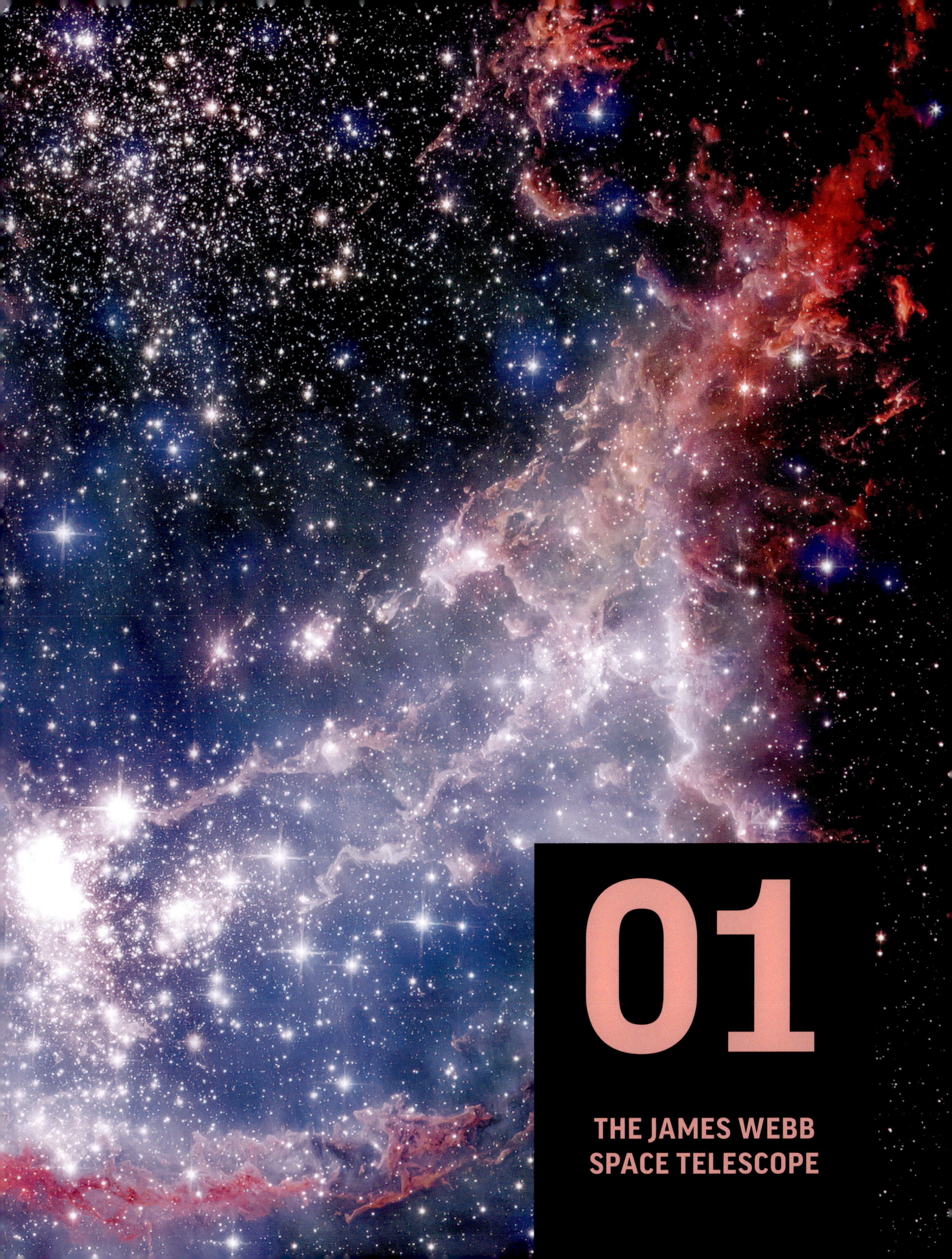

01

THE JAMES WEBB SPACE TELESCOPE

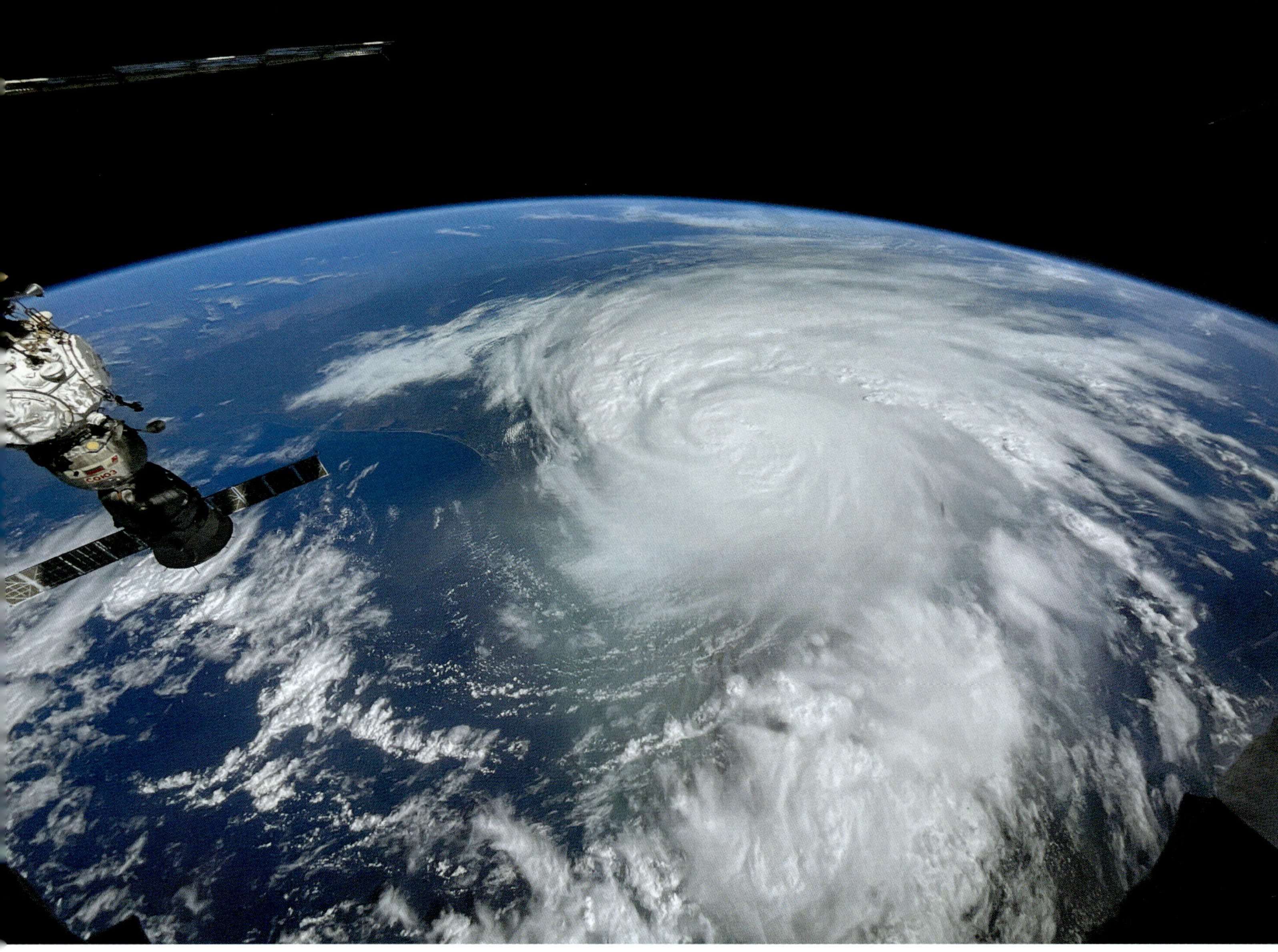

NASA Earth Observatory image shows a swirl of clouds lying between Earth and the cosmos. Earth's atmosphere blocks infrared waves.

EARTH IS NOT THE BEST place from which to view all of the cosmos. Our planet is blanketed in an atmosphere of nitrogen, oxygen and a handful of other gases. This life-sustaining shield not only allows us to breathe, but it also regulates global temperatures and protects us from the sun's intense radiation. If we didn't have an atmosphere, then life as we know it would not exist on our planet. The sun's rays would burn into the parched ground, and there would be no security blanket exerting pressure to keep water in its liquid form.

Yet, for all its life-protecting qualities, the atmosphere isn't particularly thick. Most of it is encased within the troposphere, its lowest layer, which is about 12 kilometres (7.5 miles) thick. In its entirety, the atmosphere extends about 100 kilometres (62 miles) above the ground. Compare this to the Earth's diameter at just under 13,000 kilometres (8,000 miles) and we can understand how very thin it is.

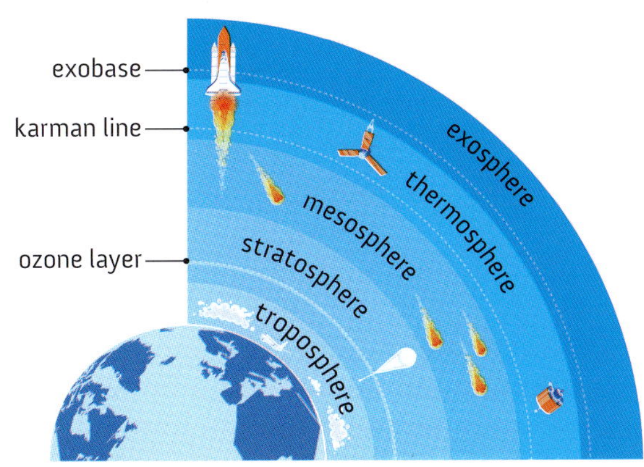

The atmosphere extends about 100 kilometres (62 miles) from Earth's surface and protects our planet from solar radiation.

But this atmosphere, thin though it is, dramatically affects incoming signals from space – from the sun and all other celestial bodies – by absorbing, dampening and distorting many of them. We see some of this distortion with our eyes, but it also effectively obscures large swathes of the universe around us. This is particularly true for wavelengths outside of the visible light range. It's a bit like trying to gaze into the distance with grimy glasses on. In addition to this natural distortion, our lights, radios and various other technologies often overwhelm the relatively weak signals coming from space. It is difficult to detect a weak torch light in a brightly lit room.

Telescopes on Earth have become more sophisticated and impressive – their gazes more sensitive, their computing increasingly powerful – but there is only so much they can do to see through this atmospheric blanket. That is why some of the most exciting astronomy projects in the last few decades, such as Hubble and Webb, sit away from Earth's blinding veil.

SPACE TELESCOPES

American astronomer Lyman Spitzer first proposed an 'extra-terrestrial observatory' in 1946.

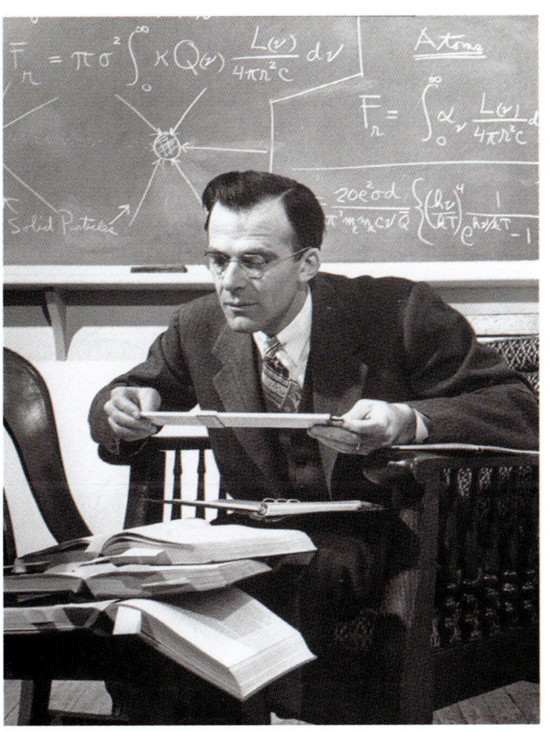

Space telescopes don't struggle with Earth's atmosphere or light interference. They sit above our protective blanket and are at a relatively safe distance from us humans and our noisy gadgets. American astronomer and theoretical physicist Lyman Spitzer first proposed an 'extra-terrestrial observatory' in 1946, arguing that it would overcome many of the challenges facing ground-based telescopes.

It was an ambitious idea. At that time, we hadn't even put a satellite in space. It would be another two decades before we were able to deploy a telescope beyond Earth's atmosphere, but Spitzer's idea formed the backbone of the many observatories which humanity has since put into space.

This pristine view of space, however, comes at a cost. Space telescopes tend to have a hefty price tag. They also have a limited number of functions, as every kilogram launched costs a lot of money – both to construct and to send into orbit. This means that when space agencies launch a space telescope, they must carefully weigh (usually both literally and figuratively) the pros and cons of every instrument and component that they include on board.

But without eyes in the sky, we would remain blind to many of the wonders of the universe and many of our questions about life on Earth and beyond, including the nature of matter itself, could remain unanswered. Webb, with its tailor-made instruments, can see into many of the veiled parts of the cosmos, allowing us to look further into it than ever before.

By means of our advancing technology, we are now able to see far beyond our own galaxy. Hubble – seen by many as the predecessor of Webb – enabled us to estimate the number of galaxies in the universe. The conclusion was that there are around 200 billion, the large majority of which are moving away from us as part of a vast, expanding universe. But with Webb, a new chapter of discovery is upon us.

ON THE OUTER EDGES OF THE RAINBOW

Radiation permeates our universe, some of which we can see and some of which we can't. The radiation emitted by stars and some other celestial bodies comes in a variety of forms. All of it comprises part of what is known as the electromagnetic spectrum of waves that travel through the vacuum of space. Radiation is one of the few things that can do this, and it is therefore important to us astronomers: to understand the universe, we analyse radiation.

The waves of the electromagnetic spectrum vary in energy, some being more energetic than others; and their energy is governed by their wavelength.

Our local star, the sun, for example, emits a range of wavelengths. Short wavelengths have higher energy levels, like gamma rays and X-rays. These waves can cause serious damage to human cells, so we keep interaction with them to a minimum. Around the middle of the spectrum, we have the less energetic ultraviolet (UV), visible and infrared radiation (IR). We can detect visible light with our eyes, feel infrared radiation on our skin, and we protect our skin

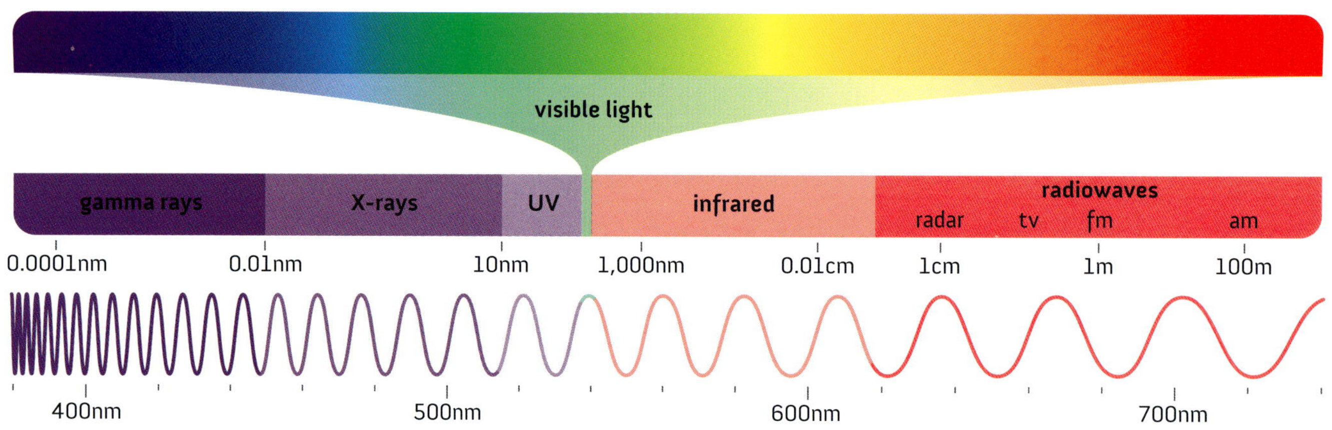

Visible light occupies a tiny portion of the electromagnetic spectrum.

from ultraviolet radiation by using suncream. At the lower energy end of the electromagnetic spectrum, we have microwaves and radio radiation. All of these different types of radiation permeate the universe.

The wide range of wavelengths that make up the electromagnetic spectrum enables us to see the universe in different ways. In the past, the main wavelengths used for astronomy were those that make up visible light – which makes sense, as we can see these with our eyes. But as technology has improved, we have been able to create instruments that can pick up other wavelengths. However, not all of these waves penetrate our atmosphere; so, to get the full range of electromagnetic waves, some detectors need to sit in space, up above the atmosphere.

Many of the hundred or more observatories, satellites and explorers launched into space have looked at the parts of the electromagnetic spectrum that are hard to observe from Earth due to atmospheric absorption. Putting visible light telescopes up in space above the atmosphere also helps counter atmospheric distortion, making it possible to gain a much clearer view of the universe.

Webb is one such example. It is detecting infrared light, in a wavelength range that cannot entirely penetrate the Earth's atmosphere. The infrared light received from space is very, very faint, so we must also contend with the fact that, on the Earth's surface, such signals would be swamped by infrared sources on our planet and from the sun. Webb therefore sits 1.5 million kilometres (just under a million miles) from our planet for the telescope to be effective, and many precautions were taken to prevent it becoming overwhelmed by these other sources of infrared radiation.

This image was captured after Webb was released into space. Once it left the spacecraft, scientists had no way of fixing any mechanical problems that arose.

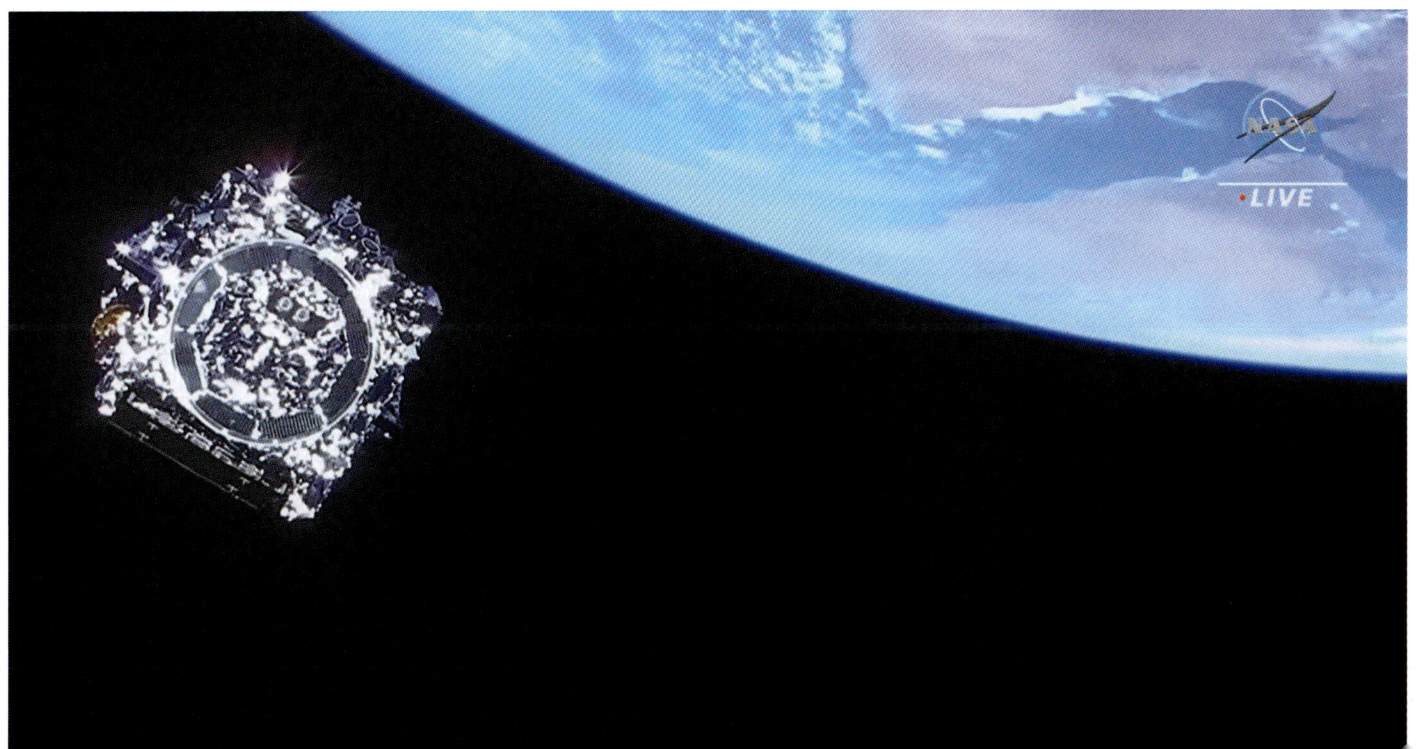

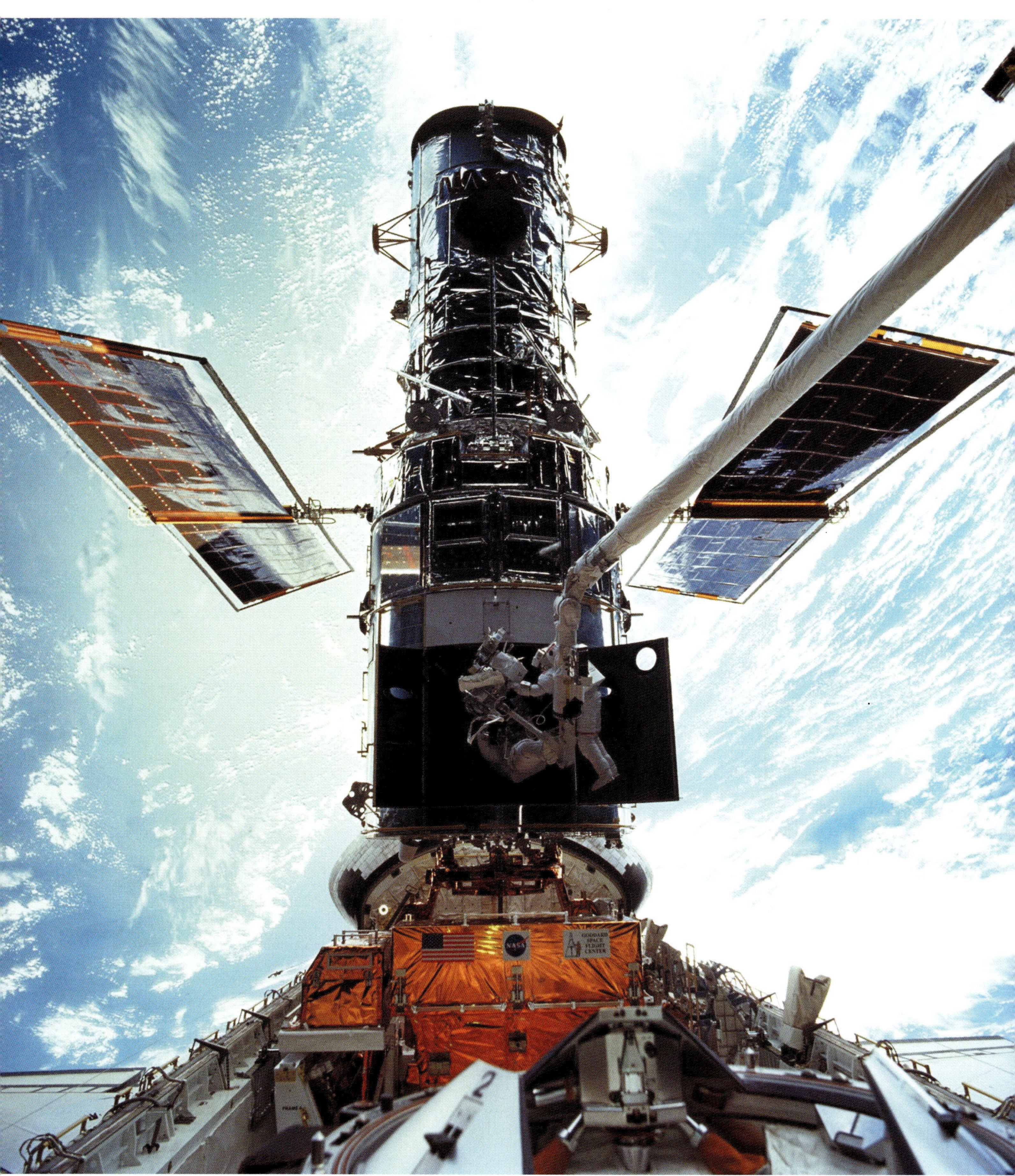
NASA

WEBB'S PREDECESSORS

Before Webb, there were a few critical space missions that laid the groundwork for the science it was designed to conduct.

HUBBLE SPACE TELESCOPE

NASA's Hubble Space Telescope entered orbit in 1990 and continues to beam glorious images back to Earth. It was named after influential American astronomer Edwin Hubble, whose work helped us understand both the scale of the universe and the fact that it seems to be expanding.

Like Webb, Hubble was decades in the making. The first scientific working group to discuss the telescope, then known as the Large Space Telescope, was held in 1974, with funding approved in 1977.

Eventually launched on 24 April 1990, and sitting in low Earth orbit (around 500 kilometres/310 miles above sea level), its viewing life did not start smoothly. Two months after deployment, NASA announced that there was a flaw in the telescope's primary mirror. Its curve was off by about two microns, or 1/200th the thickness of a human fingernail. It was a tiny defect, but it meant that the incredibly sensitive Hubble couldn't see clearly. The images it produced were blurry and no better than those obtained on the Earth's surface.

It wasn't possible to replace the 2.4-metre (7.5-feet) mirror in space, so astronauts attached a new instrument to the telescope. About the size of a telephone booth, the Corrective Optics Space Telescope Axial Replacement (COSTAR) instrument is effectively like a pair of glasses, correcting Hubble's flawed sight. In its more than thirty years of operation, the telescope has seen many upgrades, with astronauts servicing it *in situ* and adding new instruments during five missions. Unfortunately, no further upgrades have been possible since the demise of NASA's space shuttle programme in 2011.

Hubble observes from ultraviolet wavelengths through the visible range and into the near-infrared, and has made more than 1.5 million observations so far. Many of these observations have fundamentally changed our understanding of the universe.

In the early 1990s, Hubble witnessed material being pulled into a black hole, which up until then had been a theoretical possibility but had not been observed. Black holes are regions of space where gravity is so strong that not even light can escape, and can form when very massive stars collapse in on themselves. As a result

Because the Hubble Space Telescope is in low Earth orbit, astronauts have gone on several missions to maintain and upgrade the observatory. These missions ended with NASA's space shuttle programme in 2011.

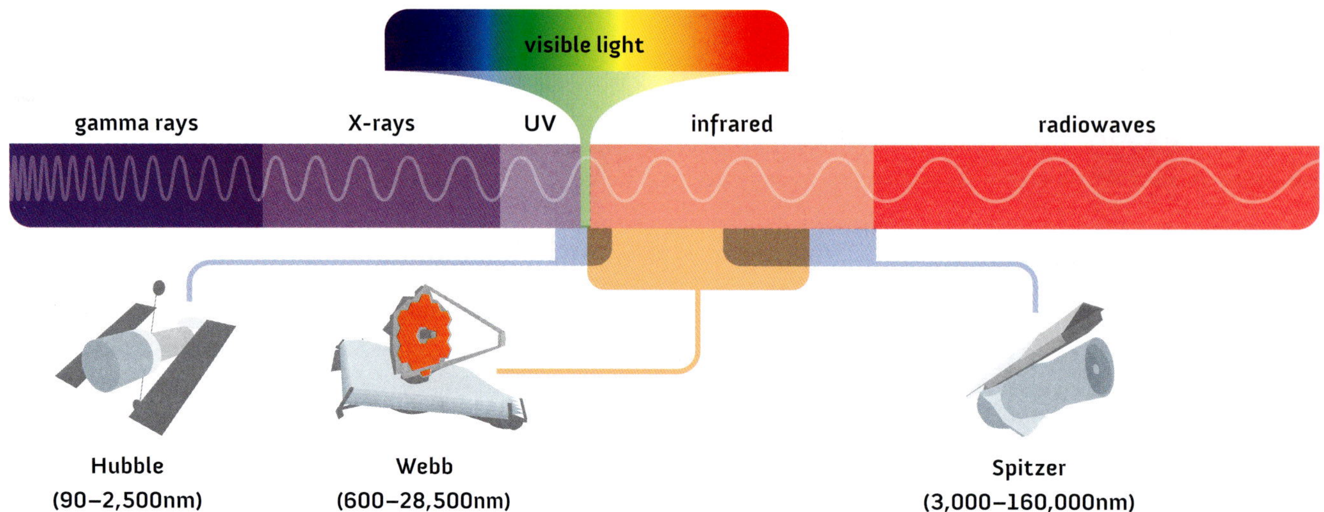

Space telescopes observe different portions of the electromagnetic spectrum. For example, Hubble looks at the visible and ultraviolet range; Webb and Spitzer both specialize in seeing infrared signals, but different parts of the range. Webb focuses on slightly more energetic forms of infrared radiation.

of this, it is not possible to image them directly – because light is sucked into a black hole, not emitted by it – so astronomers identify them by the way in which matter around them behaves. Hubble later confirmed the presence of supermassive black holes at the centre of most galaxies. Supermassive black holes are, as the name suggests, gigantic and can be up to a billion times larger than our sun. Hubble's observations have also revealed the shapes and structures of many, many galaxies; and astronomers using its data have showed that the universe's expansion is accelerating, not slowing down. The telescope was instrumental in finding the first direct proof of dark matter in the universe too.

In some ways, Webb is Hubble's successor, but the two telescopes were designed to look at different wavelengths. In fact, many of the discoveries by Hubble defined the scientific goals of Webb, including its wavelength range. Scientists are now getting the best results by combining the data from both telescopes, making formidable joint observations.

RIGHT: NASA and the European Space Agency released this image of the young star cluster Westerlund 2 and its surroundings on 23 April 2015 to celebrate Hubble's twenty-fifth year in orbit.

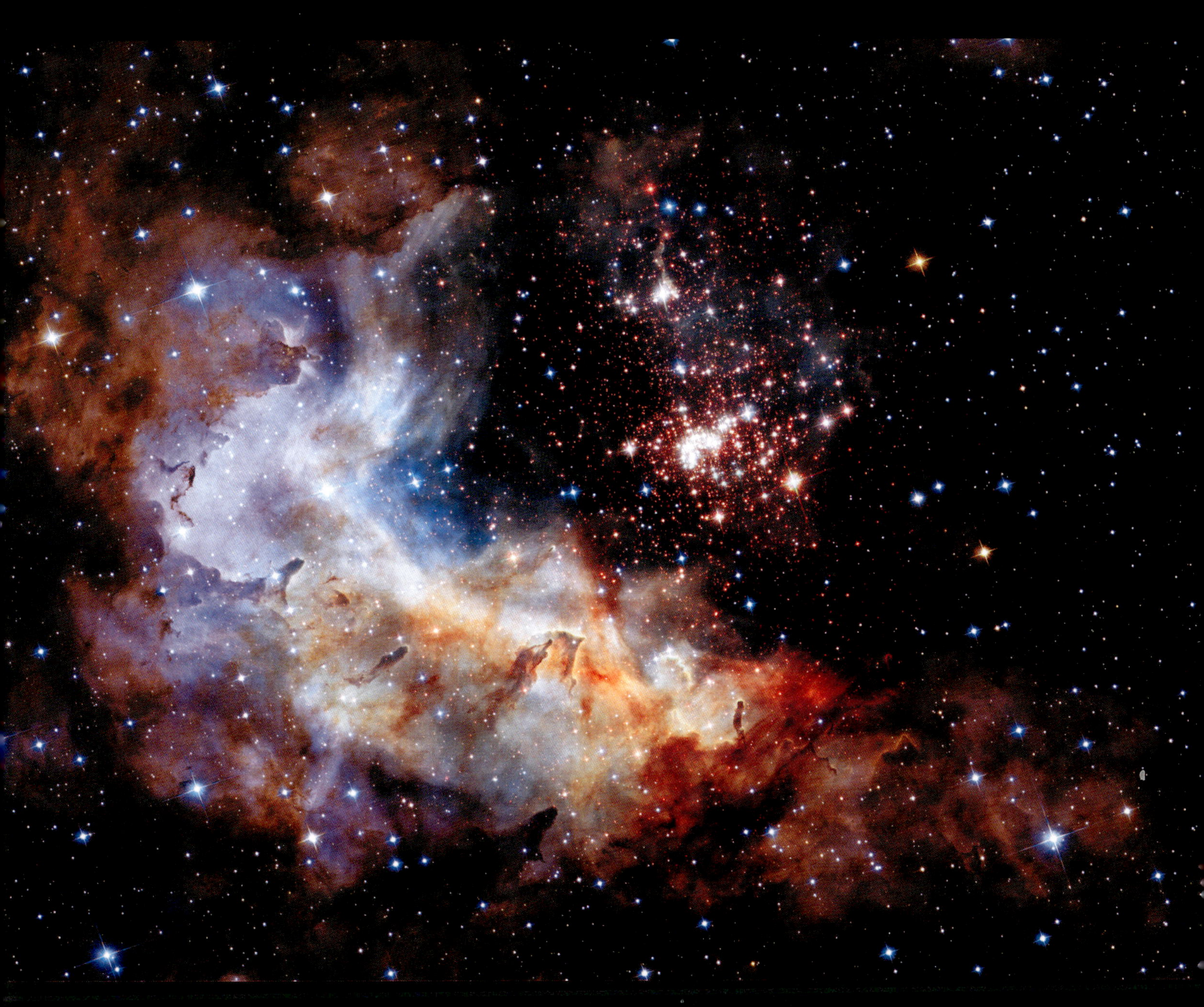

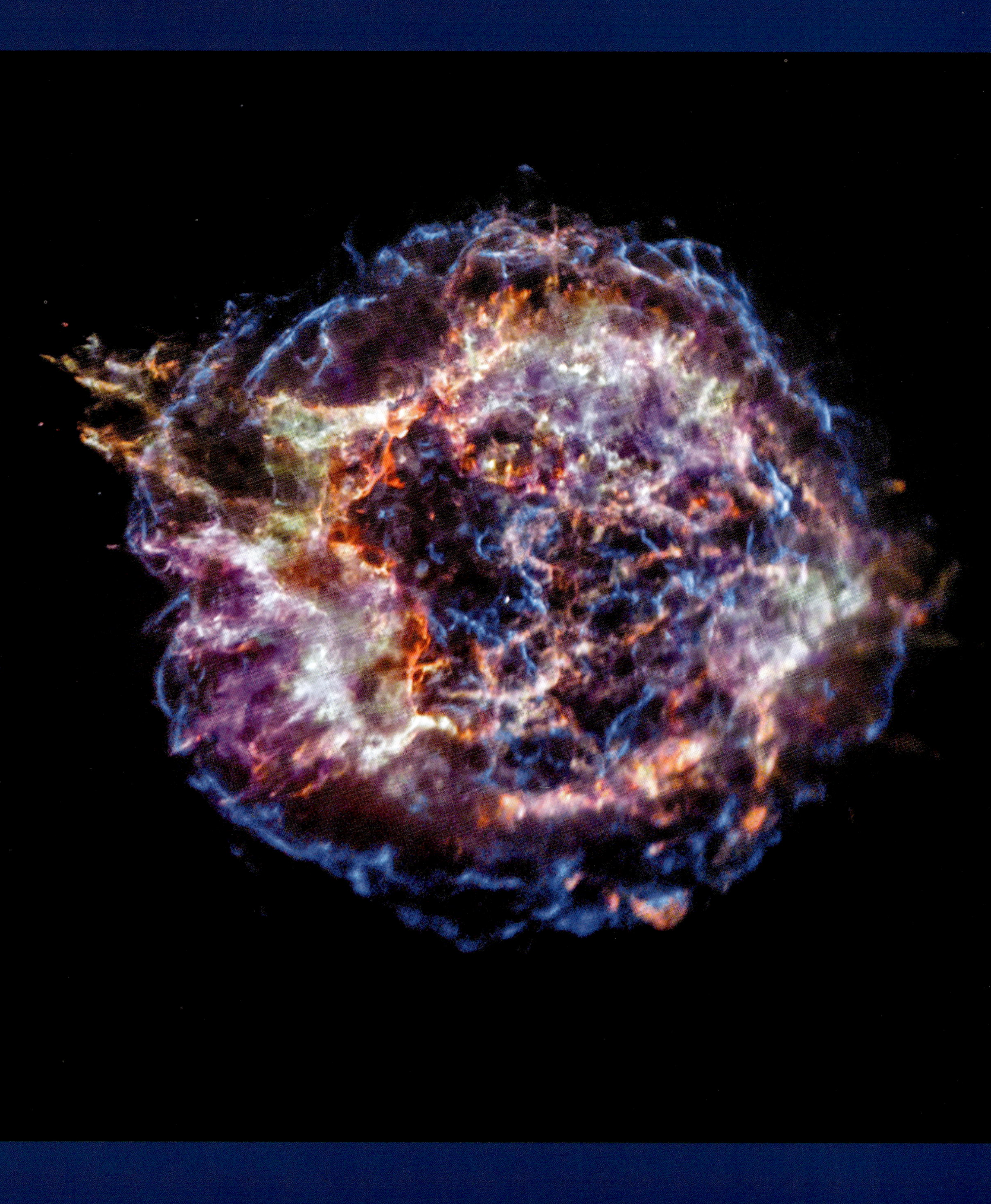

LEFT: Chandra's X-ray snapshot of Cassiopeia A shows the many elements jettisoned from the famous supernova remnant. Silicon (red), sulphur (yellow), calcium (green) and iron (purple) – as well as carbon, nitrogen, phosphorus and hydrogen – are hurtling into the space around the former supernova.

CHANDRA X-RAY OBSERVATORY

Chandra X-ray Observatory was originally known as the Advanced X-ray Astrophysics Facility, but was renamed after Nobel Prize-winning Indian-American astrophysicist Subrahmanyan Chandrasekhar. In Sanskrit, Chandra means bright, shining or glittery, and is used as a name for our moon.

Launched in 1999, the Chandra Observatory detects X-rays, energetic emissions generated by very hot parts of the universe where object temperatures run into the millions of degrees, making them very bright. This includes celestial bodies such as pulsars, galactic supernovae remnants and exploding stars, as well as disks of matter around black holes.

As Earth's atmosphere absorbs most incoming X-rays, Chandra orbits Earth above its atmosphere. Unusually, Chandra travels around our planet in a highly elliptical orbit, completing one orbit every sixty-four days.

Chandra has observed the central region of our galaxy around its supermassive black hole, and also identified other black holes. Its X-ray gaze is often layered over Webb data, allowing astronomers to gain insights across broad swathes of the electromagnetic spectrum.

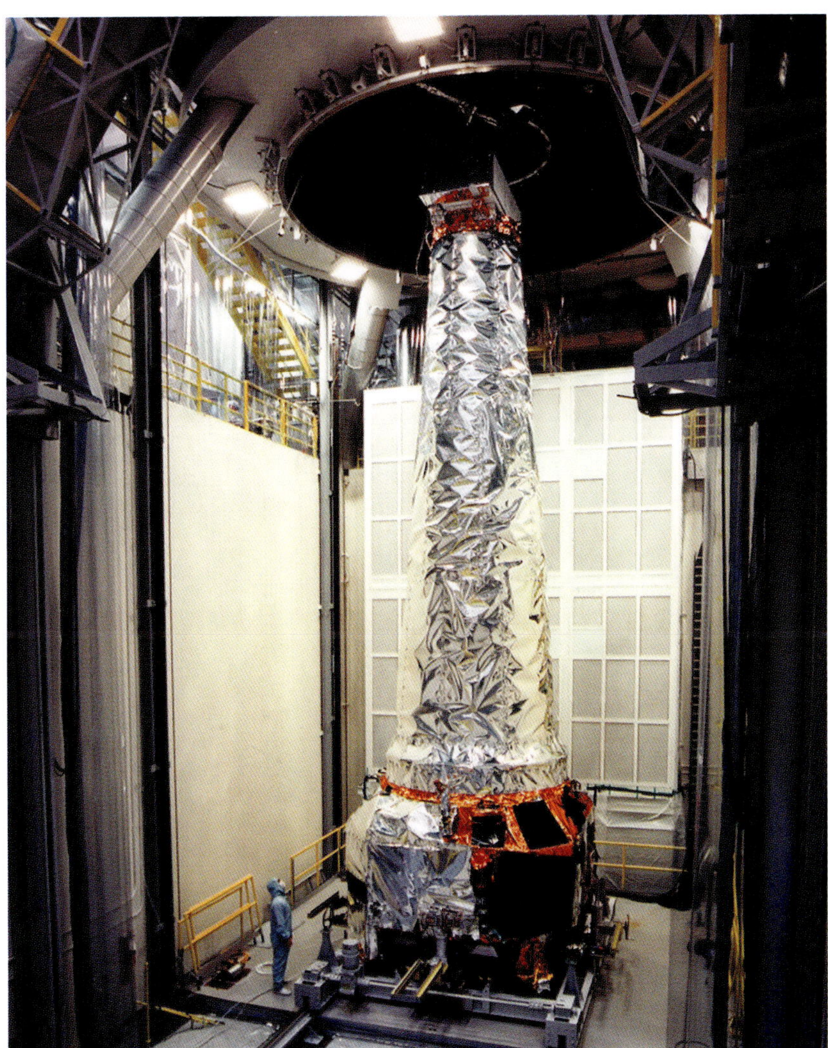

RIGHT: Before its launch, scientists tested Chandra in a giant thermal vacuum chamber to mimic the harsh conditions in space.

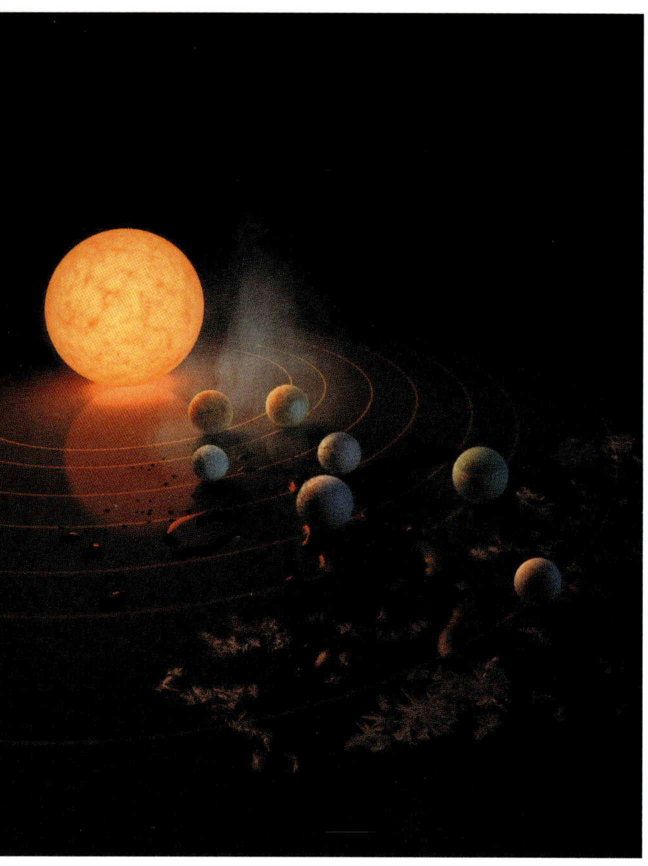

An artist's impression of Trappist-1 star and its seven orbiting planets – one of Spitzer's great discoveries.

SPITZER SPACE TELESCOPE

Launched in 2003, NASA's Spitzer Space Telescope was meant to be operational for only two and a half years, but scientists were able to continue operations until 2020. Spitzer, named after Lyman Spitzer, one of the first people to propose the idea of using telescopes in space (see page 17), was the third space telescope dedicated to infrared astronomy. It can therefore certainly be seen as a forerunner to Webb.

Spitzer's telescope mirror was just 85 centimetres (33.5 inches) in diameter, about the size of a hula hoop, which is small for an infrared telescope. The mission had three cryogenically cooled scientific instruments, namely an array camera, a spectrograph and a multiband imaging photometer. So these instruments could operate effectively, they were cooled to a chilly -268°C (-450°F) using 360 litres (79 gallons) of liquid helium. This cryogenic system ensured that the cameras could detect the low levels of infrared radiation without being affected by the heat of other onboard systems. But with a limited supply of helium, this cooling could not last forever and it eventually ran out in 2009 – having lasted much longer than its engineers had expected it to. But the mission continued with just the infrared array camera, in what the mission scientists dubbed the Spitzer 'warm phase'.

During its surprisingly long life, the Spitzer Space Telescope enabled scientists to make a number of groundbreaking observations. One of its greatest discoveries was the star system known as Trappist-1. Using Spitzer's onboard instrumentation, it was discovered that this star has seven exoplanets (planets that orbit the distant stars we see in the night sky) in orbit, four of which are located within the habitable zone around it.

Spitzer was also one of the first telescopes to image light from an exoplanet. A hot Jupiter-like planet in close orbit to its star was able to reflect enough infrared radiation to be detected by Spitzer's cameras.

In 2007, Spitzer's spectrometer enabled researchers to use its data to detect the first molecules in the atmosphere of an exoplanet. The spectrometer data was also used to identify the first carbon-rich planet, known as WASP-12b, orbiting a star.

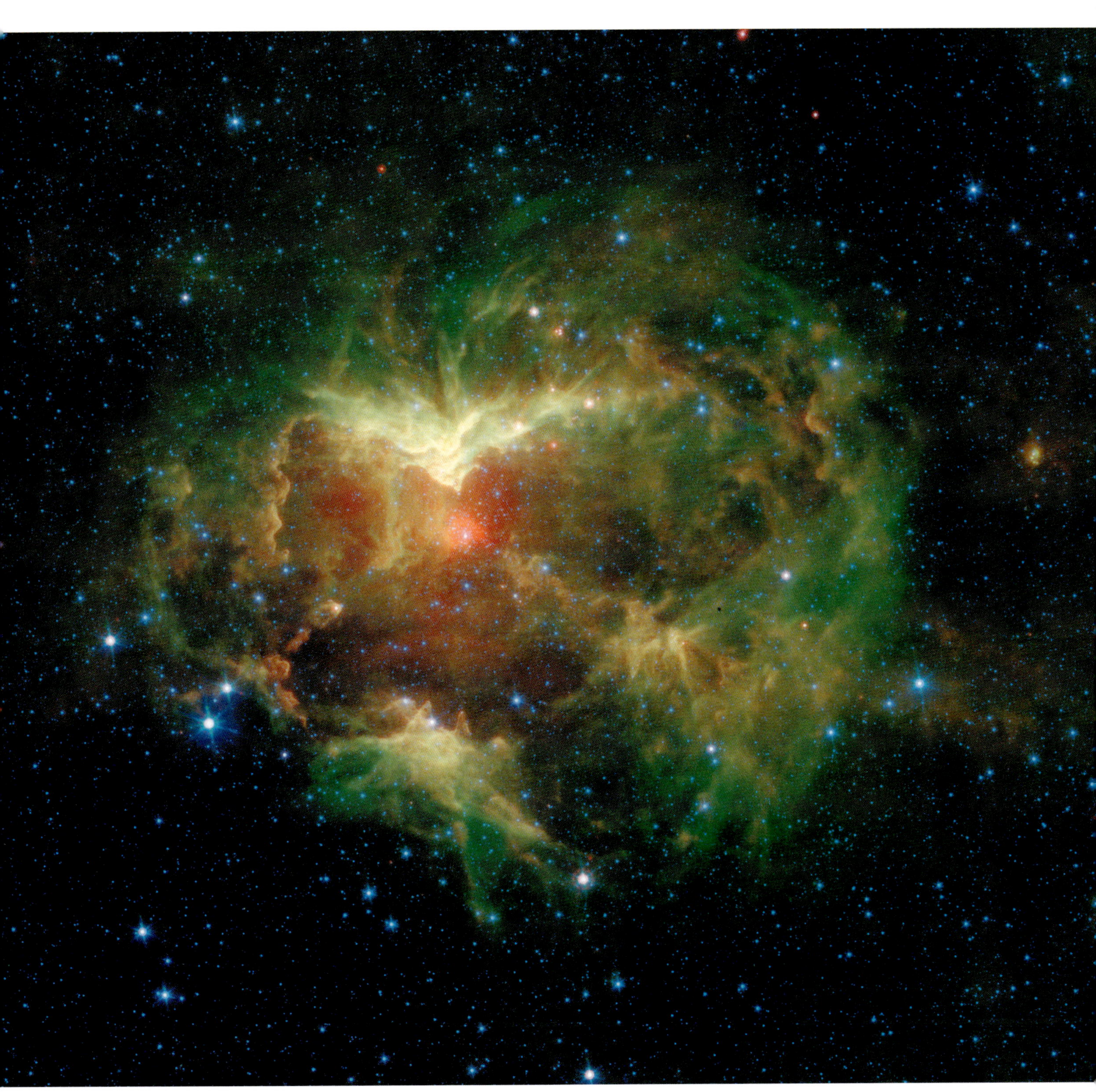

The Jack-o-Lantern Nebula, discovered by the Spitzer telescope. It was called that because the hollowed out dust cloud resembles a Halloween pumpkin.

KEPLER SPACE TELESCOPE

The Kepler Space Telescope, named after German scientist Johannes Kepler, one of the founders of modern astronomy, was launched by NASA in 2009. It was designed with the sole aim of hunting for exoplanets. In particular, its aim was to discover planets that were Earth-like – between half and double our planet's size, and located in or near habitable, so-called Goldilocks zones around the stars they orbit. In these locations it's neither too hot nor too cold, but just right for liquid water to exist.

Kepler's only instrument was a photometer, a device that monitored the optical brightness of stars. It detected exoplanets by using what is called the transit method. Most planets do not emit radiation, but we can see the planets of our solar system from Earth because the planets orbiting the sun reflect some of the sun's light.

The closest star to us within our galaxy is Proxima Centauri, which lies approximately 40 trillion kilometres (24,854,800,000,000 miles) away from Earth. At this distance it is very hard to detect the starlight reflected off any exoplanets orbiting the star, so a different technique for detection is needed. The transit method monitors the star's brightness and can detect when an exoplanet moves in front of its star. The transit of the exoplanet in front of the star periodically dims the star's light, alerting scientists to the possible existence of an exoplanet. To do this the star or planet system must be at the right orientation to detect the transit.

Despite having only one instrument on board, the Kepler mission was remarkably successful – using its data, scientists confirmed the existence of more than 2,700 exoplanets. In 2018, Kepler ran out of fuel and retired, but left behind an amazing legacy.

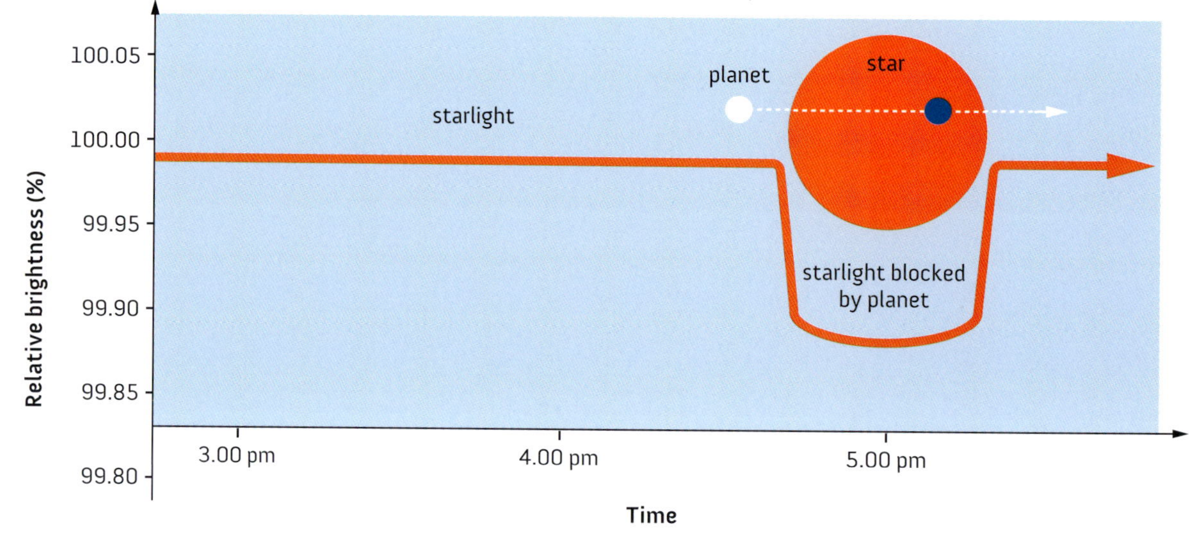

Using the transit method, scientists can detect an exoplanet moving in front of its star.

ENTER THE JAMES WEBB SPACE TELESCOPE

When I first started to work on Webb, it was called the Next Generation Space Telescope. I loved this: it sounded to me as if it had come straight out of a *Star Trek* episode. This was in the early 2000s, but the original concept for the telescope was proposed by NASA scientists and engineers much earlier, in 1996.

Hubble was the first major telescope in space that observed specifically in the visible wavelength range; buoyed by its success, scientists were looking to the next big space project.

They decided to develop a telescope that observed in the red and infrared range. Unlike visible light, infrared light can travel through dense clouds of dust and gas, revealing newly born stars and planetary systems hidden within. Infrared light has a

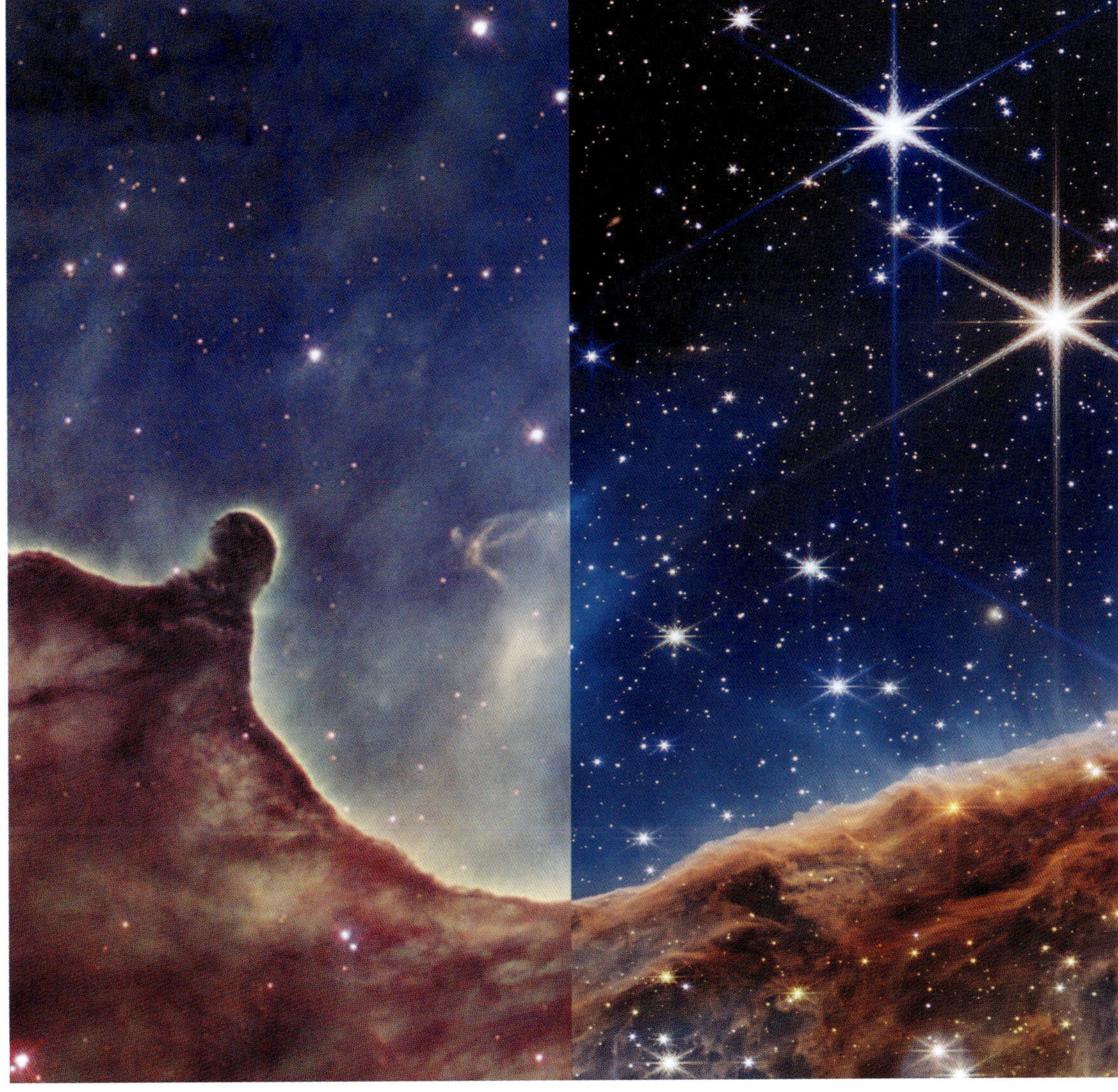

By comparing Hubble (left) and Webb (right) images of the Carina Nebula, it is possible to see the giant leap in the crispness and clarity of Webb's gaze compared to its predecessor.

significantly longer wavelength than visible light, meaning that it can more easily pass through the dust clouds without interacting with the particles contained within them. Visible light has a shorter wavelength, which is on a par with the size of particles within these dense clouds. Consequently, when visible light passes through the clouds, it is more likely to interact with the particles and so get scattered or absorbed by them.

Infrared can also reveal some of the secrets of the early universe. When we observe radiation reaching us from space we can literally look back in time. This is because even when travelling at the speed of light, it takes a measurable amount of time for light to travel through space. Even light from our local star, the sun, takes 8 minutes and 20 seconds to reach us. If we are looking at something further away it can take a lot longer for the light to get to us.

The emissions from the first stars and galaxies – those that formed soon after the Big Bang – still permeate the universe today. However, because the universe is expanding, the radiation emitted by these early bodies has been stretched, lengthening their wavelengths. So, radiation that was originally emitted as visible light can be detected by us today as infrared radiation due to the expanding universe. It therefore makes sense to detect infrared radiation for a detailed study of the early universe.

It took scientists and engineers from different space agencies around the world over two decades to design and make JWST. It involved the work of more than 10,000 people, and years of development and problem-solving to design an infrared space telescope with the desired accuracy and sensitivity. I am immensely proud to be one of those people who made a small contribution to the project.

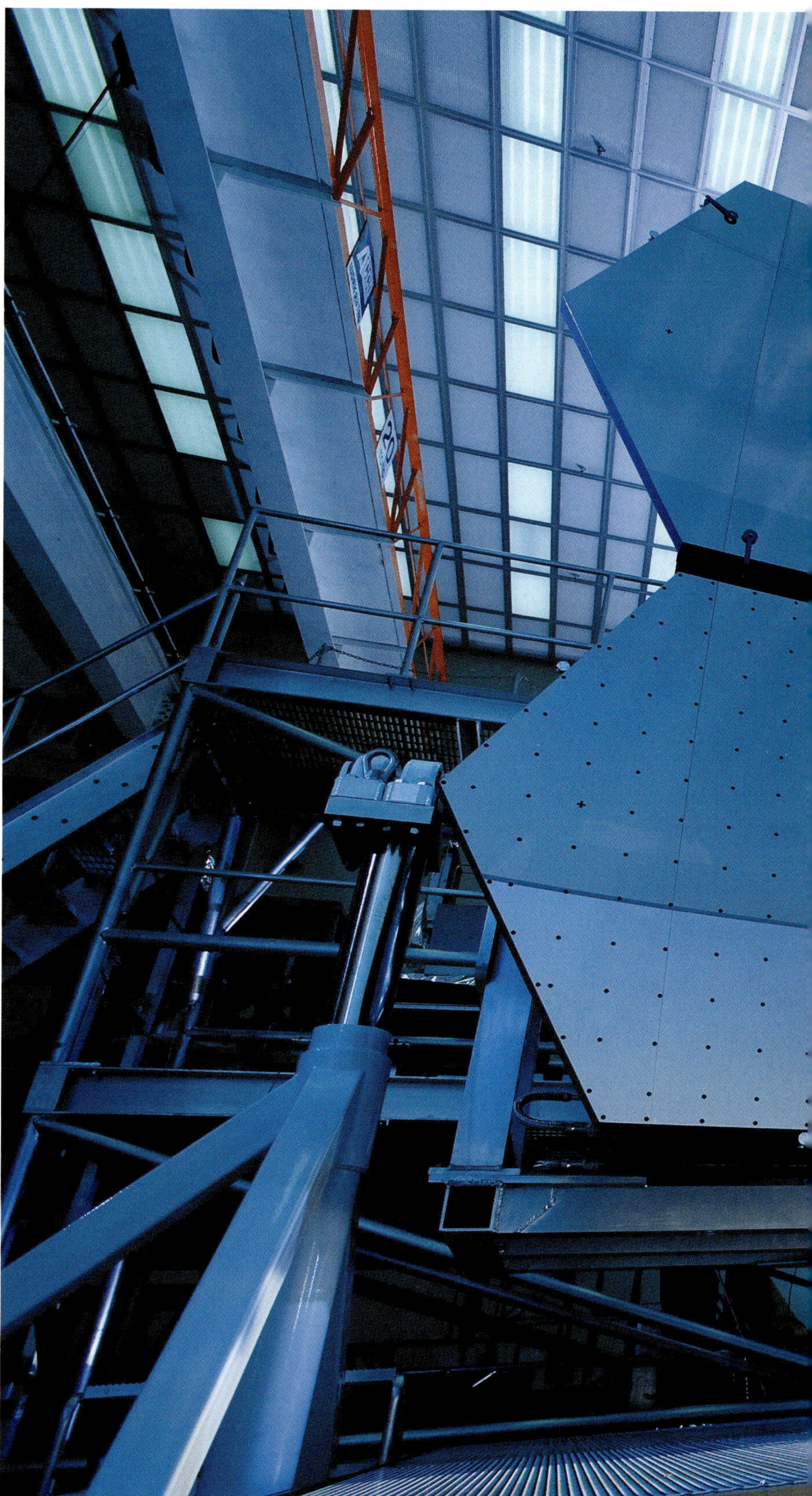

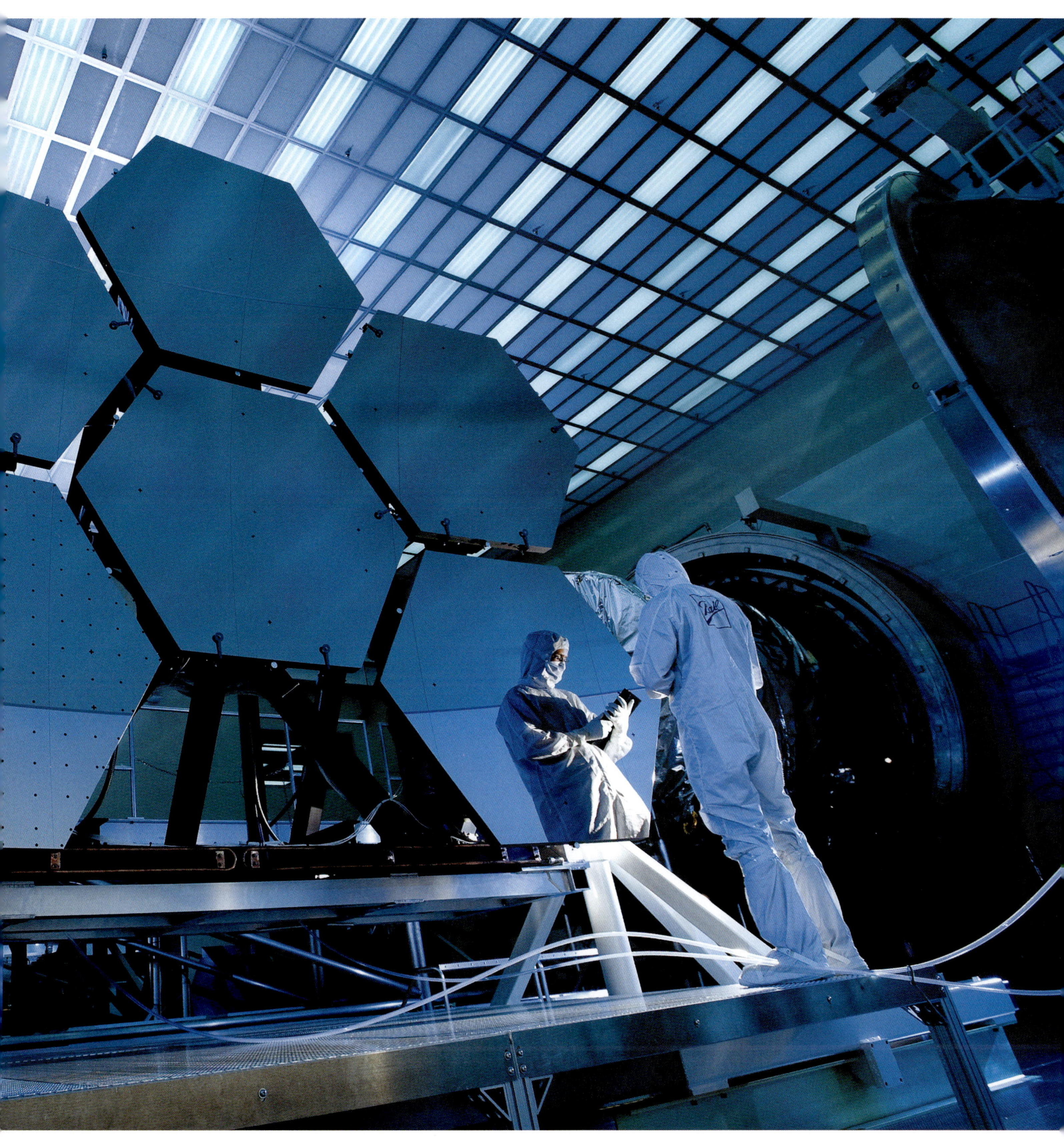

Webb's giant primary mirror is made up of eighteen hexagon-shaped mirror segments. During testing, the mirrors were subjected to temperatures as low as -248°C (-415°F).

TRICKY INFRARED TELESCOPES

Infrared telescopes, by their very nature, are complicated to build. Constructing one and then launching it into space is even more difficult. Webb's designers ran into two main design challenges.

First, the telescope needed to have a gigantic mirror to capture a good amount of infrared radiation. The larger the mirror, the more radiation it can collect and the more details it can capture. So, when it comes to telescope mirrors, bigger is always better. Also, the longer the wavelength of radiation being captured, the larger the telescope needs to be – which explains the huge size of radio telescopes.

But large single mirrors are heavy, breakable and cumbersome to launch into space. In the end, the designers decided to make a large mirror out of lots of segments. Webb's primary mirror contains eighteen hexagonal mirrors, each of them 1.32 metres (4.5 feet) across. When lined up together, the hexagons form a rough circle; they are very precisely aligned – to 1/10,000th the thickness of a human hair.

China's FAST radio telescope is one of the largest single-dish radio telescopes in the world. Its diameter is 500 metres (1,600 feet).

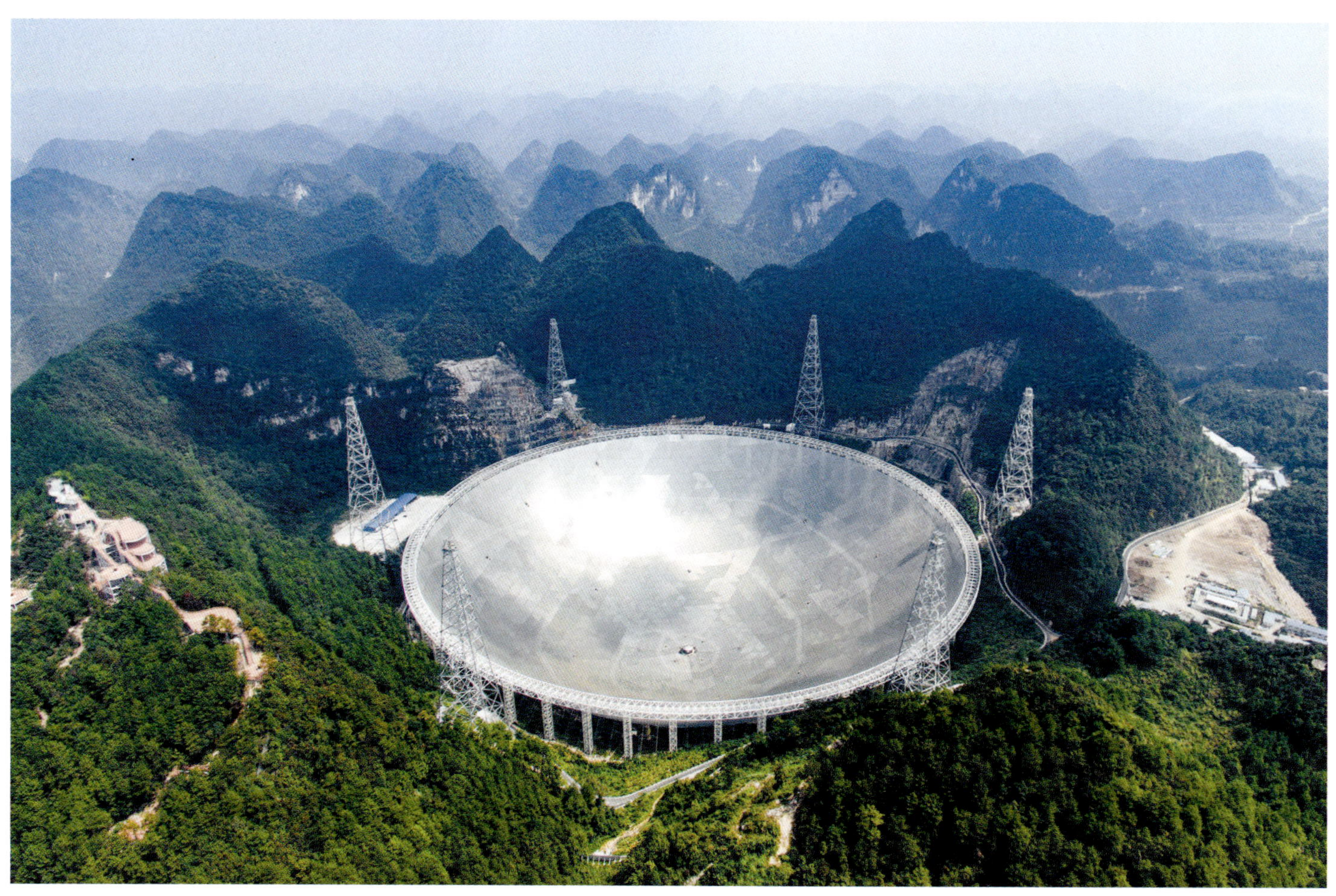

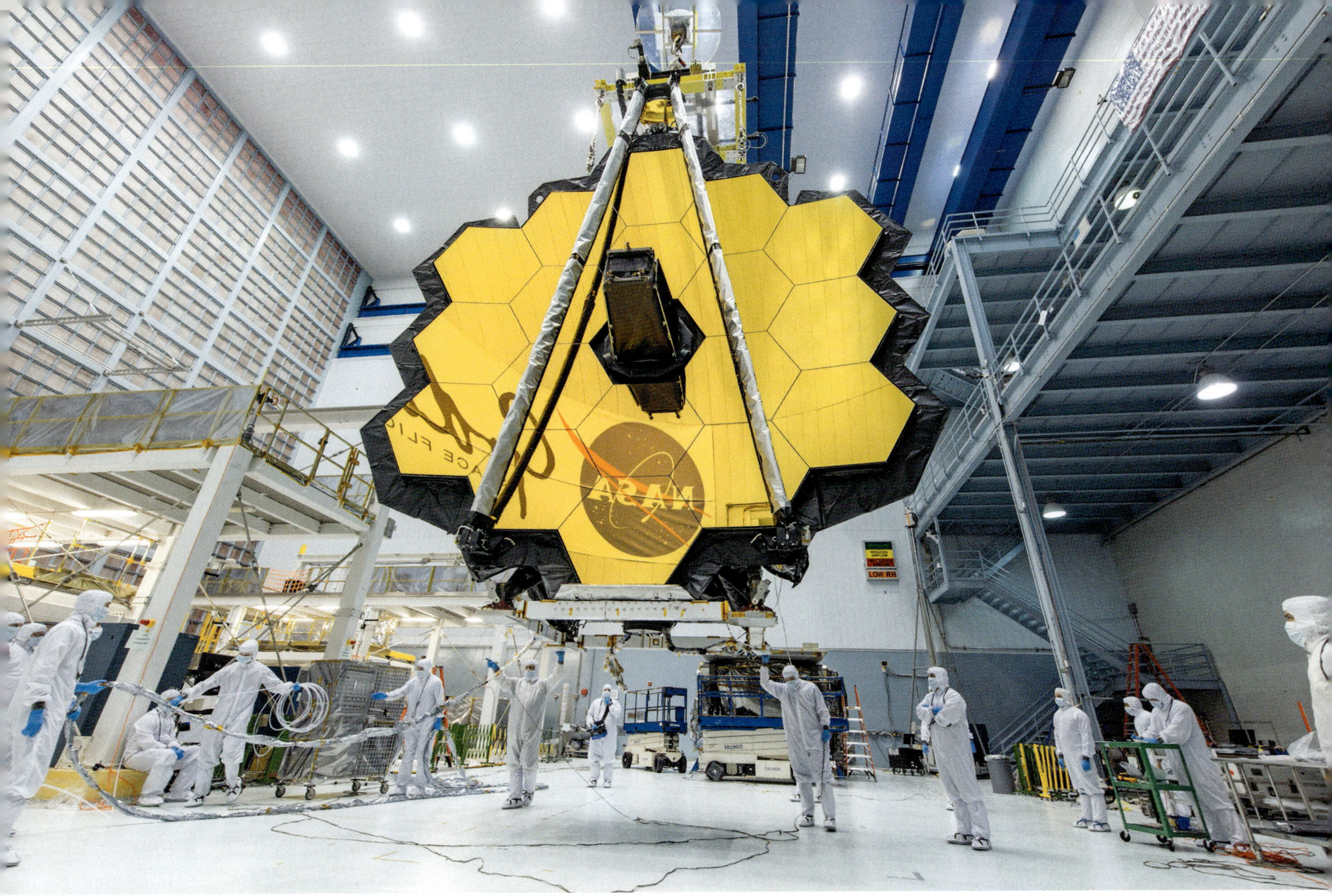

ABOVE: Webb's mirror array was coated in gold to improve its signal collecting.

BELOW: Webb's sunshield membrane protects its instruments from solar radiation and interference from the spacecraft.

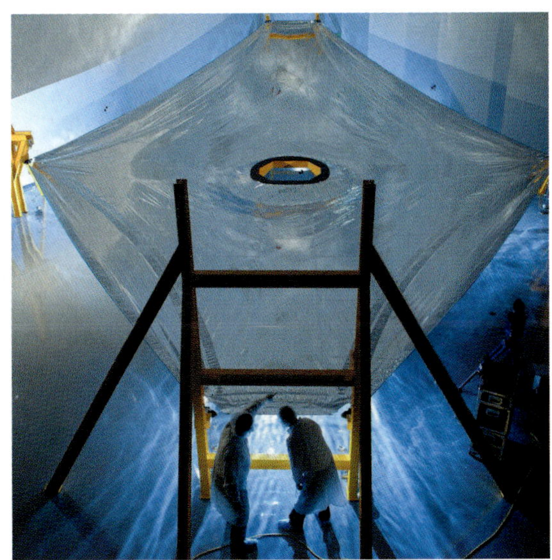

Before deployment, these mirrors were folded together, like an origami flower waiting to blossom. Unfurled in space, they span 6.5 metres (21.5 feet) at their widest point. The primary mirror has a total collecting area of about 25 square metres (269 square feet). The light collected in this mirror is reflected off a secondary mirror, which is in the telescope's 'nose', before being bounced into a gap in the middle of the primary mirror and sent to the instruments. All of the telescope's mirrors have a thin coating of gold to improve their reflection of infrared radiation. Its larger primary mirror and superior sensitivity mean that Webb can detect objects up to a hundred times fainter than Hubble can, as well as objects that existed much earlier and closer to the Big Bang.

The second major challenge was to keep the telescope cold. Infrared energy is heat energy, so if the telescope gets warm it will be swamped by its own radiation and unable to see the radiation gathered from space. Protection was therefore needed from radiation from the sun, the moon and Earth, as well as from its own electronics, which also dissipate heat. With this in mind, Webb was designed with a five-layer sunshield the size of a tennis court. This shield was able to screen both the spacecraft and its instruments from surrounding radiation.

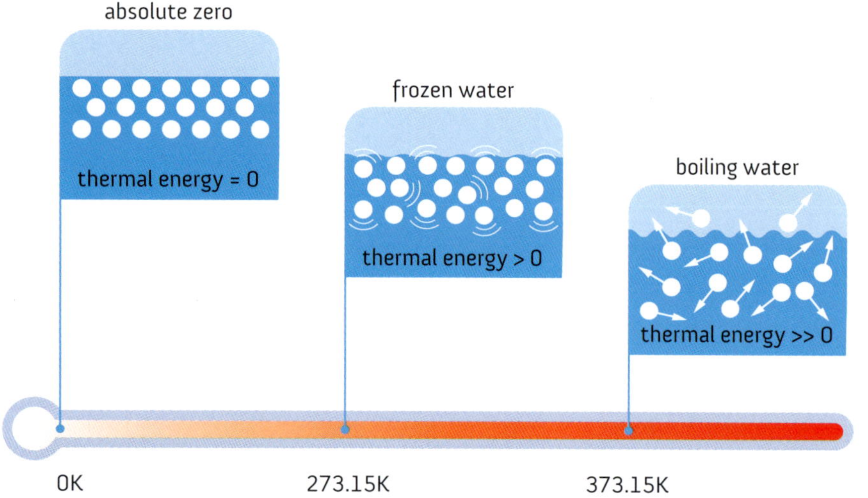

The energy of water molecules on the Kelvin temperature scale.

But that is not enough. The infrared detectors on Webb's instruments also need to be kept very cold. If not cold, they exhibit something called 'dark current', which means that they produce a signal even if there is no radiation reaching the detector. So, each detector needs to be cooled, and their operating temperature is dependent on the infrared wavelength they are detecting. To address this, Webb has an onboard cryocooler. This allows the detectors to work efficiently as well as preventing the emission of radiation that overwhelms the signals from space.

Three of Webb's instruments, the Near-Infrared Imager and Slitless Spectrograph (NIRISS), the Near-Infrared Camera (NIRCam) and the Near-Infrared Spectrograph (NIRSpec), have detectors that operate at 40 Kelvin, which is -233.15°C (-387.67°F). On the Kelvin temperature scale, zero is a deficit of all heat energy and called absolute zero (the equivalent of -273.15°C/-459.67°F): the coldest that anything in the universe can get. At this temperature, atoms and molecules no longer move due to the lack of heat energy. So, 40K is extremely cold.

However, for the Mid-Infrared Instrument (MIRI), radiation is detected in the mid-infrared range, as the name suggests. The cryocooler for MIRI's detector needs to be even colder and is the coldest instrument on board the spacecraft – a gobsmacking 6K, which is -267.15°C (-448.87°F), close to absolute zero.

Scientists tested MIRI's thermal shield in a thermal vacuum chamber.

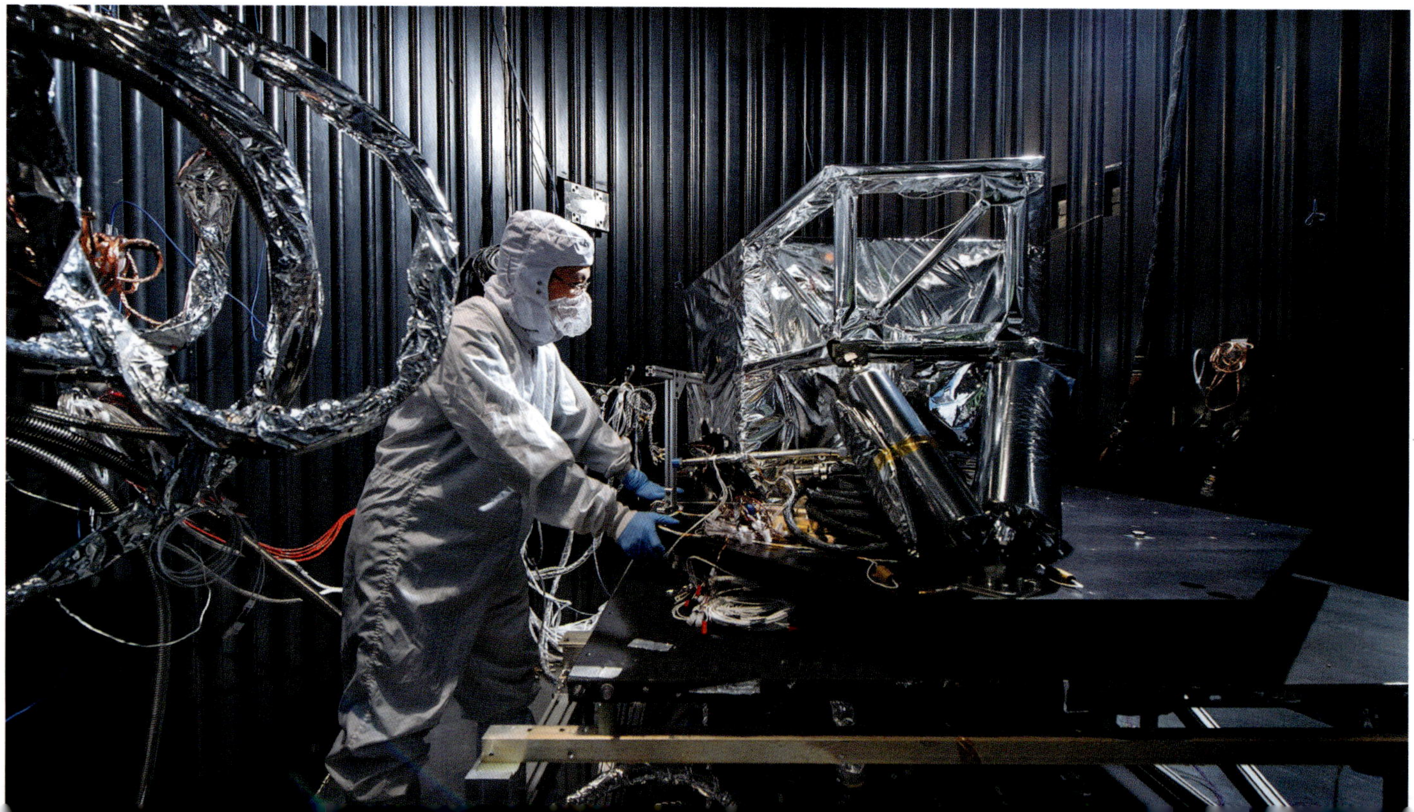

BLOSSOMING IN SPACE

It's difficult to get a giant space observatory onto a rocket. Due to its revolutionary design, the Webb Observatory, when stowed in the rocket, was folded in on itself like a flower in a bud. After it detached from the rocket, Webb began the careful process of unwrapping itself. First, it deployed its solar array, then its high-gain antenna, before releasing its sunshield and other parts. Finally, it stretched out its mirror wings. Unlike Hubble, if anything went wrong with Webb there was no rescue mission available. Hubble sits a mere 550 kilometres (341.5 miles) above sea level, so it was possible to send a team of astronauts to tweak any faults. Webb's final destination was 1.5 million kilometres (932,056 miles) away from our planet – a location that no astronaut could get to. This added a whole layer of jeopardy to the already complex proceedings.

In total, there were around fifty major deployments and 178 release mechanisms that needed to be successfully activated before the telescope was ready for action.

Webb took more than two weeks to completely unfurl, but it took many more days to reach its final orbit.

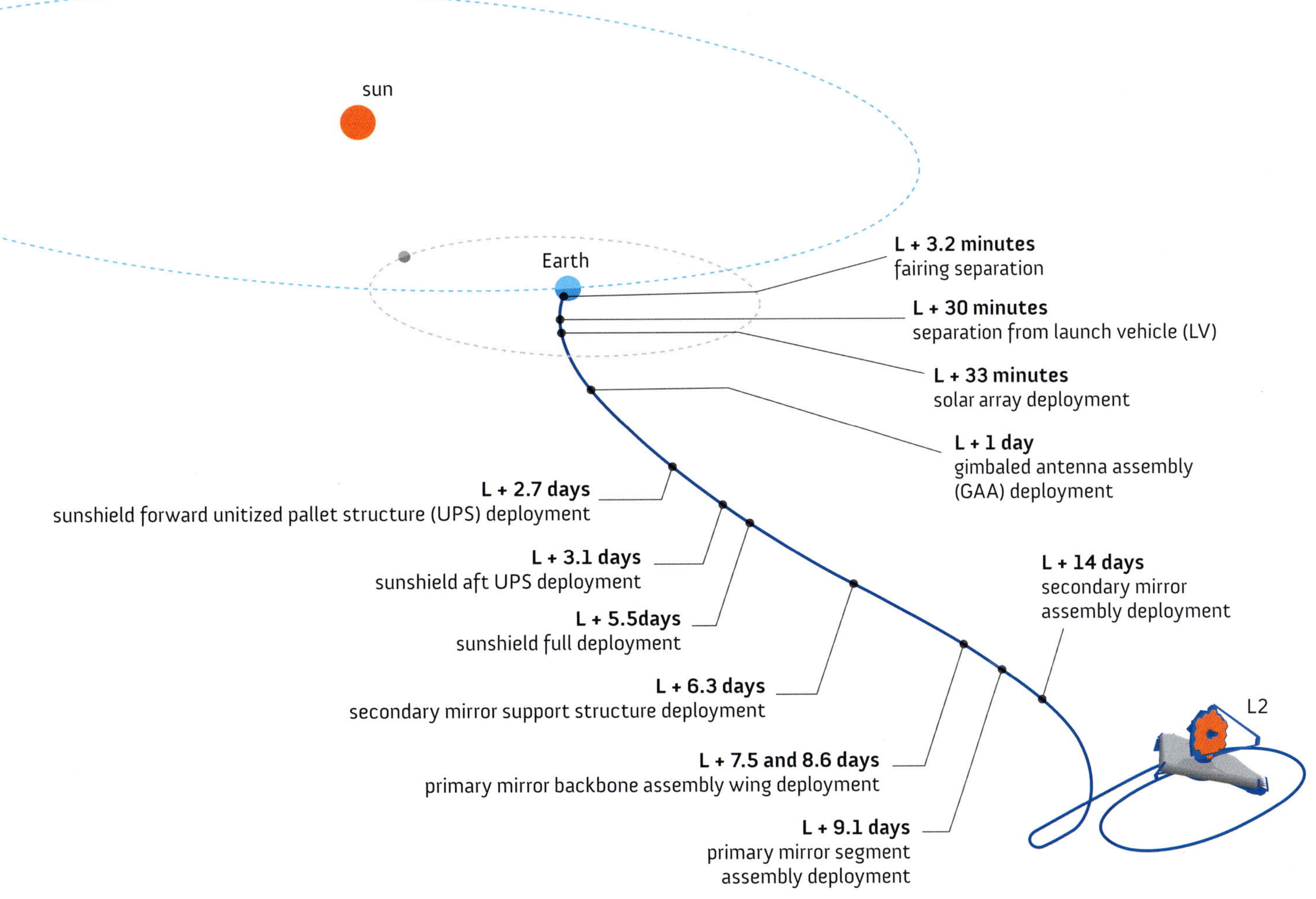

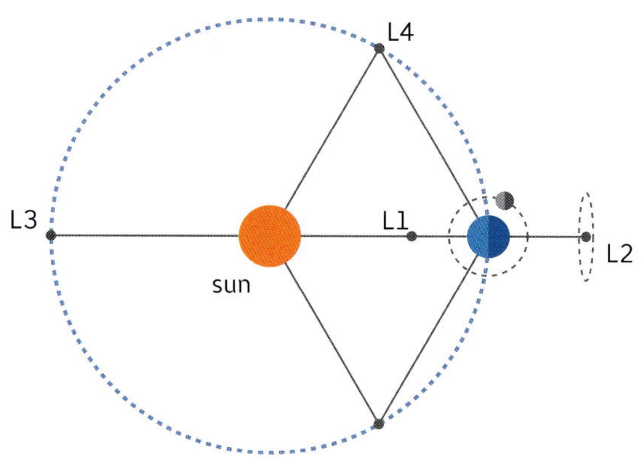

Webb orbits around the Lagrange Point L2, which is about 1.5 million kilometres (1 million miles) from Earth.

UNSTABLE LOCATION

Webb is located at a point in space known as the Lagrange point L2, which lies approximately 1.5 million kilometres (just under 1 million miles) beyond the Earth's orbit around the sun. At this point in space an object is kept in its place by the gravitational force of other celestial bodies, so Webb can stay in this position with minimum consumption of fuel. Webb moves in a halo orbit, a three-dimensional orbit circling this L2 point, and its position varies between 250,000 and 832,000 kilometres (155,000–516,980 miles) from it. L2 is not a fixed point but follows Earth around the sun. The issue with the Lagrange positions, however, is that they are not entirely stable, and so the spacecraft requires a small amount of fuel to maintain its position around them.

Webb was launched with enough fuel to maintain this orbit for about ten years, but it is thought that the launch went so well that it consumed less fuel than was allocated to get it into position. This means that the observatory now has enough propellant to stay in its orbit for around twenty years – twice the length of time scientists had planned for and hopefully, just as with Hubble, a lot more amazing science and images.

AN INFRARED SPACE OBSERVATORY

Webb's design and function provide a platform for groundbreaking science, but it is the instruments on board that really do the heavy lifting.

Webb has three components: its mirrored eyes (known as the Optical Telescope Element, or OTE), the Spacecraft Element (mainly made up of the spacecraft bus, which provides the infrastructure for the spacecraft including power, telemetry and an appropriate thermal environment), and then the Integrated Science Instrument Module (ISIM), which is the main payload and heart of the telescope.

The telescope feeds the signal information into the ISIM, which contains four custom-designed instruments. These include infrared cameras and spectrographs, which detect the wavelengths of specific molecules.

At launch, Webb weighed around 6,500 kilograms (14,330 pounds) – slightly heavier than a male African elephant – and it is, to date, the most expensive and most advanced telescope ever built. About six months after launch, NASA revealed the first image taken with the telescope, a deep-field camera image of galaxy cluster SMACS 0723. The cluster is about 4 billion light-years from Earth in the constellation of Volans and teeming with thousands of galaxies, including some of the faintest objects observed in the infrared range. Since then, Webb has yielded dozens of images and terabytes of data, which are helping us to understand our vast and largely unknown universe.

Released on 12 July 2022, this near-infrared image of galaxy cluster SMACS 0723 is the first full-colour Webb image.

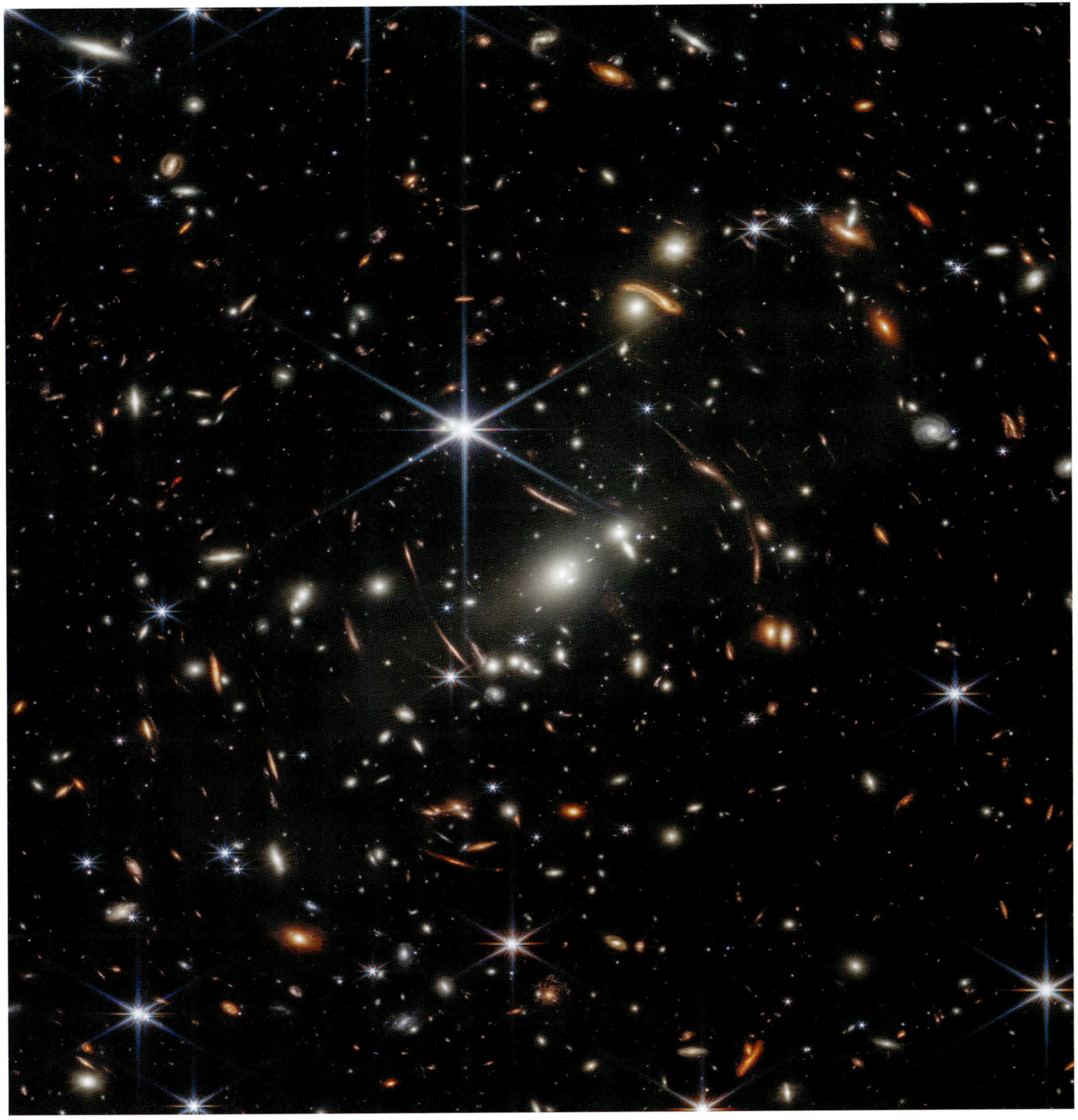

INTEGRATED SCIENCE INSTRUMENT MODULE (ISIM)

The ISIM houses Webb's cameras and science instruments. It makes up about a quarter of Webb's weight, and is the heart of its science capabilities. Due to the sensitivity of the instruments, the entire observatory is engineered to protect the ISIM from the sun's extreme radiation and the low-level interference of Webb's machinery. The ISIM's electronics compartment, which contains the instruments' computing and electrical components, is nestled in a thermally wrapped box, which is itself mounted on a cryogenic cooler. The electronics' heat is specifically directed into space, where it cannot interfere with the instruments or mirrors.

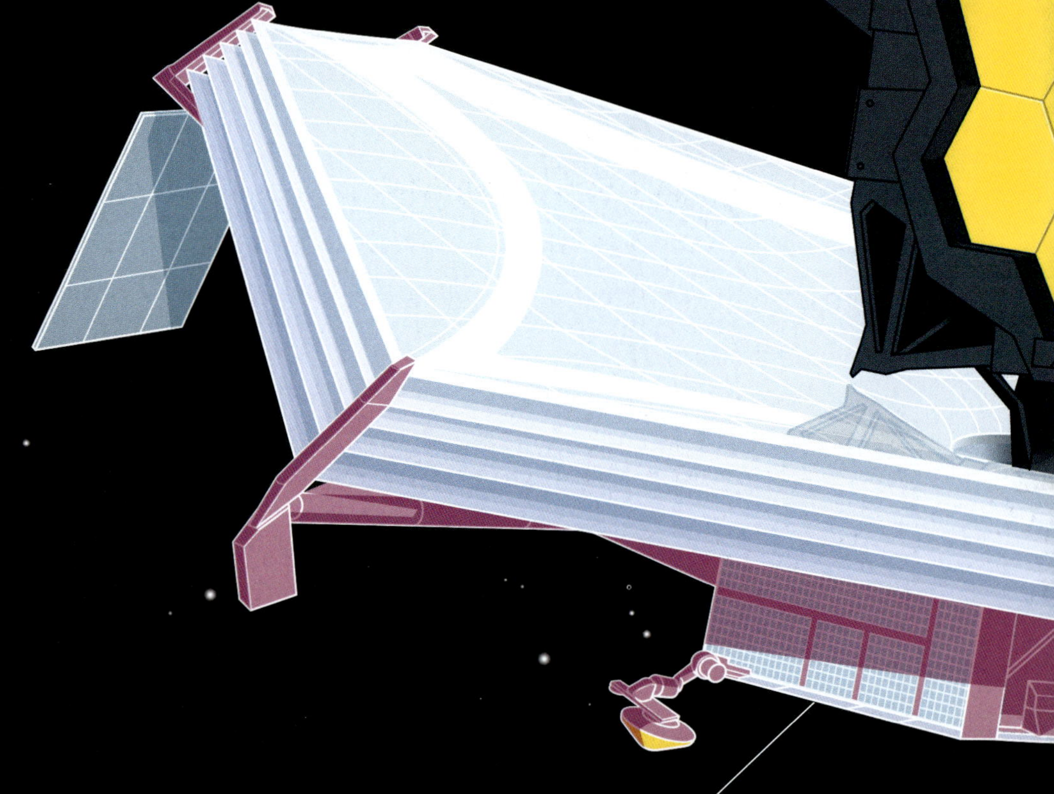

SPACECRAFT BUS

The spacecraft bus is where the observatory's steering, communication and control machinery live. It contains six important subsystems that are needed to keep Webb running, namely the electrical power, attitude control, communication, command and data handling, propulsion, and thermal control subsystems. The electrical power subsystem converts sunlight from the solar panels into electricity to run the instruments and other subsystems. The attitude control subsystem makes sure that the observatory is in the correct orientation and facing the right direction. The communications subsystem beams information to the scientists and engineers back on Earth, while the command and data handling subsystem is the brain of the bus, acting as a go-between for all the different subsystems. The propulsion subsystem contains the fuel and communicates with the attitude control unit so that the observatory can change its position if it needs to. The thermal control subsystem keeps the whole bus operating at an optimal temperature.

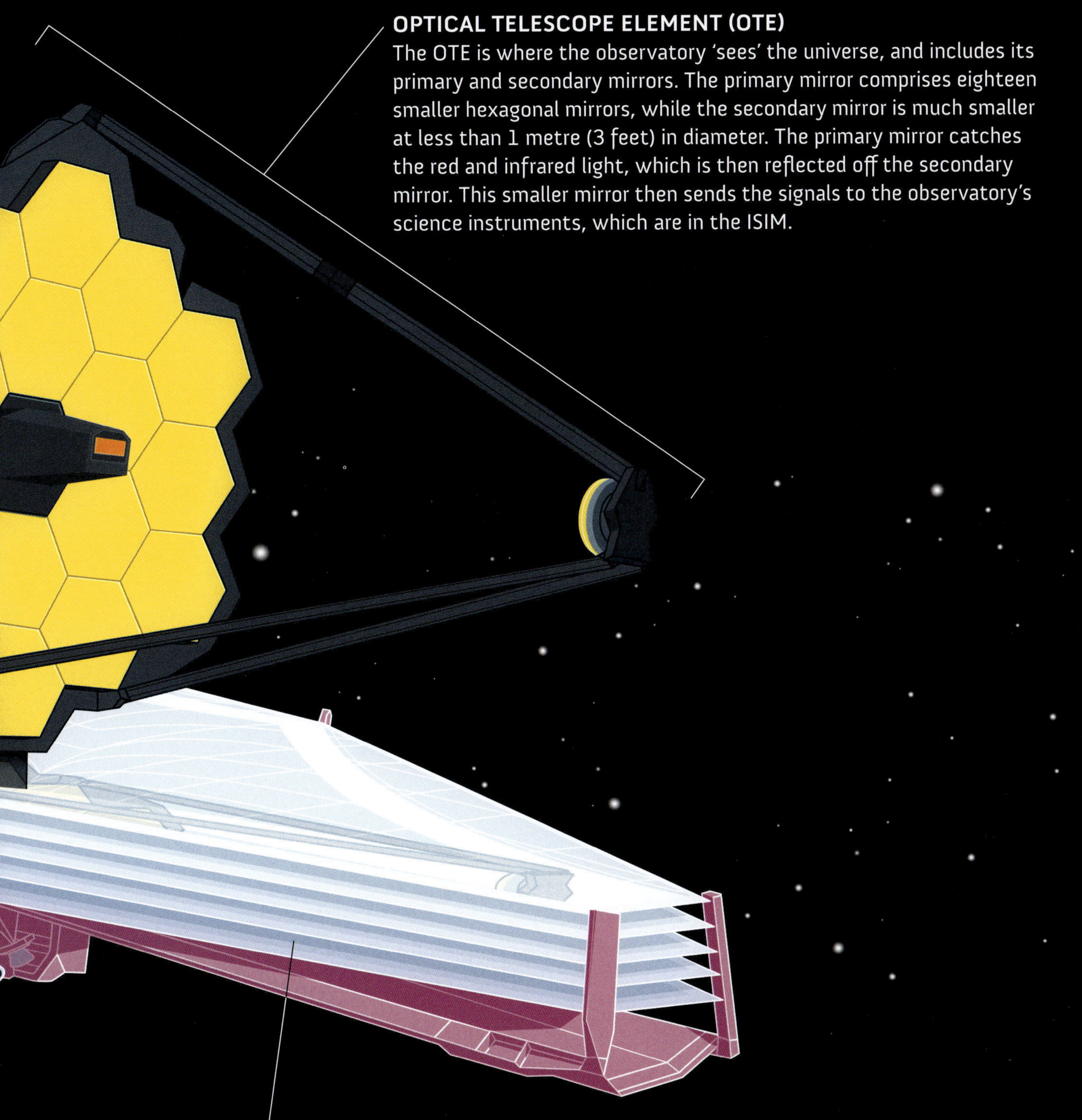

OPTICAL TELESCOPE ELEMENT (OTE)
The OTE is where the observatory 'sees' the universe, and includes its primary and secondary mirrors. The primary mirror comprises eighteen smaller hexagonal mirrors, while the secondary mirror is much smaller at less than 1 metre (3 feet) in diameter. The primary mirror catches the red and infrared light, which is then reflected off the secondary mirror. This smaller mirror then sends the signals to the observatory's science instruments, which are in the ISIM.

SUNSHIELD SUBSYSTEM
The sunshield subsystem divides the toasty sun-facing side of Webb from the incredibly cold science-enabled side. The sunshield has five layers. It has a warm side, which faces the sun and reaches about 110°C (230°F) – hot enough to boil water – and a cold side. Its cold side, which faces away from the sun and looks into deep dark space, can reach -237°C (-394.6°F). The telescope itself operates at around -223°C (-369.4°F), close to the coldest natural temperature ever recorded in our solar system, on the planet Uranus

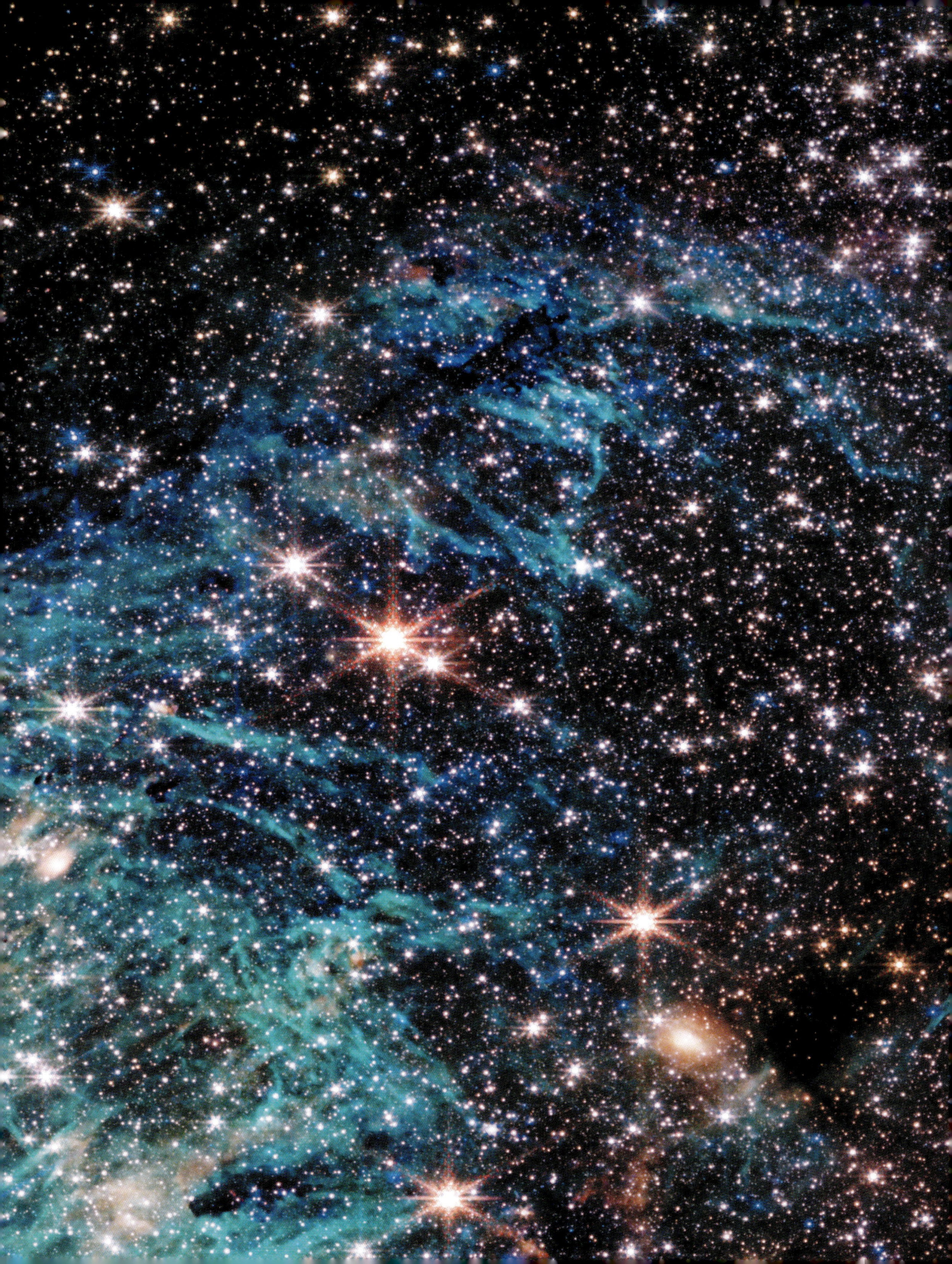

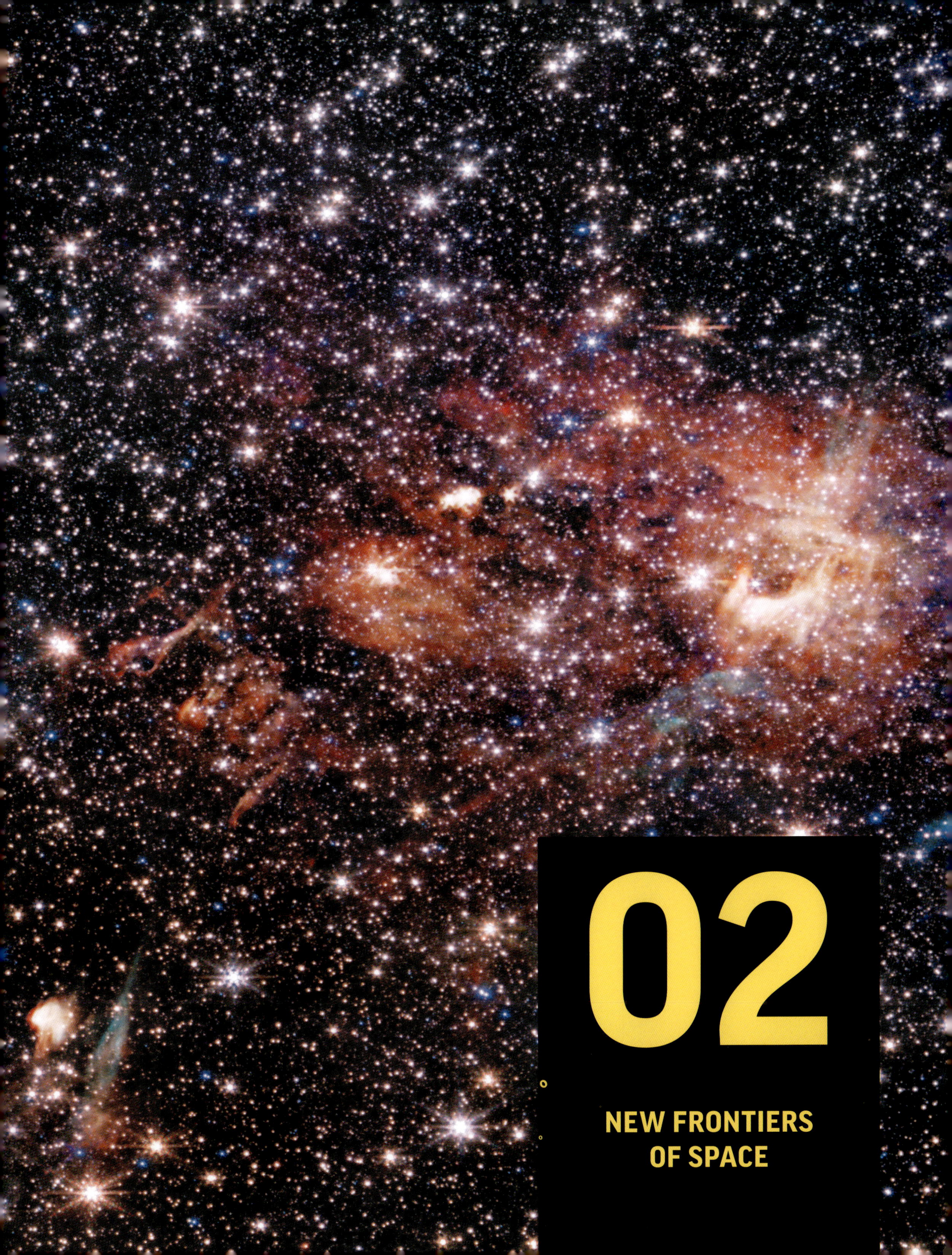

02

NEW FRONTIERS OF SPACE

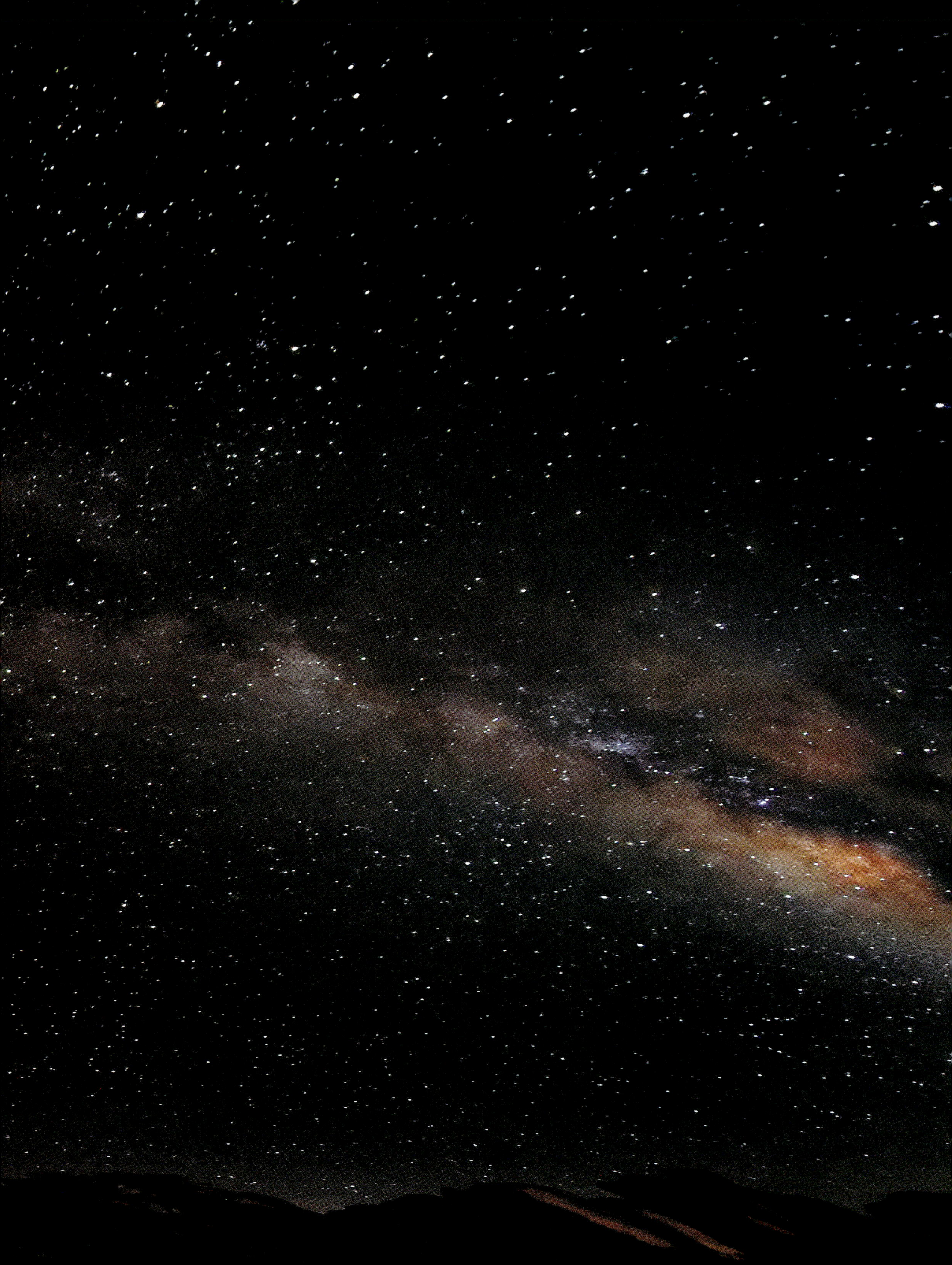

THE STORY OF OUR ASTRONOMICAL understanding dovetails neatly with the trajectory of our technology, and our stargazing abilities are tied up with our development as a species – we have developed more sophisticated tools to investigate the universe as our understanding of it has grown. And of course, the better our instruments, the more we have been able to discover. This cycle of curiosity and technological advancement continues today; Webb is our latest and most sophisticated attempt to understand the diverse and wondrous universe around us.

NEW TECHNOLOGIES AND APPROACHES

Our ability to take precise observations and measurements to explain the movement of the stars has increased over time. By the early seventeenth century, astronomers were using telescopes to observe far away objects that were invisible to the naked eye. By using two lenses in a tube, they could magnify objects up to three or four times their original size.

But, if you are trying to unlock the secrets of the universe, that sort of magnification gets you only so far. In 1609, Italian scientist Galileo Galilei, often regarded as the father of observational astronomy, developed a telescope that could magnify objects even further, forming the foundation of optical astronomy.

Visible light is only a tiny portion of the electromagnetic spectrum, however. By the early twentieth century, we knew that other wavelengths existed – though we'd yet to detect them coming from space. This all changed in the 1930s.

MULTIWAVELENGTH ASTRONOMY

Karl Jansky, an American physicist and radio engineer working at Bell Laboratories in New Jersey in the US, designed and built a radio antenna capable of detecting radio waves. He had joined the company in 1928 and set about detecting the source of an unknown interference that had been plaguing their transatlantic shortwave radio communications – a faint hissing sound that persisted day and night and that seemed to be coming from a fixed point in the sky. After years of painstaking observations,

Signals from our own galaxy, the Milky Way, interfered with transatlantic communications, which ultimately led to the establishment of multiwavelength astronomy.

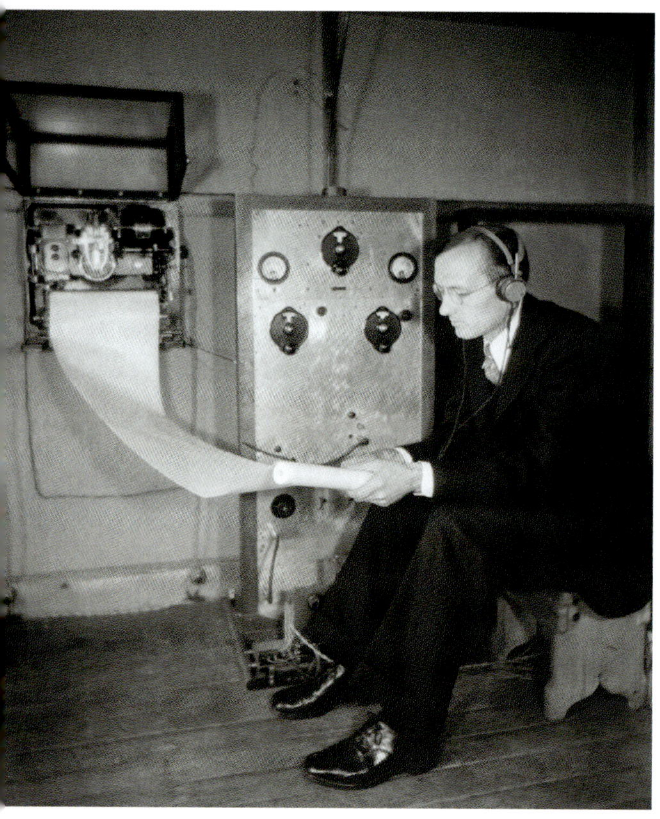

American radio engineer Karl Jansky established the field of radio astronomy.

in 1933 Jansky wrote two papers, in which he concluded that the interference was in fact coming from our galaxy, the Milky Way. With this conclusion he levered open the new discipline of radio astronomy. We were no longer restricted to observe the universe just using visible light: now, with the detection radio waves too, the new era of multiwavelength astronomy was born.

Objects within our universe emit and reflect a spectrum of electromagnetic radiation. Over time, engineers and astronomers have designed and built instruments to investigate these different portions of the spectrum. As we have seen with Webb and its specialist infrared capabilities, telescopes and instruments can be optimized to observe specific wavelengths.

Something special happens when different instruments observe the same object in the sky using different parts of the electromagnetic spectrum: the resulting data is more than the sum of its parts, and we learn more about an object than we could ever hope to by observing it in only one segment of the spectrum. Some of Webb's most interesting insights and glorious images come from combining its data with that from other telescopes, such as visible information from the Hubble Space Telescope and X-ray insights from Chandra.

Newton's reflecting telescope.

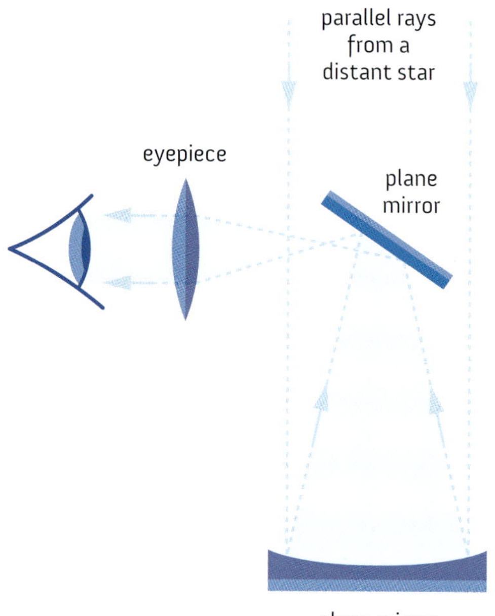

SPECTROSCOPY

Sir Isaac Newton, a key figure in the Scientific Revolution in the seventeenth century, did not build the first reflecting telescope – Scottish mathematician and inventor James Scott had come up with this design a few years earlier. But Newton made important improvements to it that helped to popularize its use. One of the main benefits of the reflecting telescope was that light did not pass through the glass, so the quality of the glass used was not so critical. Also, unlike a lens, where two surfaces need to be shaped to get the magnification, with a mirror telescope only one surface needs to be fashioned.

Some people suspect that Newton's new telescope design was probably inspired by his investigations into the properties of light, such as the prism experiment he conducted. Here, Newton passed a narrow beam of sunlight through a glass prism and saw that the light split into a spectrum of colours. This experiment demonstrated that white light is actually composed of different colours of light, each with a different wavelength and corresponding colour.

Newton's prism experiment was the forerunner of a technique called spectroscopy – the study of different wavelengths of electromagnetic radiation to better understand what happens when it interacts with matter. Depending on the type of material and the wavelengths involved, matter can absorb or reflect radiation. Under some conditions, matter can even emit radiation. In spectroscopy, we study the emission, absorption and reflection characteristics of radiation to understand the object that the radiation has interacted with – whether that is whole galaxies or the atmosphere surrounding a distant planet.

Spectroscopy enables us to do a remote analysis of the chemical composition, temperature, density and motion of a celestial body billions of kilometres away. This technique has played a pivotal role in the development of our understanding of the universe, including the discovery of new elements and the confirmation of key measurements that led to the establishment of the Big Bang theory, the best theory we have to explain the origins of the universe.

During my career, I have managed the design, build and implementation of several spectrographs, for ground- and space-based telescopes, including, of course, the subsystems for Webb's NIRSpec instrument that we explored in Part 1. These spectrographs have all been used to understand the movement and chemical composition of celestial objects that sit out there in the cosmos.

On 31 July 2023, the European Space Agency's Euclid space observatory sent back its first image using its near-infrared spectrometer and photometer.

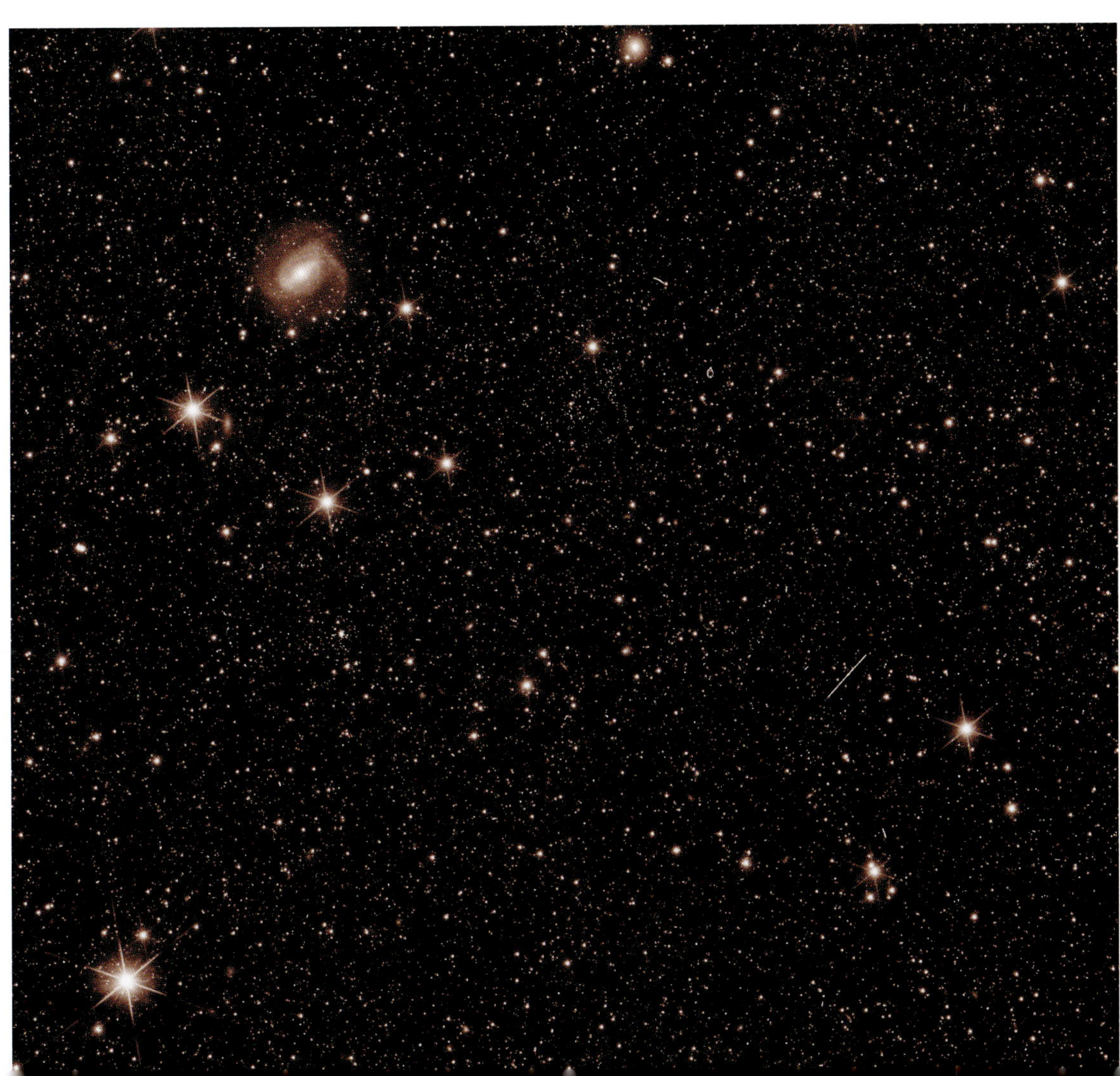

WEBB'S SCIENCE GOALS

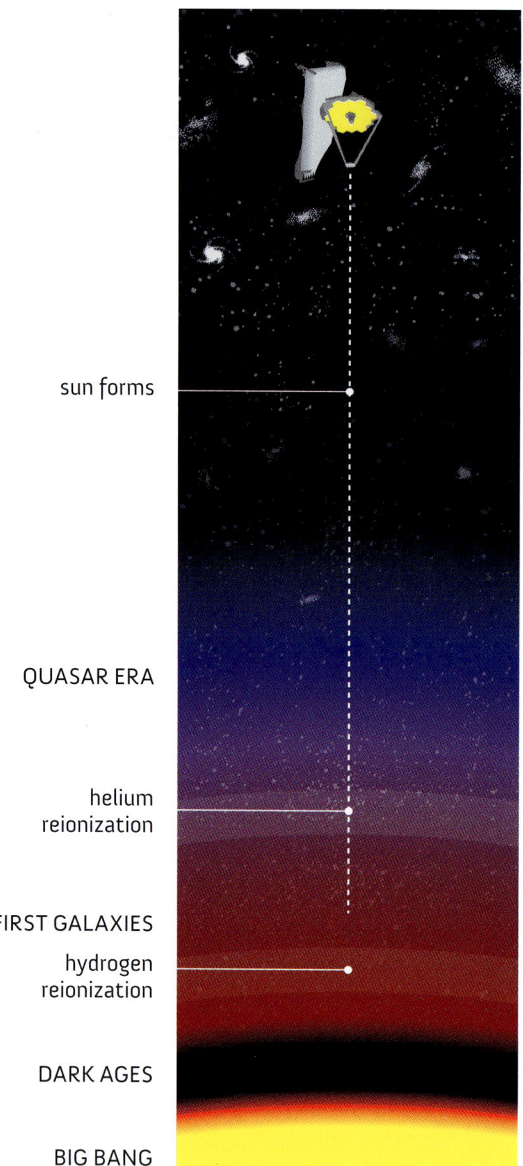

ABOVE: A timeline of the universe, from the Big Bang to Webb observing the signals from the first galaxies.

BELOW: Signals from the earliest galaxies and stars stretch as the universe expands.

The Big Bang theory suggests that around 13.8 billion years ago the universe came into being from a single, unimaginably hot and dense point, known as a singularity. Everything we can see in the universe (and many things we can't) expanded out of that tiny pinprick, the beginning of everything – including space and time. We don't believe that this expansion occurred in an already existing space but rather that it initiated the creation and expansion (and cooling) of space itself.

Hundreds of millions of years after that universe-creating explosion, all matter and energy were mixed in a dark and energetic primordial soup. But as this soup cooled, the particles began to coalesce into atoms, specifically hydrogen – the most basic atom is, in its simplest form, made up of a single proton and an electron. Up until this point, the soup was so dense that light could not pass through it, the particles so close together that the light would bounce between them. But once atoms had begun to form, the light particles could finally move freely through the lumpy soup of the universe. Looking back in time, as Webb enables us to do, we can see this early light from the first stars and galaxies. It is as though they were first 'switched on', like lights in a dark room, and their ancient light still reaches us across the cosmos today.

By the time this radiation reaches our telescopes, it has been travelling for billions of years across the universe. What began as visible light has become less energetic and has shifted towards the 'red' end of the electromagnetic spectrum. Astronomers call this phenomenon *redshift*. Blueshift of light can also occur. This happens when an object is moving towards an observer. In this case, the movement of the object causes the light to get bunched up and 'shifted' towards the blue end of the spectrum.

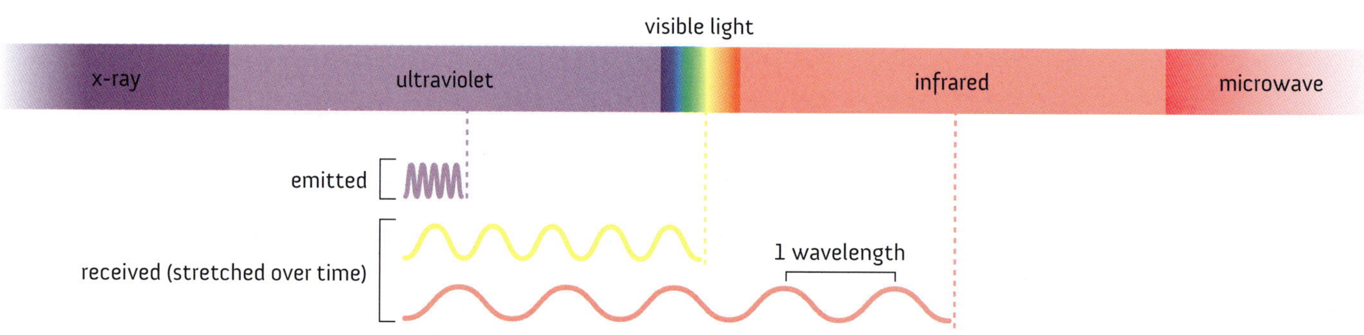

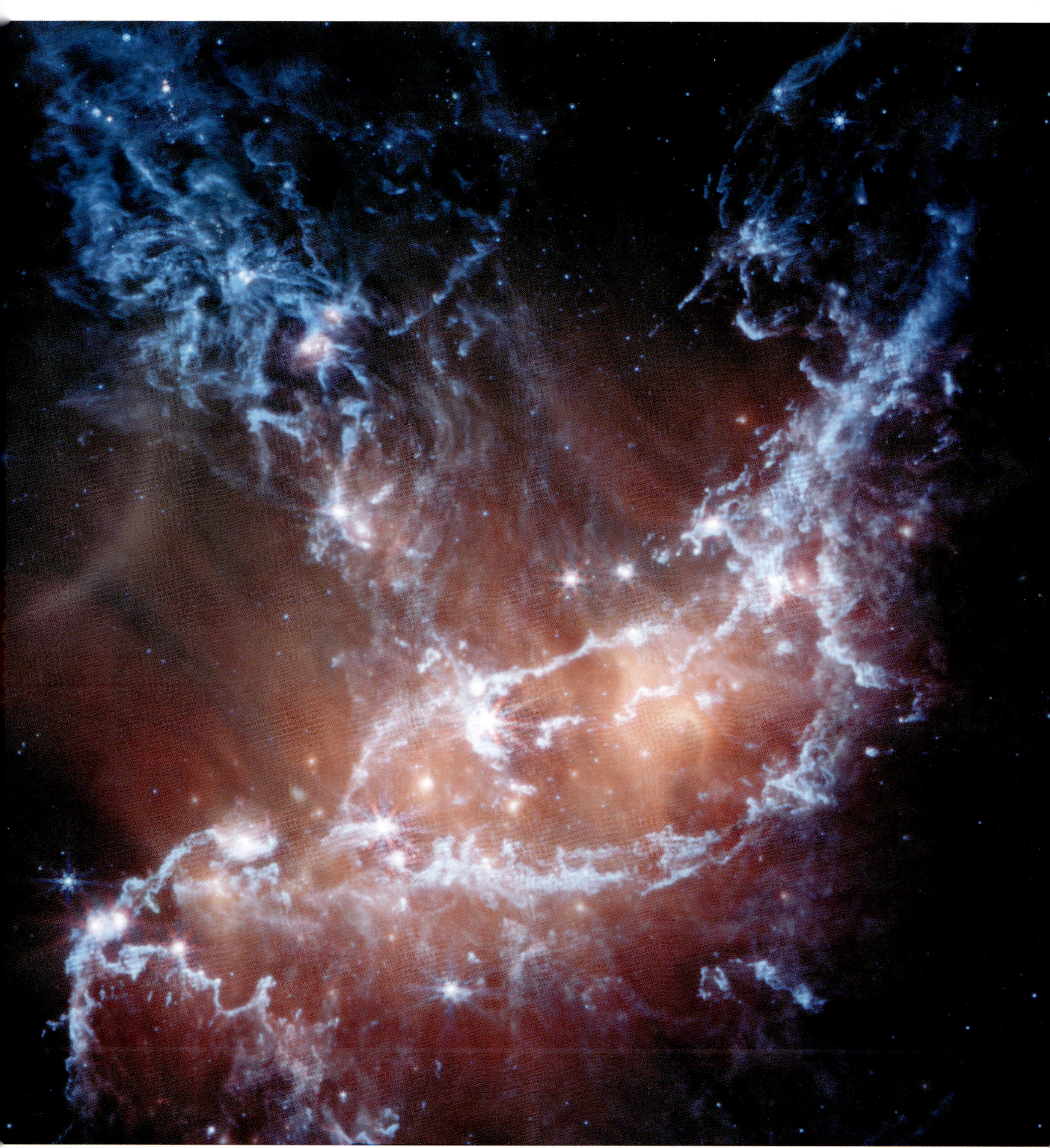

Released on 10 October 2023, Webb's MIRI instrument has captured emissions and dust within NGC 346, as well as a number of protostars.

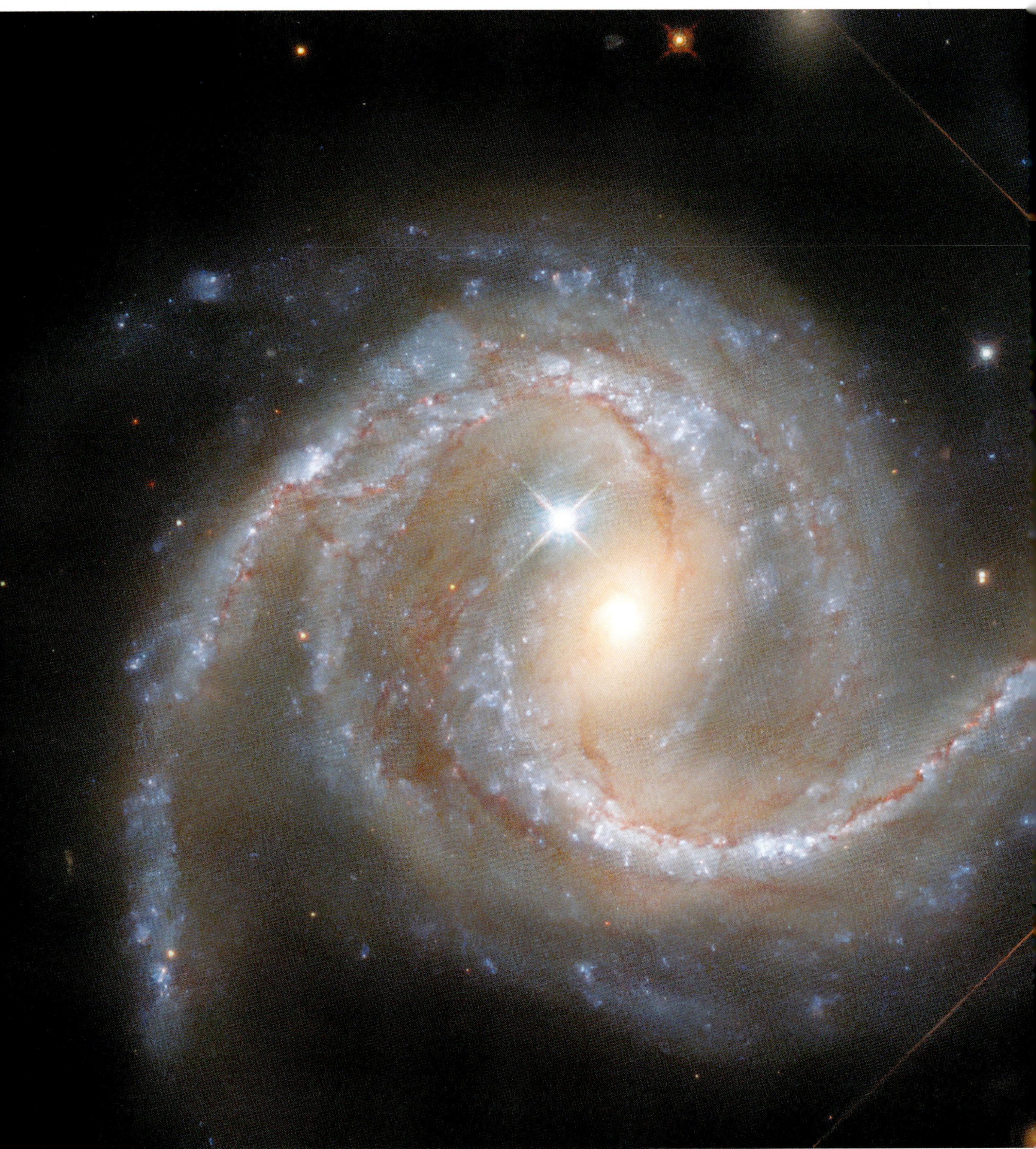

A Hubble image of NGC 5495. This spiral galaxy is about 300 million light-years away in the constellation Hydra. The two bright stars in the image – to the top left of the galactic centre and the very bright star on the right – are not part of NGC 5495 and are actually in the Milky Way.

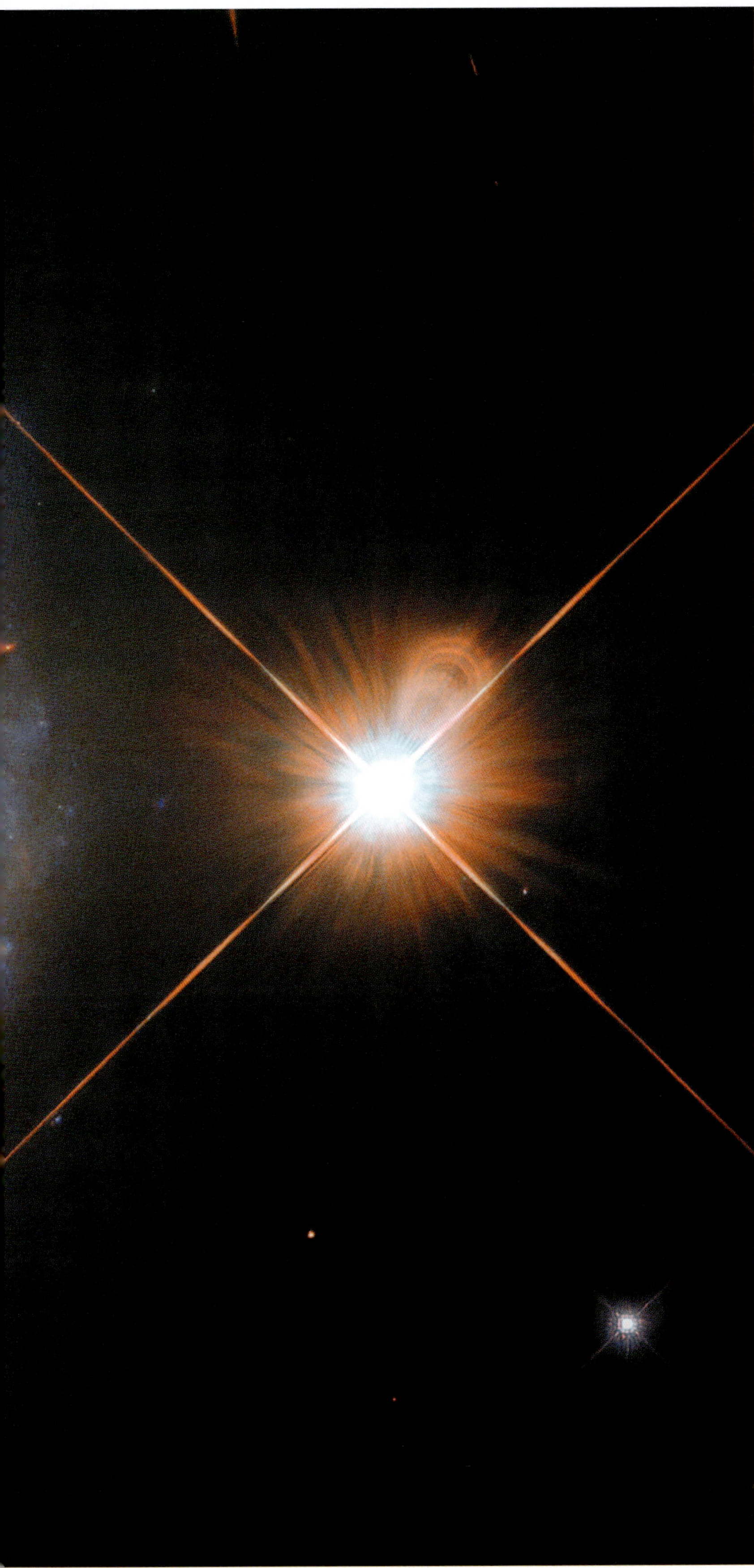

Webb, with its infrared capabilities, is a powerful time machine that can detect light from the early stages of universe. It is performing ultra-deep surveys of the cosmos, analysing some of the observed radiation with spectroscopy to give even more detailed information on the objects being observed.

Understanding these early years of the universe and the first light sources is important because they influenced the formation of later objects, such as galaxies. The conditions of the nascent universe are impossible to recreate on Earth, so this is also one of the few ways in which we can investigate this high-energy transformative physics. With Webb, we can probe these ancient epochs in a way that has not previously been possible.

A COSMOS OF GALAXIES

Images taken by Hubble showed us that in the observable universe there are hundreds of billions of galaxies – giant structures containing stars, planets, dust and gas. But it took us millennia of observations and technological development to reach this understanding.

In 964 BCE, Persian astronomer 'Abd al-Rahmān al-Sūfī recorded the first galaxy outside of our own Milky Way. He described the Andromeda galaxy, which is more than 2.5 million light-years away, as a 'little cloud' – a title it held for many centuries. Since then, we have created an inventory of galaxies and are continually adding more.

Today, galaxies are fundamental to our understanding of cosmology. They show us how matter can behave on gigantic scales, and their current arrangement suggests what previous epochs of the universe may have looked like – how earlier galaxies appeared and how they interacted with each other.

We now also know that most large galaxies have supermassive black holes at their centres. Some scientists theorize that the universe's first stars may have been the progenitors of these black holes, which formed when they exploded, but we are still not sure. There remain many, many things that we do not know about galaxies, such as how they form, how they generate stars and why there is such a staggering variety of them.

In part thanks to the discerning eyes of Hubble, we are familiar with beautiful spiral galaxies that look like a whirlwind of stars, and elliptical collections of stars and dust that look like a smudge on the black backdrop of space. But we believe that truly ancient galaxies looked a bit different. Very distant galaxies tend to be smaller, with roughly shaped clusters of stars rather than spirals, although we're not sure why. Also, the rate of star formation in these early galaxies is much higher than we see in today's galaxies. Webb is already shining light on these questions.

Webb's sensitive infrared gaze, coupled with its sophisticated spectroscopy, allows us to probe the chemical composition of galaxies, among other characteristics, and fill in some of the blank spaces in our knowledge of the universe.

Both of these infrared and optical views of Andromeda galaxy were created using data from the Spitzer Space Telescope in 2004. They highlight how objects' appearances vary when viewed with different wavelengths.

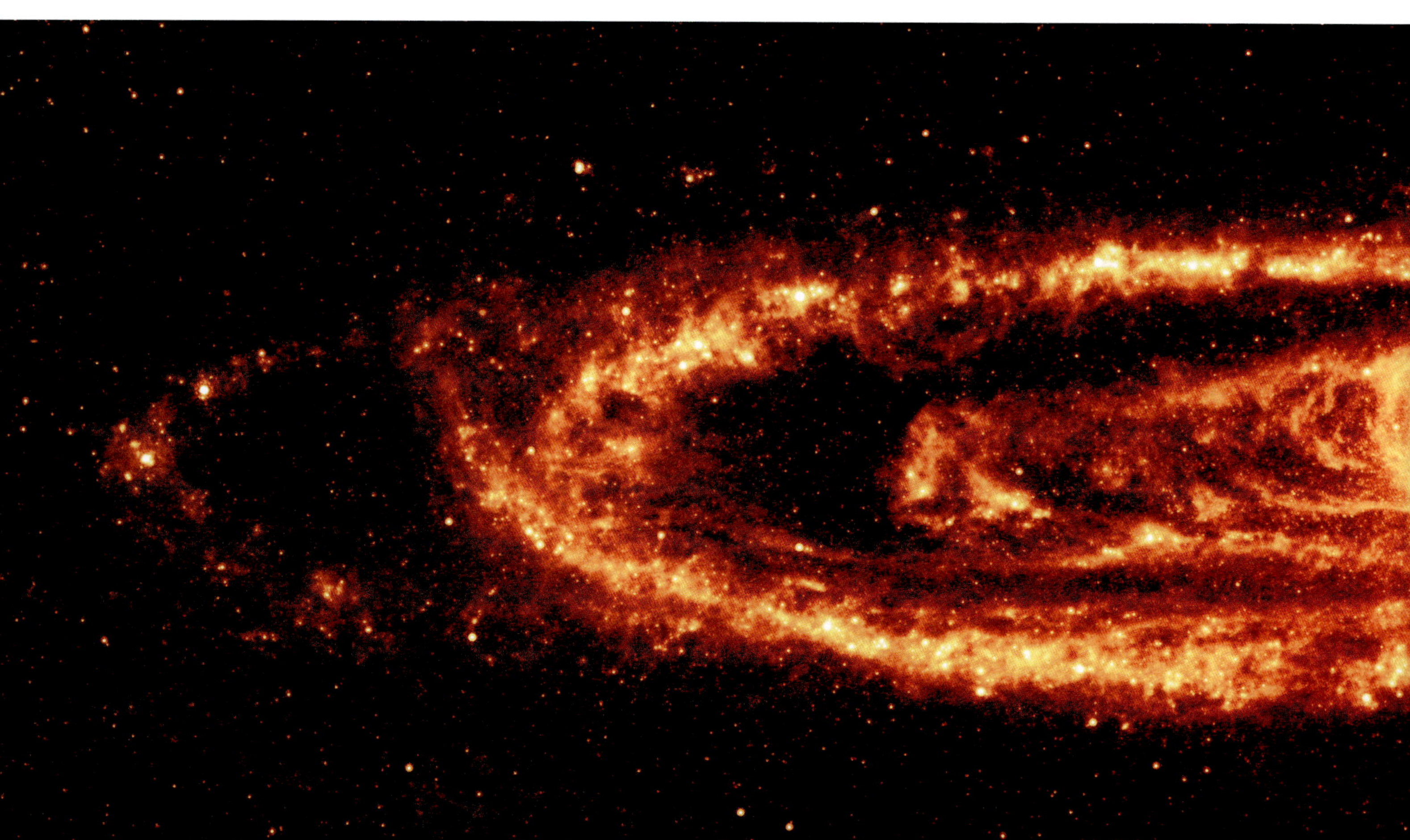

ENIGMATIC DARK MATTER

Perhaps one of the biggest questions in cosmology – and physics in general – relates to dark matter. This substance permeates the universe: we know it is there because of the effect it has on the matter we can see, but we can't see dark matter itself. It doesn't interact with radiation, so we are not sure what exactly it is. In the 1930s, astronomers saw that galaxies seemed to be spinning faster than they should, indicating that there was much more mass contained in them than could be observed with the technology available at the time.

By the 1980s, most astronomers were convinced that dark matter existed, and we now know that it accounts for more than a quarter (26.8 per cent) of the universe. Dark energy – an unknown energy that also permeates the entire universe and drives its expansion – accounts for about two thirds (68.3 per cent) of the universe. The matter that we can see – the planets, stars and galaxies that interact with electromagnetic radiation – is amazingly only about 4.9 per cent of the cosmos.

Some describe dark matter as the 'scaffolding' of the universe on which the 'visible' matter (such as atoms) collects, and suggest that this is how stars and galaxies may form. Webb, with its sensitivity and ability to probe ancient stars and galaxies, will help us to understand dark matter's role in the universe.

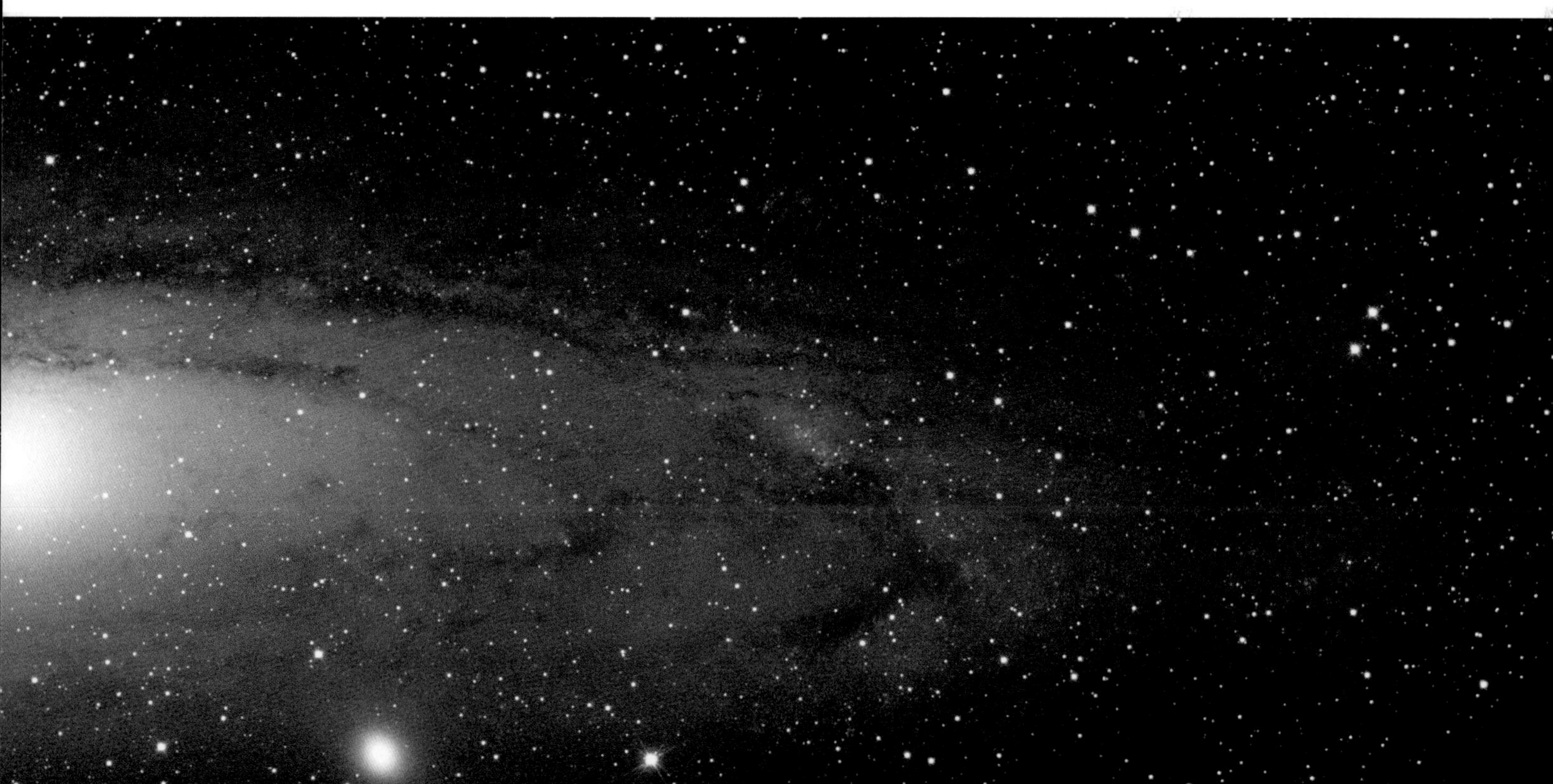

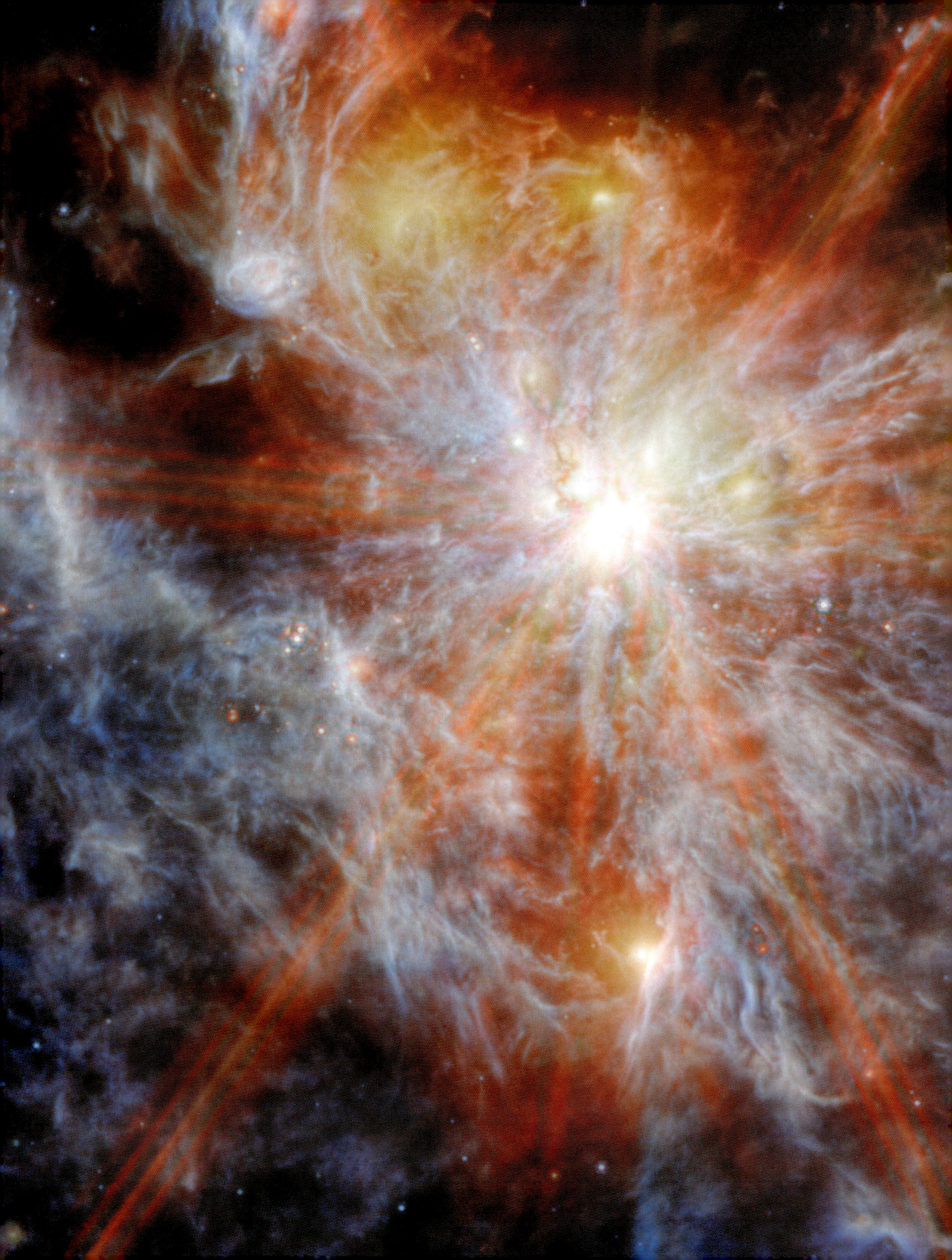

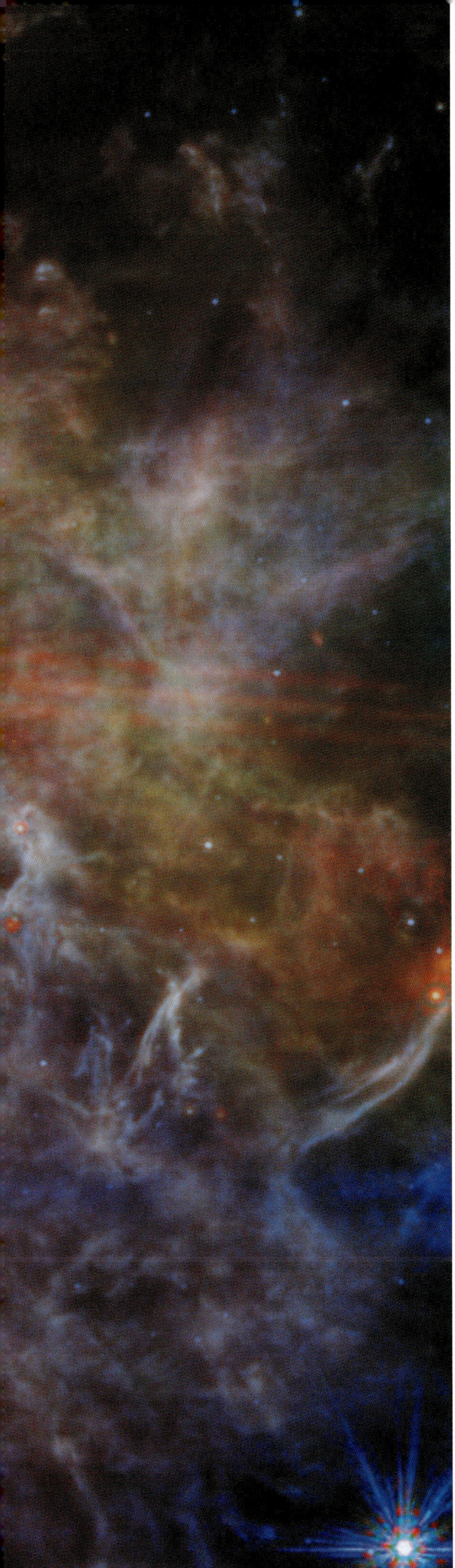

THE BIRTH OF STARS

Within giant swirling clouds of dust and gas called nebulae, new stars are being born right now. Occasionally these whorls will start to collapse in on themselves as the gravitational attraction acting on the matter within them causes the coalescence of the material. At the centre of this collapse, pressure builds and things begin to heat up. As more mass gathers at the centre, so the gravitational force increases and attracts even more matter. When a critical mass is reached a protostar is formed – and within this pre-star, temperatures and pressure reach levels where fusion is possible.

This body continues to suck gas and dust from the nebula around it. To better understand this process, we often run computer simulations. These show that these spinning knots of glowing matter can sometimes form not just one star, but in fact split into two or three stars in the same vicinity. This is reflected in what we see in the universe, where binary and ternary star systems are often observed. But there are still many outstanding questions about how this process works. We have been limited mainly because the visible light generated by these processes cannot escape the heart of these nebulae where new stars are being born – and so these areas have remained unseen by visible-light telescopes.

Webb's highly tuned stare will detect the infrared radiation that can escape from these clouds, picking up the signals that come from these energetic processes and their environment. With this information, astronomers can study the complete lifecycle of stars and observe how they evolve, interact and ultimately seed the next generation of stars and planets with the heavy elements they jettison back into space when they expire.

Such knowledge could also help us to understand how our own star and planet formed. Once upon a time, the sun was a protostar born in a nebula and drew matter into its centre. Astronomers look at other stars in the universe to better understand how our own sun and planets evolved. But although our solar system has lots in common with other stars and planet systems out there, our planet is, so far, the only example we have of a system able to support life. Our stellar evolution could point to the conditions that allowed life to burgeon on our comparatively tiny blue dot in space – and could therefore show us where in the universe we should look to find other planets capable of supporting living organisms.

N79, as seen here by MIRI, is a giant star-forming region in the Large Magellanic Cloud. The distinct starburst pattern is characteristic of Webb due to the design of its collecting mirrors, and indicates a very bright object.

THE ORIGINS OF LIFE

Earth is a special place. This tiny cosmic speck of dust is the only place in the vast expanse of space where life definitely exists. There are many interwoven reasons why we have life. Earth's location in the Goldilocks zone means it is the right distance from its local star, the sun, to have liquid water.

Our atmosphere also contributes to Earth's ability to have liquid water on its surface. Our tiny atmospheric blanket also gives us a protective shield against the solar radiation that would otherwise fry us. It also helps the regulation of the global temperature by allowing the transport of heated gases around the planet. Our planet is rocky, too – unlike the gas and ice giants of the outer solar system – which means that we have solid ground under our feet.

NASA's Jupiter probe, Juno, imaged the swirling jets of wind on the gas giant's surface.

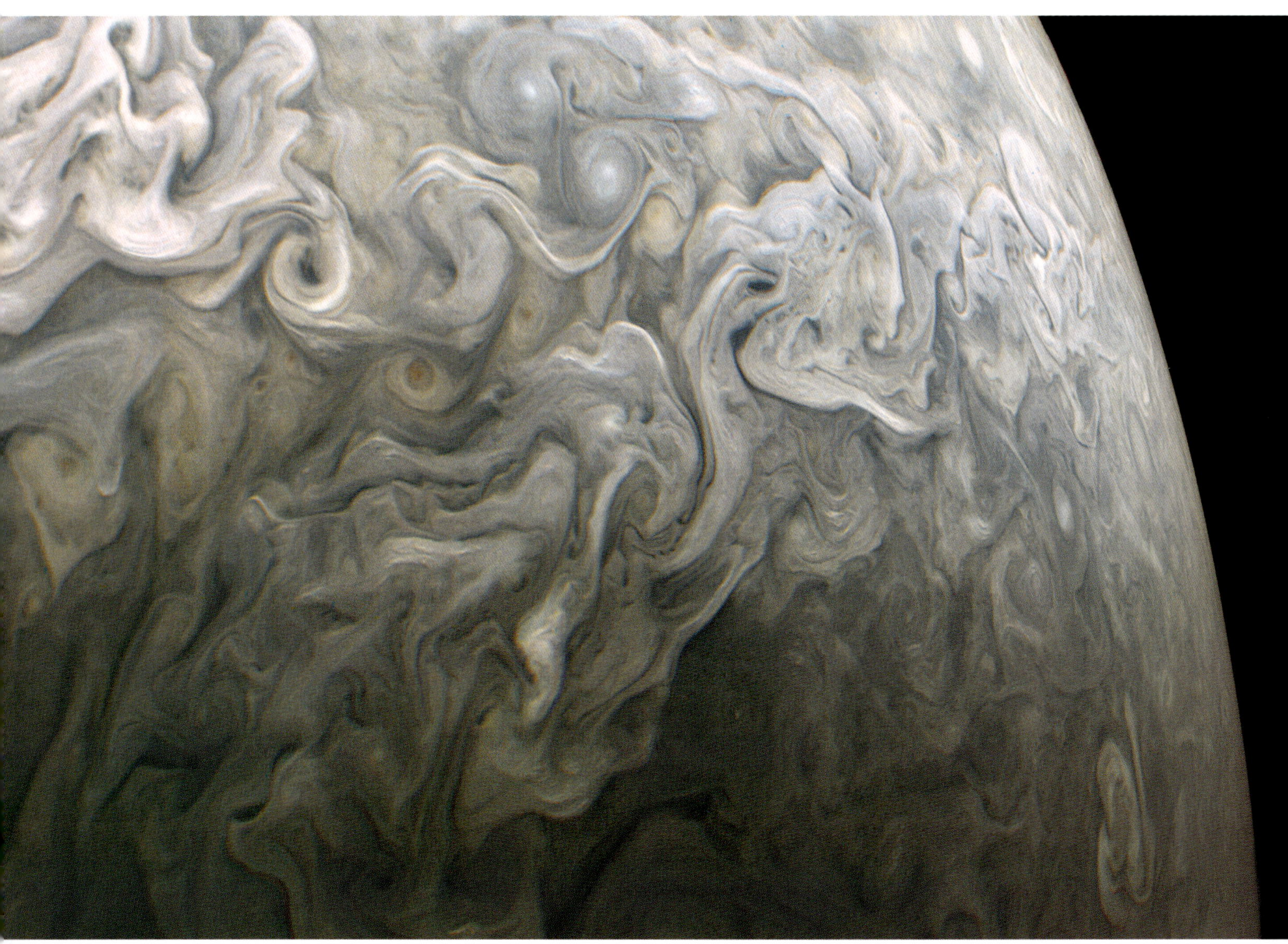

For most of our existence, we thought that Earth and the other planets in the solar system were unique. Then, in 1992, the first exoplanet was discovered. We have found many thousands more since – and there are certainly many others yet to be discovered. We can thank the planet-hunting Kepler space mission for much of our exoplanet knowledge (see page 87).

Multifunctioning Webb will also add to our catalogue of exoplanets using the transit method (see page 28). When an exoplanet's orbit brings it between us and the star, astronomers can infer information about the exoplanet, such as its size and period of orbit. In some cases, if we have the right orientation, we can analyse the small portion of the star's light that passes through the thin band of atmosphere that sits around an exoplanet. If we can do this, the use of spectroscopic techniques can then enable us to understand the chemical composition of the exoplanet's atmosphere. Webb also has a coronagraph on board, which enables it to observe exoplanets directly, even though they are situated next to bright stars. This is because coronagraphs block the glare from host stars, allowing astronomers to see the dim exoplanets orbiting them.

There are still some mysteries regarding planetary formation. As scientists, we expect that Webb's observations of known exoplanets will help us to better understand how these planets and planetary systems formed. One of the most prevalent types of exoplanet is a 'hot Jupiter'. In tight, scorching orbits, these worlds speed around their stars. From our perspective on Earth, a 'year' on such worlds can last only a few days, if not hours.

This begs the question as to whether the chilly, distant Jupiter of our solar system is unique, or do 'hot Jupiters' arise in the outer solar system and then travel inward? One theory is that young planets meet ripples of dusty debris, leftover material from early in their solar system's history, and that this dust then causes the planets to lose momentum, with the drag forcing them to migrate inward towards their sun. Could this be the long-term future for our Jupiter?

In the past, in our hunt for worlds that could possibly support life, we scientists have focused on stars with planets located in the habitable zones. But the atmospheres of exoplanets also play a very important role in temperature regulation. An extra thick blanket of atmosphere could allow habitable planets to exist outside of the habitable zone. Webb will be exploring the cosmos looking for these indicators of life and the places in which we think it could possibly flourish. But it's also important to remember that we are basing this search on 'life as we know it' on Earth. We're

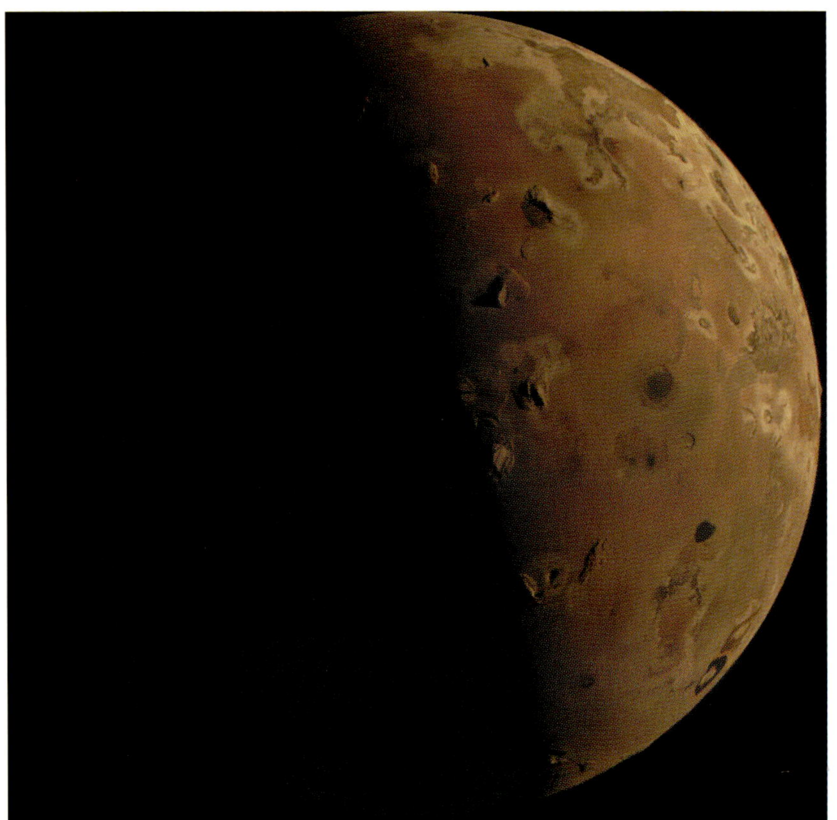

looking for what we already know: the hallmarks of carbon-based life. It may be that life can exist in other conditions that we're currently unfamiliar with. There is a whole area of research called astrobiology that envisages what different forms of life could have possibly formed on the many varied exoplanets that we see.

And while we are interested in discovering habitable planets and life in the wide reaches of space, we are also very eager to unpack the secrets of planets closer to home. Webb will be able to search for the molecules associated with life in our solar system and to support other science missions currently underway – such as the rovers on Mars.

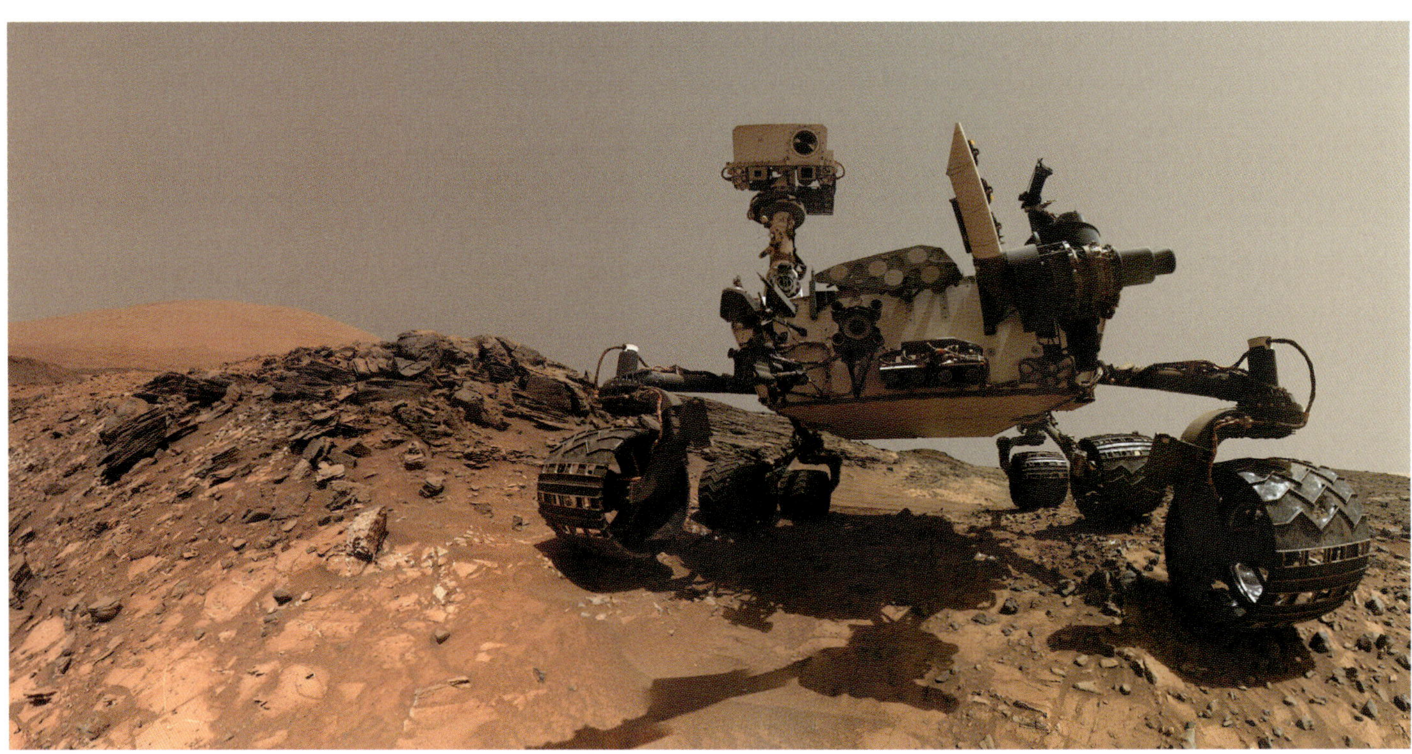

ABOVE: An image of Jupiter's volcanic moon, Io, taken by NASA's Juno probe.

BELOW: NASA's Curiosity rover is exploring the Martian surface. Here is its famous selfie. The rover is looking for evidence of life on the Red Planet's surface.

WEBB'S INSTRUMENTS

Webb has four custom-built science instruments on board. The rest of the telescope – the light gathering elements, the spacecraft bus and the sunshields – works to provide these instruments with high-quality radiation to undertake science observations that would not be possible on any other observatory.

NEAR-INFRARED CAMERA (NIRCAM)

NIRCam is Webb's primary imager and detects near-infrared radiation. It has three components: a camera, a spectrograph and a coronagraph. It is best suited to exoplanet hunting and can detect light from the early universe and stars in nearby galaxies as well as from our own.

NEAR-INFRARED SPECTROGRAPH (NIRSPEC)

NIRSpec takes the observed infrared radiation gathered by the telescope and spreads that radiation out into a spectrum. This spectrum is like a fingerprint. Chemical reactions happening in stars or other celestial bodies absorb radiation at very narrow and specific wavelengths. These absorptions appear as little black lines on a spectra. By analysing the spectra, it is possible to see what chemical reactions are happening within the celestial body. It's a bit like a remote chemical analysis that can indicate the object's physical composition.

A diagram of Webb's instruments and their capabilities.

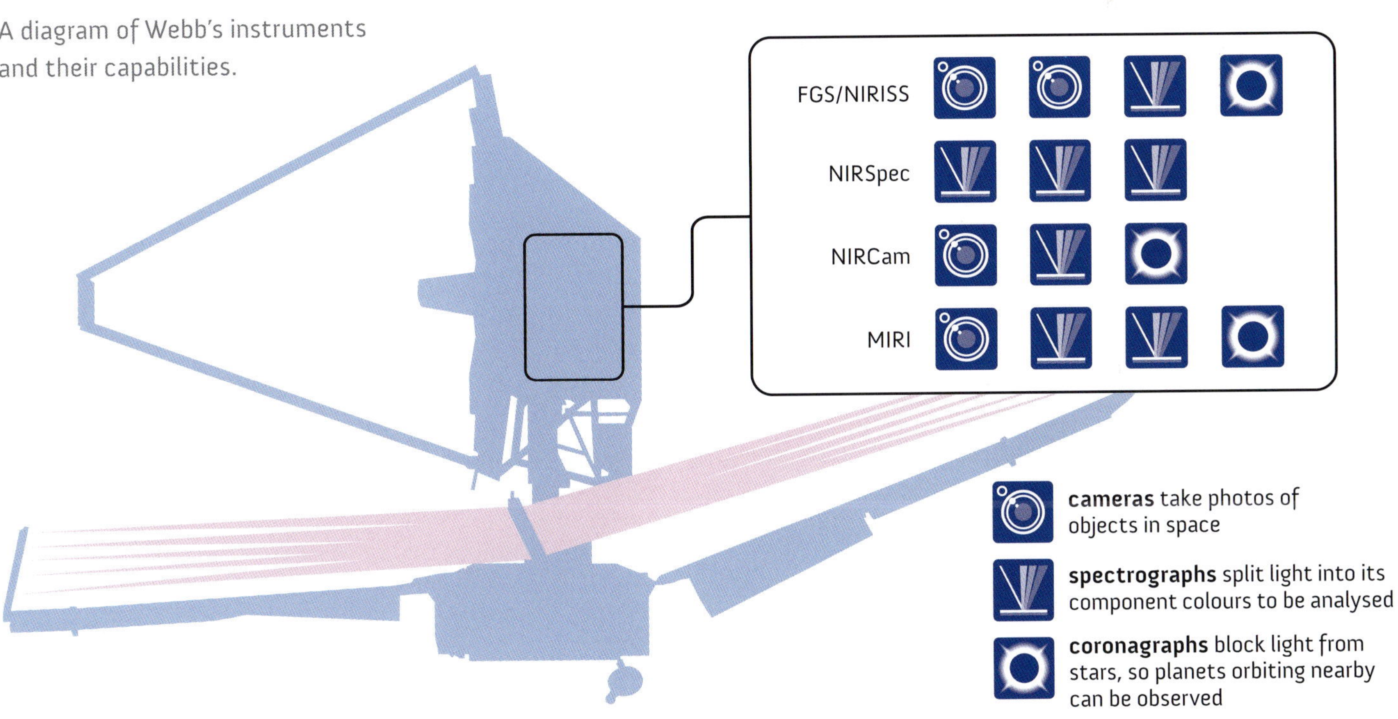

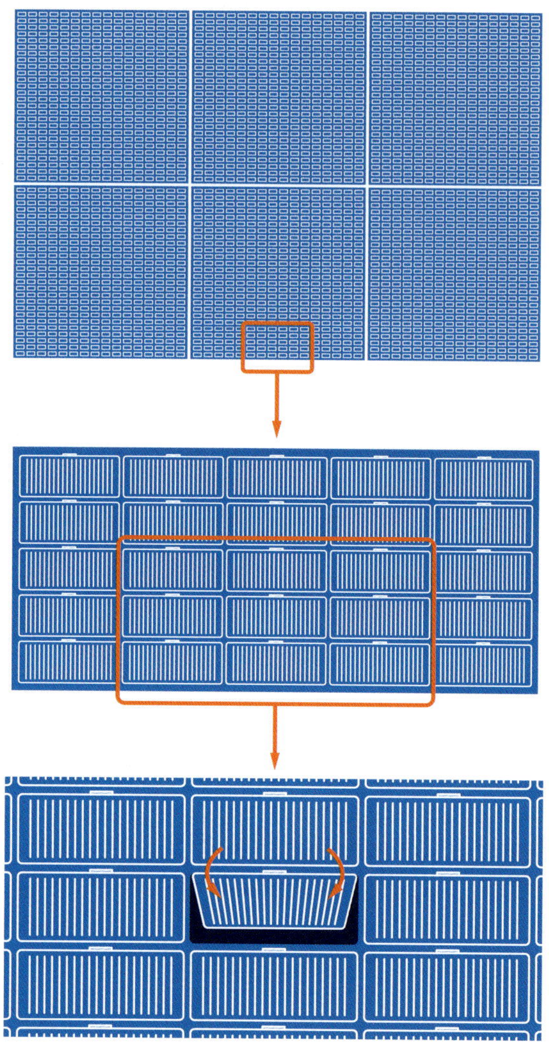

ABOVE: A diagram of NIRSpec's microshutter array, with more than 250,000 tiny flaps that open and close to focus on specific objects in the sky.

As well as a chemical analysis, spectroscopy can give an indication of the movement of the celestial body due to redshift and blueshift (see page 46). NIRSpec is designed to observe up to 100 objects simultaneously. This has been made possible through an engineering wonder called the microshutter array: an arrangement of microshutter cells, which are microscopic flaps that cover tiny windows. These flaps can be opened to let radiation pass through the windows, or kept closed if nothing of interest is falling on the flap. Each flap is just 100 x 200 microns wide. The array is made up of 250,000 of these flaps, which are individually controlled by the application of a magnetic field, and each set of 250,000 is split into four quadrants, each quadrant being the size of a postage stamp. This means that when NIRSpec is observing a field of stars, for instance, we can open and close the appropriate flaps on demand to analyse specific stars and ignore others.

Another important part of NIRSpec is the Integral Field Unit (IFU). This was the part that my team and I worked on. An IFU is a device used in astronomy spectroscopy to obtain detailed spectra of a large extended object, such as a galaxy nebula or Saturn's moon, Titan. In standard spectrographs, spectra are gathered from a single point in the sky, like a star or a planet. An IFU takes a larger extended celestial object and divides it up into a series of smaller areas called spaxels. Spectra can then be obtained from each of these spaxels. This enables us to investigate the spectra at the heart of a galaxy, for example, and to compare this with what is happening at the edge of the galaxy.

MID-INFRARED INSTRUMENT (MIRI)

As the name suggests, MIRI observes mid-infrared frequencies. It has a camera and a spectrograph, with sensitive detectors that enable it to observe redshifted light. This type of light is very important in understanding the early universe, because much of that young visible light has been stretched into longer wavelengths due to the expansion of the universe by the time it reaches Earth. MIRI also has a widefield camera that builds on the astrophotography capabilities of Hubble.

FINE GUIDANCE SENSOR/NEAR-INFRARED IMAGER AND SLITLESS SPECTROGRAPH (FGS/NIRISS)

The FGS/NIRISS comprises two components in one instrument. The FGS allows Webb to point accurately at the celestial object that is to be analysed, while the NIRISS has several different modes and can investigate different near-infrared wavelengths.

RIGHT: The NGC 3256 galaxy, imaged here using MIRI and NIRCam data, is the remnant of a cosmic collision between two galaxies.

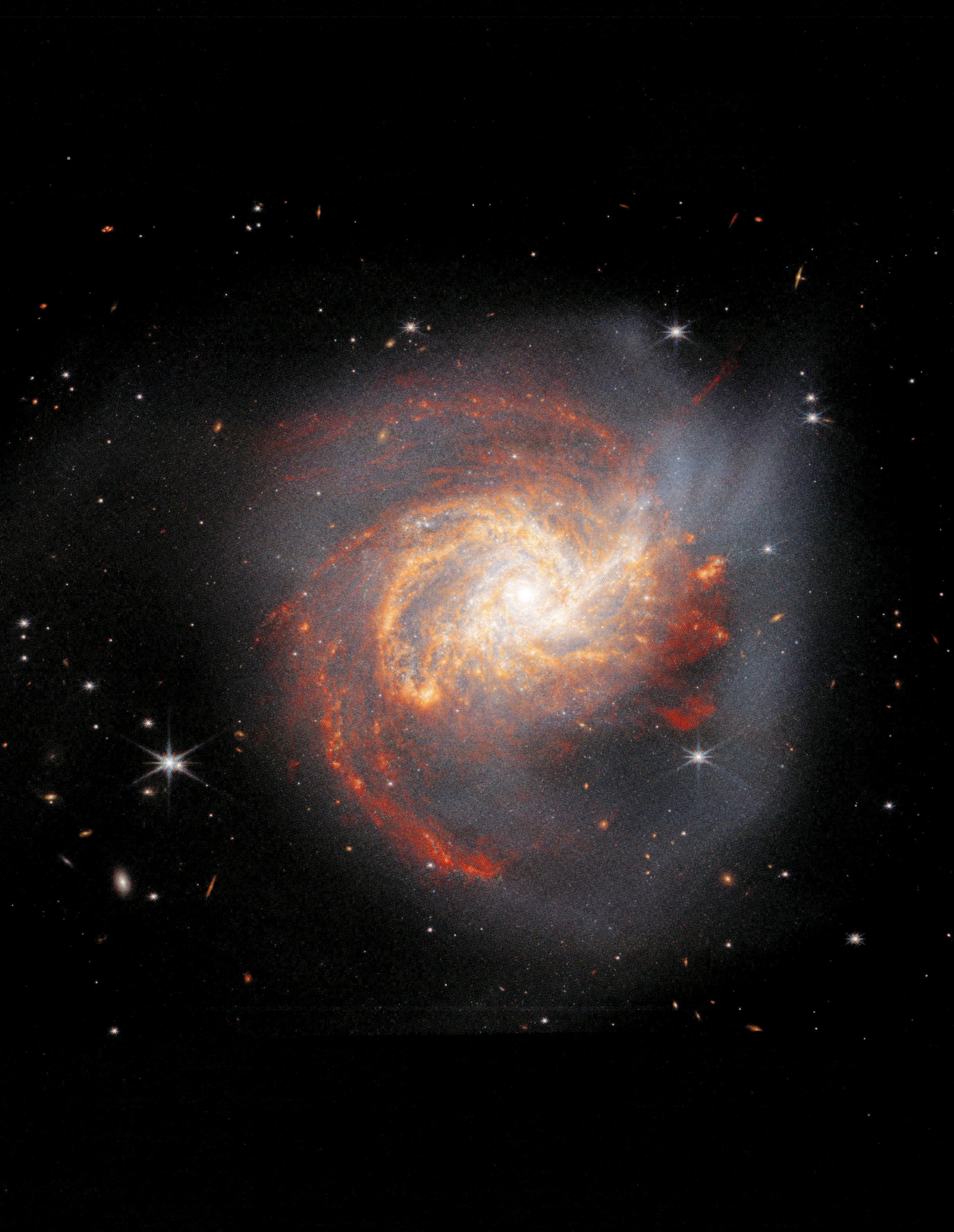

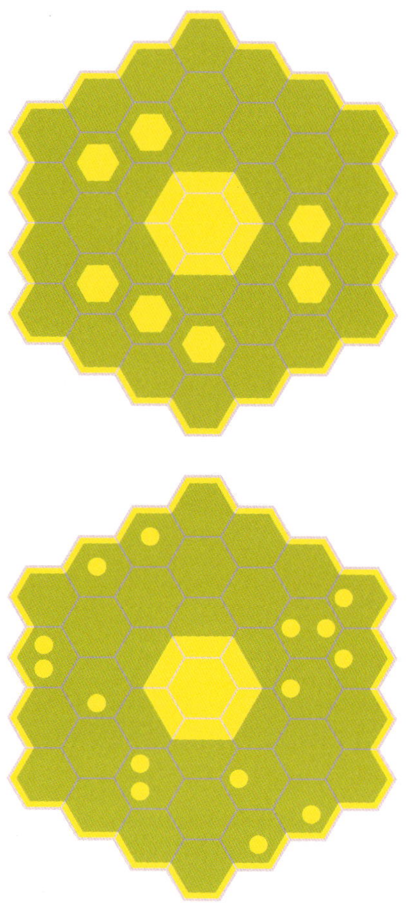

Aperture masks block out some of the light to turn the telescope into a mini-interferometer.

The NIRISS is particularly good at first-light detection (picking up light from the first stars and galaxies) and finding and analysing exoplanets. It has two spectroscopy modes. With its Wide-Field Slitless Spectroscopy, NIRISS can capture the overall spectrum of what it is looking at, whether that's a galaxy or a field of stars, while its Single-Object Slitless Spectroscopy enables it to capture the spectrum of a single object in the sky.

Its Aperture Mask Interferometry (AMI) uses a 'mask' to block out some of the light gathered by the telescope mirror that reaches the detector. This seems like an odd thing to do: as astronomers we are all about the light, usually going for bigger telescopes every time. However, by using this mask the instrument is able to do interferometry. Interferometry is a research tool that is utilized in many disciplines of science and engineering. Interferometers work by combining two or more light beams to form an interference pattern that can be measured and analysed: hence the name 'interfere-meter' or interferometer.

With its AMI mode, the different windows in the mask create multiple beams. When these different beams are combined, this results in an interference pattern made up of light and dark patches. By analysing these patches, we get better resolution despite getting less light. Resolution is the ability to see the detail in the objects observed. For instance, if two objects appear very close together in space, using this technique enables us to see them as two distinct objects rather than one large blurry one. NIRISS uses this interferometric effect to capture more detailed images than would otherwise be possible.

PAINTING THE SKY

The images that Webb is beaming back to us from its far-out position in space appear to be an artist's palette of red plumes and cyan wisps across the velvety black of space. But Webb does not have cameras that detect visible light; instead it picks up radiation in the near- and mid-infrared range – so, of course, its cameras cannot take a photo of space that we can see with our eyes. The final images presented to the public are not what the onboard cameras and instruments of this infrared mission are seeing but rather are the result of hours of painstaking work.

Webb has two cameras on board: NIRCam, which captures shorter wavelengths of infrared light, and MIRI, which observes longer ones. Images and data taken by Webb are digitized and beamed back to Earth via radio waves. If these images were processed as

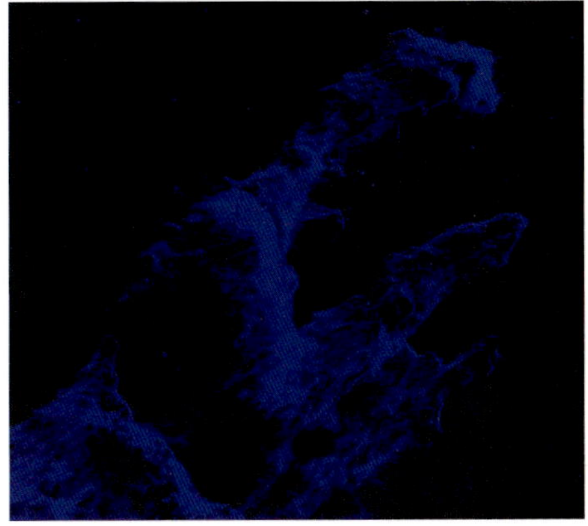

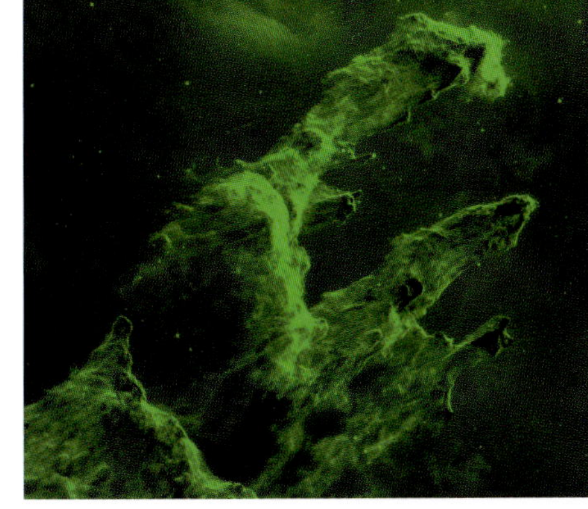

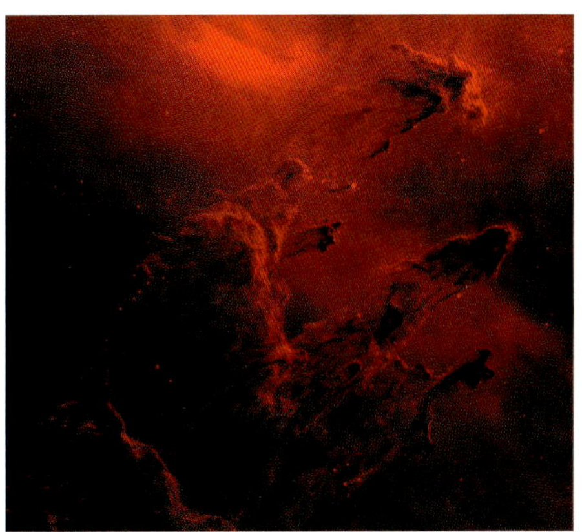

Scientists assign colours to different wavelengths. Here are the MIRI colour assignments of the Pillars of Creation in the Eagle Nebula.

infrared images, they would appear to be completely black to our human eyes because we cannot detect the infrared light they are showing. But there is a lot of information contained in the images, with each pixel potentially containing one of more than 65,000 different shades of infrared radiation.

For us to get the glorious images we can see, an imaging specialist will take the different shades of infrared light and map them onto wavelengths of visible light. The shortest infrared wavelengths are mapped to a range of visible blue wavelengths, longer infrared wavelengths to visible green wavelengths, and the longest infrared wavelengths to visible red wavelengths. This way, invisible infrared images are shifted into the visible part of the spectrum, where we can see and appreciate them.

But the process is more complex than simply assigning colours to wavelengths. Each Webb image has been pored over by professionals, tweaked and edited to be pleasing to the eye while still being scientifically accurate. The imaging specialists also consult with designers and with the scientists who made the observations to ensure that their work contains the essence of the scientific data.

Sometimes these images are useful for astronomers because they can highlight different features of a giant structure, such as a nebula, but for the most part we work with the raw data – extracting insights from a series of numbers on a screen. Many of the cameras have filters to target specific elements or molecules. For example, NIRCam has twenty-nine filters in total, some of which can detect molecules such as water and methane. MIRI, on the other hand, has nine filters. Webb images often contain combinations of these filters, and this data gives astronomers even greater insight into the mechanisms happening way out there beyond our planet.

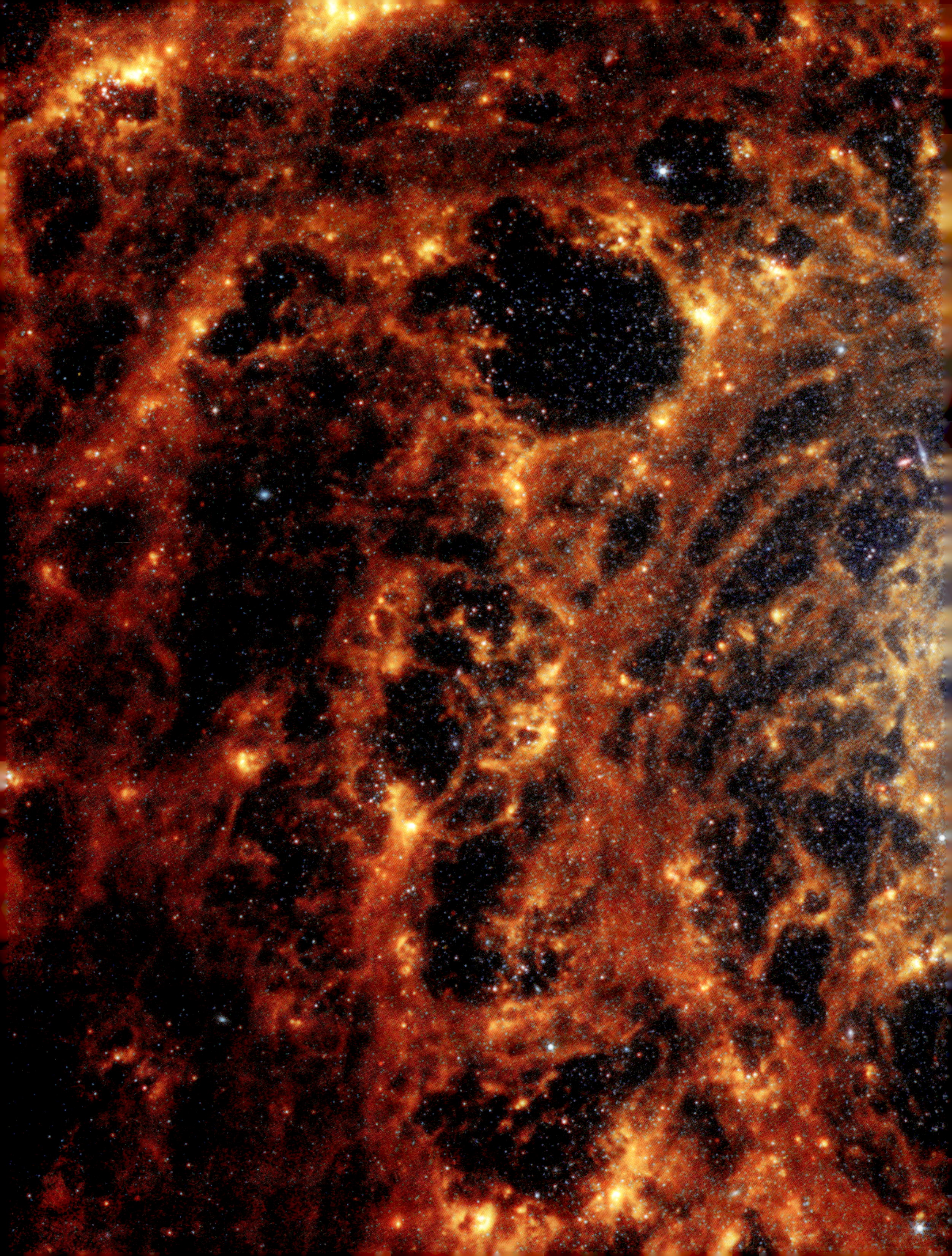

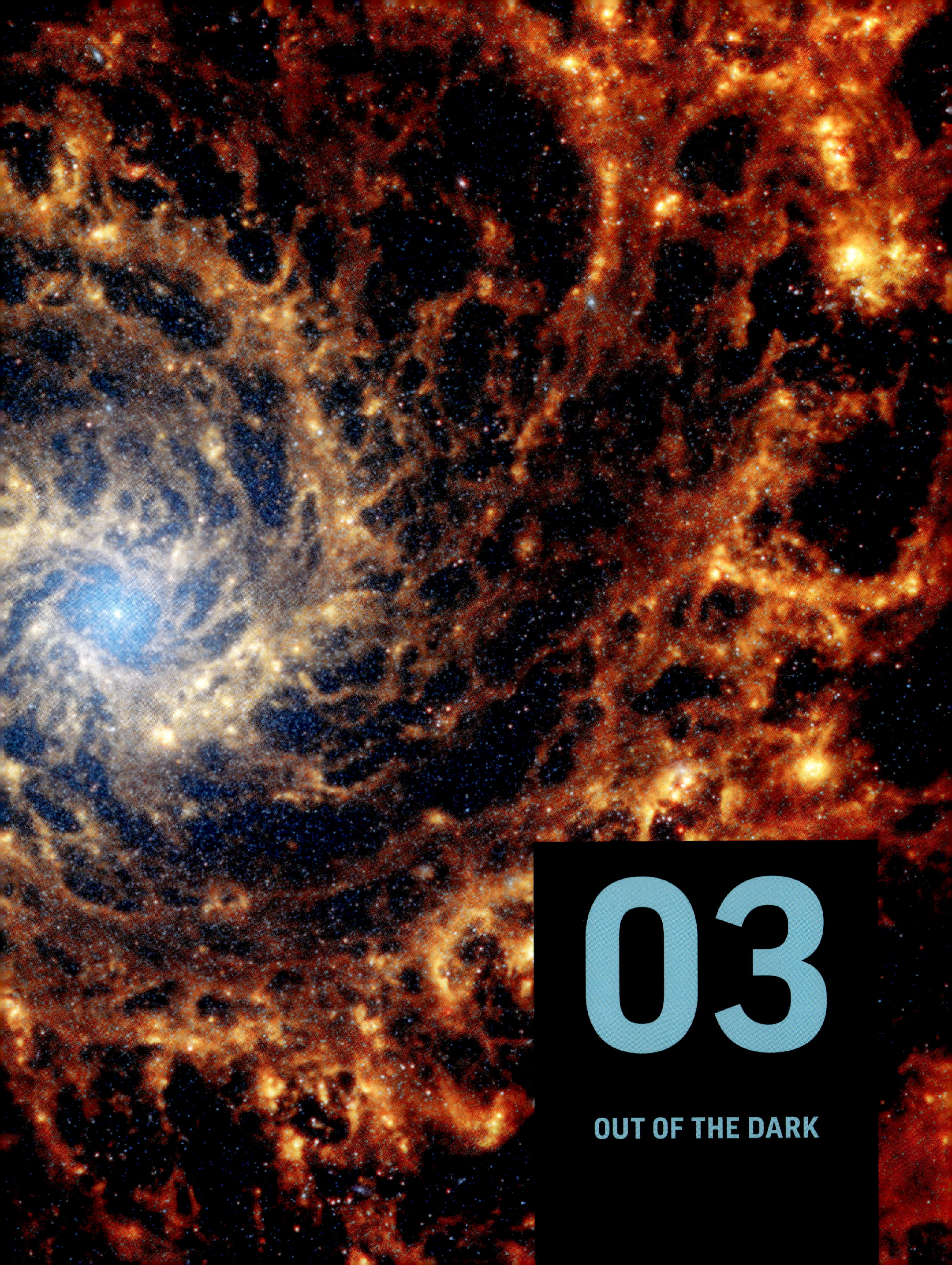

03

OUT OF THE DARK

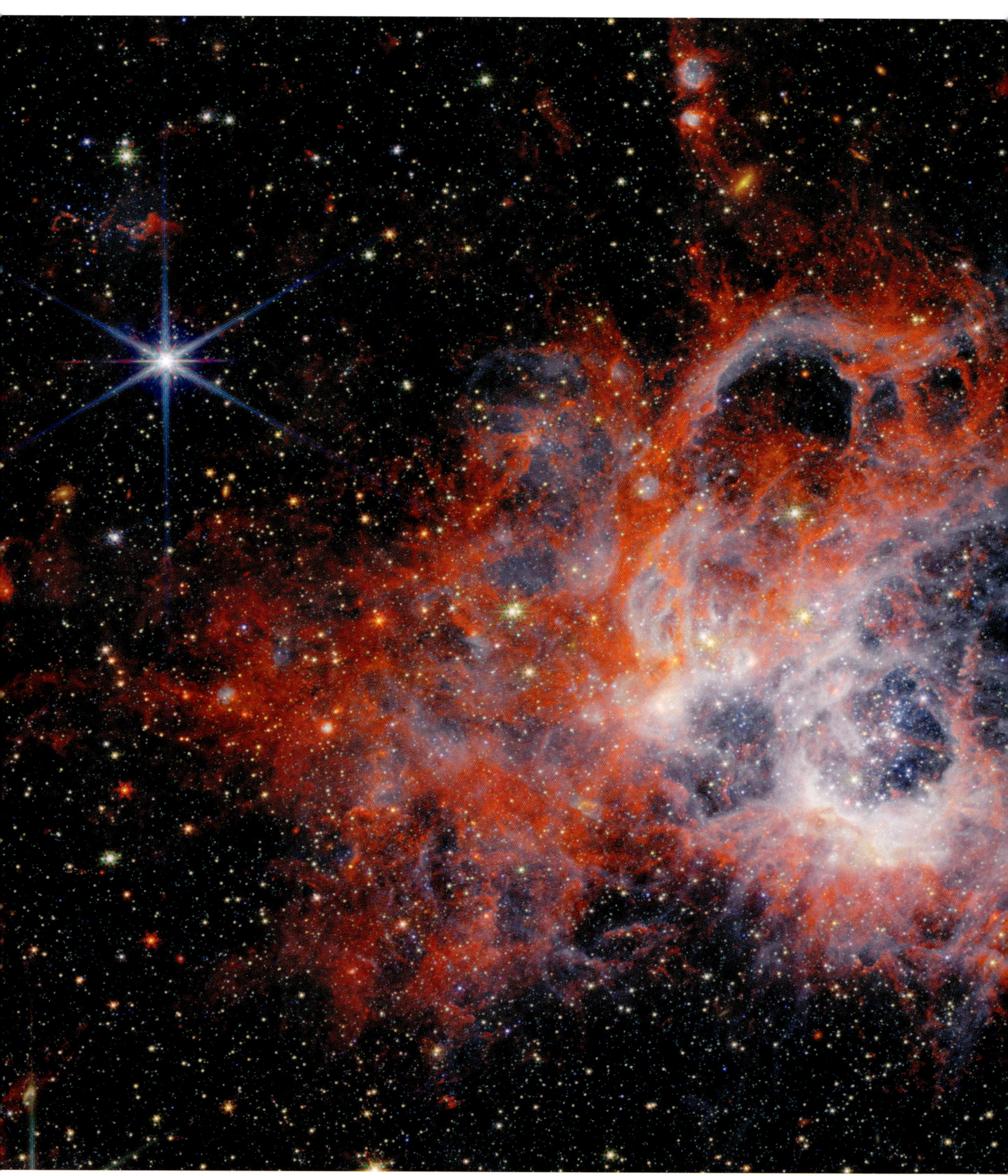

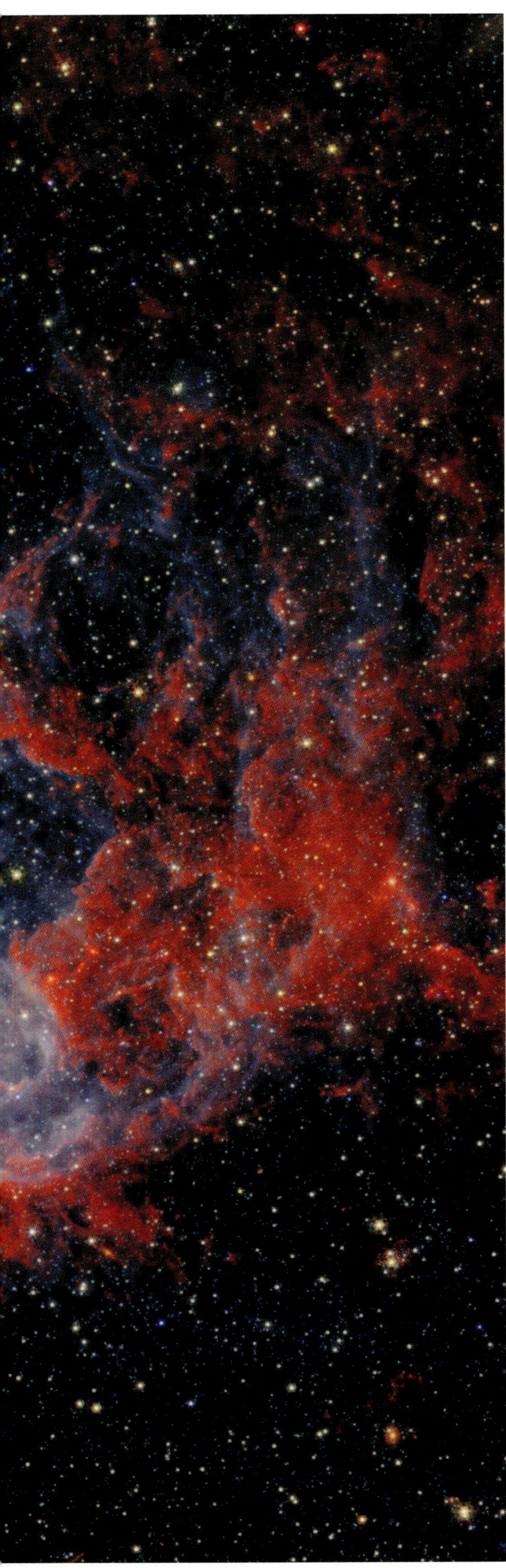

WEBB IS WELL KNOWN FOR its remarkable images – sumptuous cinematic views of the universe, revealing the secrets of previously obscure corners of the cosmos.

But the signals that Webb recognizes are not visible to our eyes. It does not detect the blue and green light we can see, nor the long, languid ripples of radio waves, nor even the energetic frissons of X-rays or gamma rays. Its giant mirror reflects infrared signals (see page 32) towards its super-sensitive instruments, and this sensitive gaze gives us an unprecedented view of the universe.

The sheer amount of data it collects each day is mind-boggling: around 57 gigabytes, which is the equivalent of about thirty full-length movies. To put this into perspective, Hubble gathers only about 1.5 gigabytes, which is about three DVDs per day.

Webb transmits its data via the Deep Space Network, which is an array of radio antennas dotted around the globe. There are antennas in Goldstone, California; Canberra, Australia; and Madrid, Spain. With these three antennas, we get full coverage of all space beyond our planet – they literally span the universe around us. That means that scientists can communicate with Webb throughout the day, with signals being transmitted or received by whichever antenna is pointing towards Webb.

The Deep Space Network then transmits the data from the antennas to the Space Telescope Science Institute in Baltimore in the United States, where it is processed and shared with the scientific community, and ultimately archived.

The images we get to see are the end result of hours of Webb observations, in addition to the many hours invested by various experts – from data scientists to astronomers to imaging specialists.

More than 200 fiery, giant stars are burning in NGC 604, a star-forming region in the Triangulum galaxy. Composed using NIRCam data, the region is about 2.73 million light-years away.

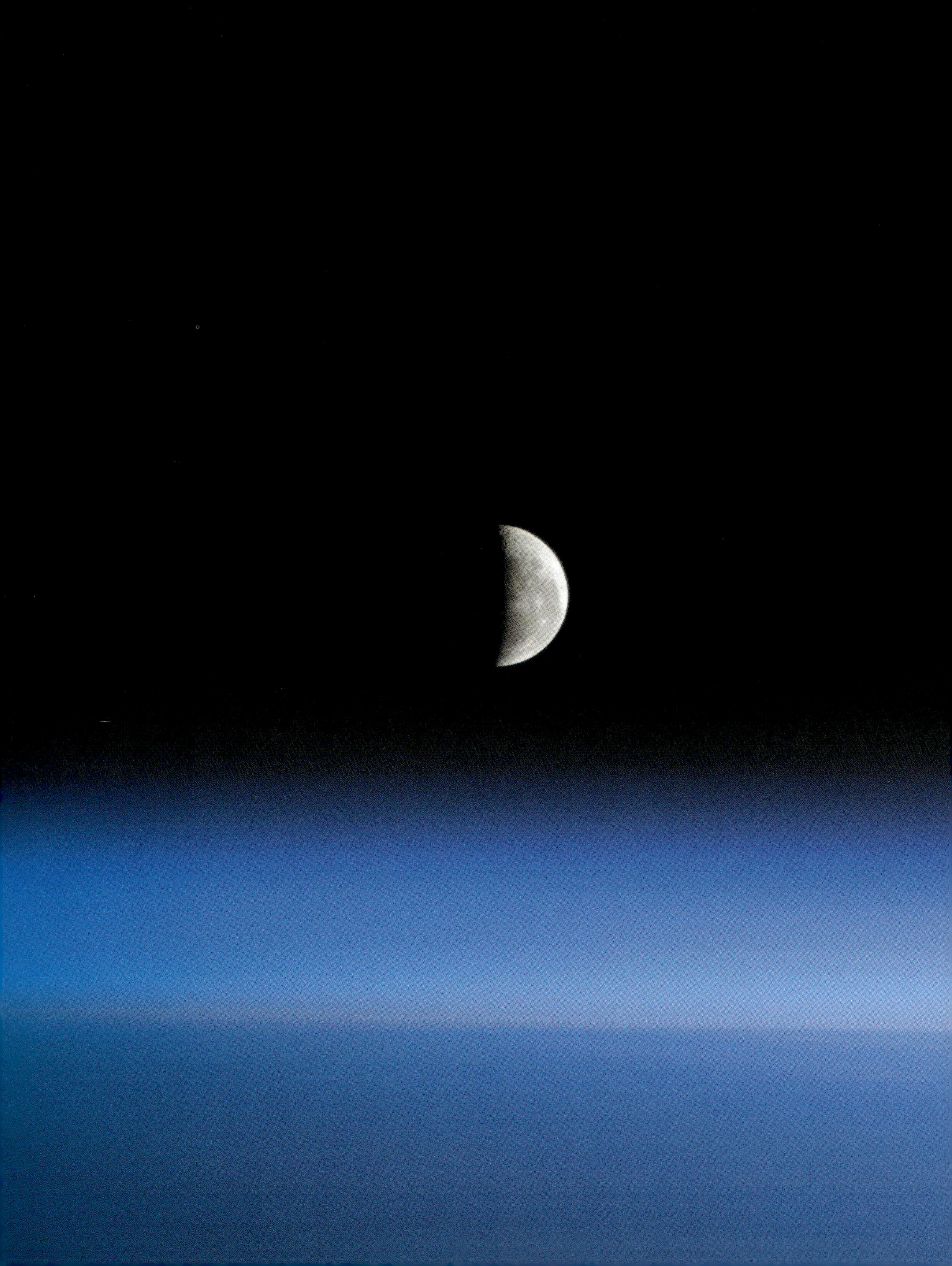

CLOSE TO HOME: OUR SOLAR SYSTEM AND BEYOND

The term 'cosmogony' is used to describe the various models that have helped us to make sense of the cosmos and how it came to be. Growing up, I only remember hearing about how the early Greek and Roman cultures tried to define the origins of the universe through their myths and legends. Yet in virtually every culture across the world, people have wondered about the same big questions: where did we come from? Why are we here?

The idea that the Earth revolves around the sun is usually credited to sixteenth-century Polish mathematician and astronomer Copernicus. However, the little-known ancient Greek astronomer Aristarchus of Samos had proposed the idea of the heliocentric universe (the sun-centred universe) hundreds of years before Copernicus. He imagined the moon travelling in a path around the Earth, which in turn revolved around the sun. Since then, we have become much more certain about the goings-on in our cosmic neighbourhood – although there is still much to be understood.

A photo of Earth and the moon from the International Space Station as it soared above the Southern Ocean.

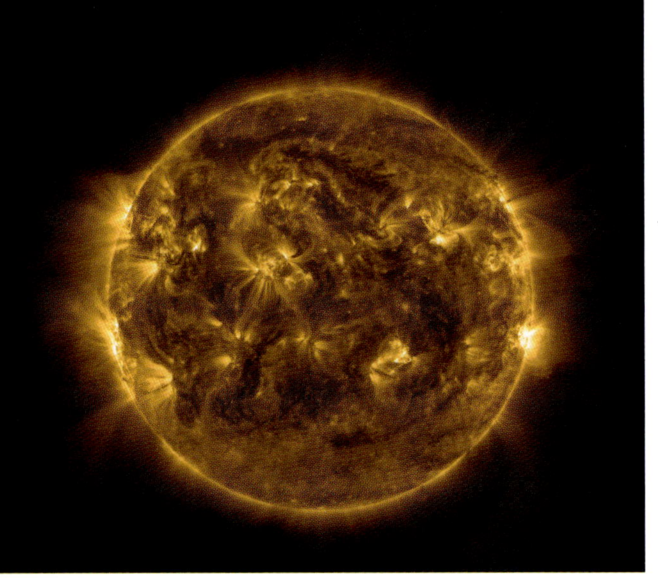

ABOVE: Our sun is constantly undergoing nuclear fusion, producing large quantities of energy.

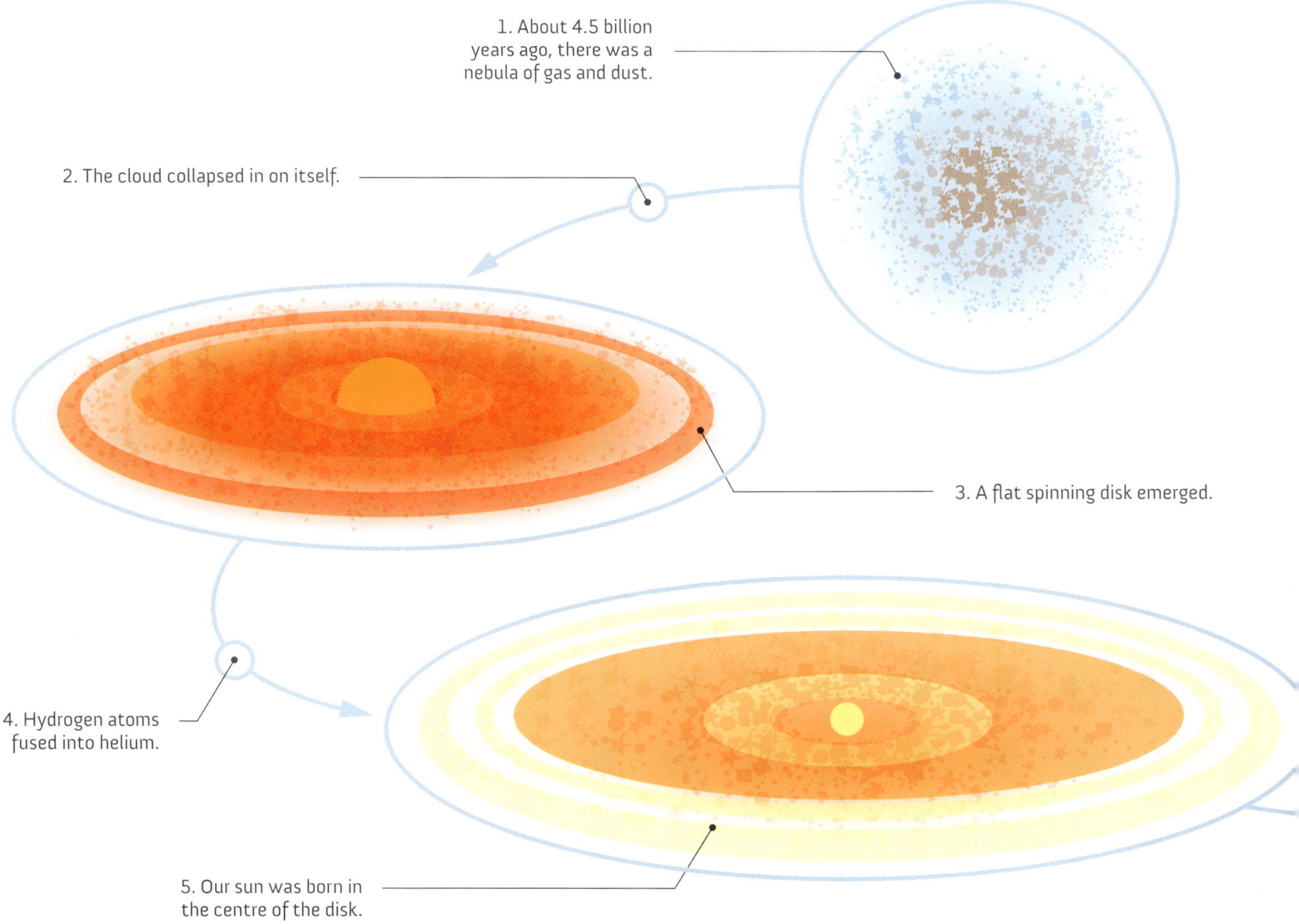

HOW OUR SOLAR SYSTEM FORMED

About 4.5 billion years ago, a nebula was disturbed, allowing material within it to start to clump together. This event marked the beginnings of our solar system. The clumping started with small particles that went on to gather into increasingly large clumps. Friction between the colliding objects caused the clumps to lose angular momentum and migrate towards the centre, forming a spinning disk known as an accretion disk. At the centre of this disk, material coalesced and became denser. Then, when enough material collected at the centre, a protostar formed.

As more material gathered, and when a critical mass was reached, intense temperatures and pressure occurred within the protostar's core. When this happened, a reaction called fusion took place.

1. About 4.5 billion years ago, there was a nebula of gas and dust.

2. The cloud collapsed in on itself.

3. A flat spinning disk emerged.

4. Hydrogen atoms fused into helium.

5. Our sun was born in the centre of the disk.

Fusion sees atoms such as hydrogen fuse together to make new elements such as helium. This fusion process, producing new elements and releasing huge amounts of energy, happens at the heart of every star and causes them to shine brightly, releasing the electromagnetic radiation that we can detect with our eyes and with telescopes.

The remaining matter in the disk that did not make the star also clumped together, forming planets, moons, asteroids and other smaller bodies. But that was not the end of the story: these objects collided with each other, broke apart and continued to jostle together for billions of years. The story of our planet – and that of the others that orbit our sun – is written in their chemical make-up, surfaces and atmospheres. With Webb, we are reading more of that story.

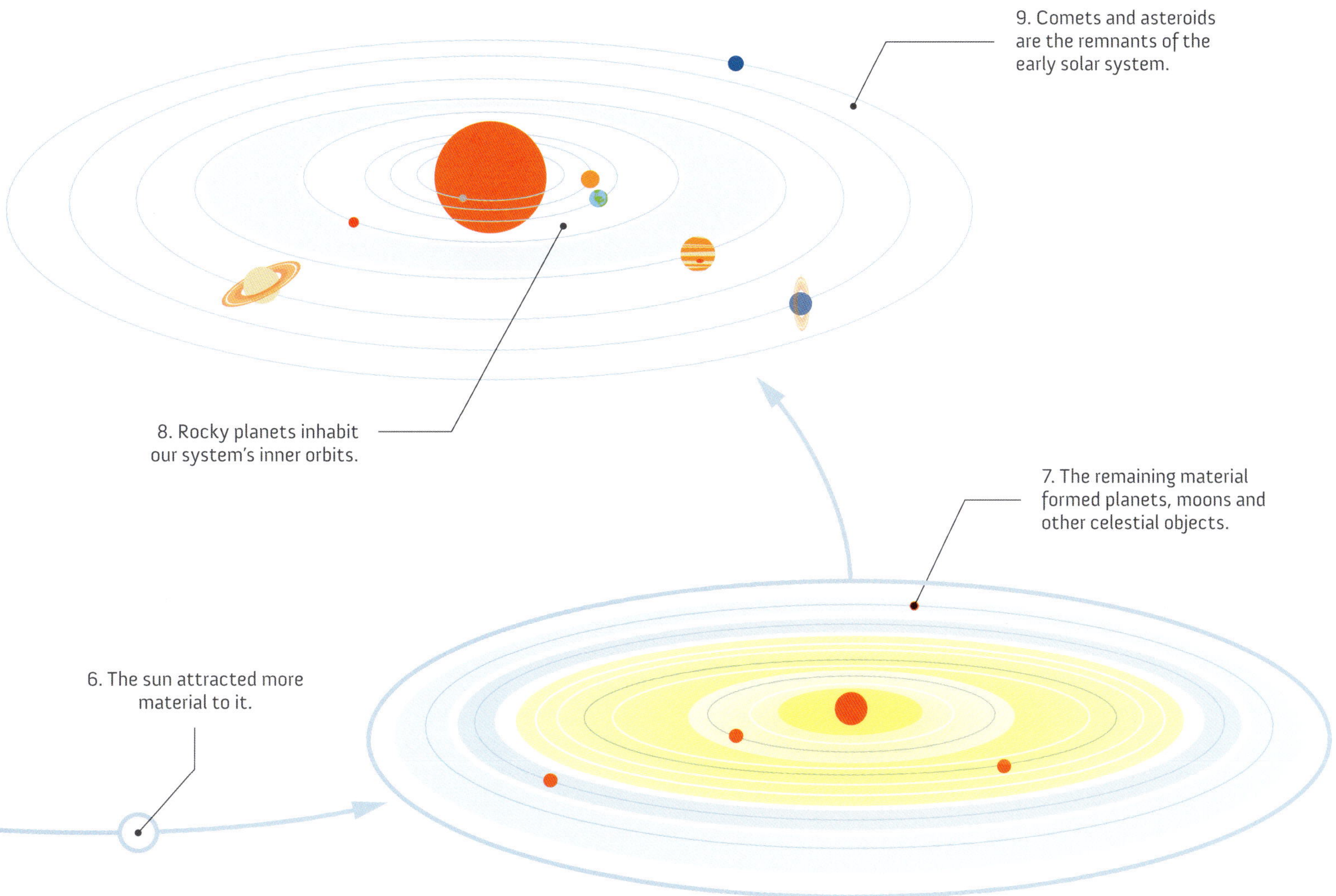

JUPITER

Distance from Earth: 588 million to 968 million kilometres (365 million to 600 million miles)
Observing: can be observed in all wavelengths

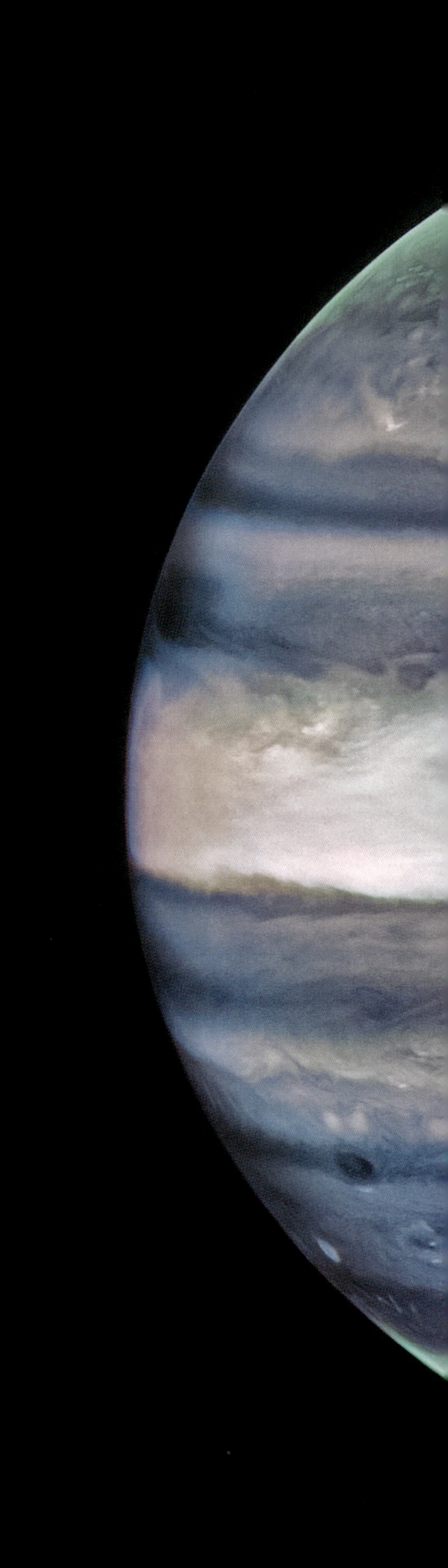

Powerful storms swirl on Jupiter's surface, while auroras glow at its poles. This image is a composite of several Webb images.

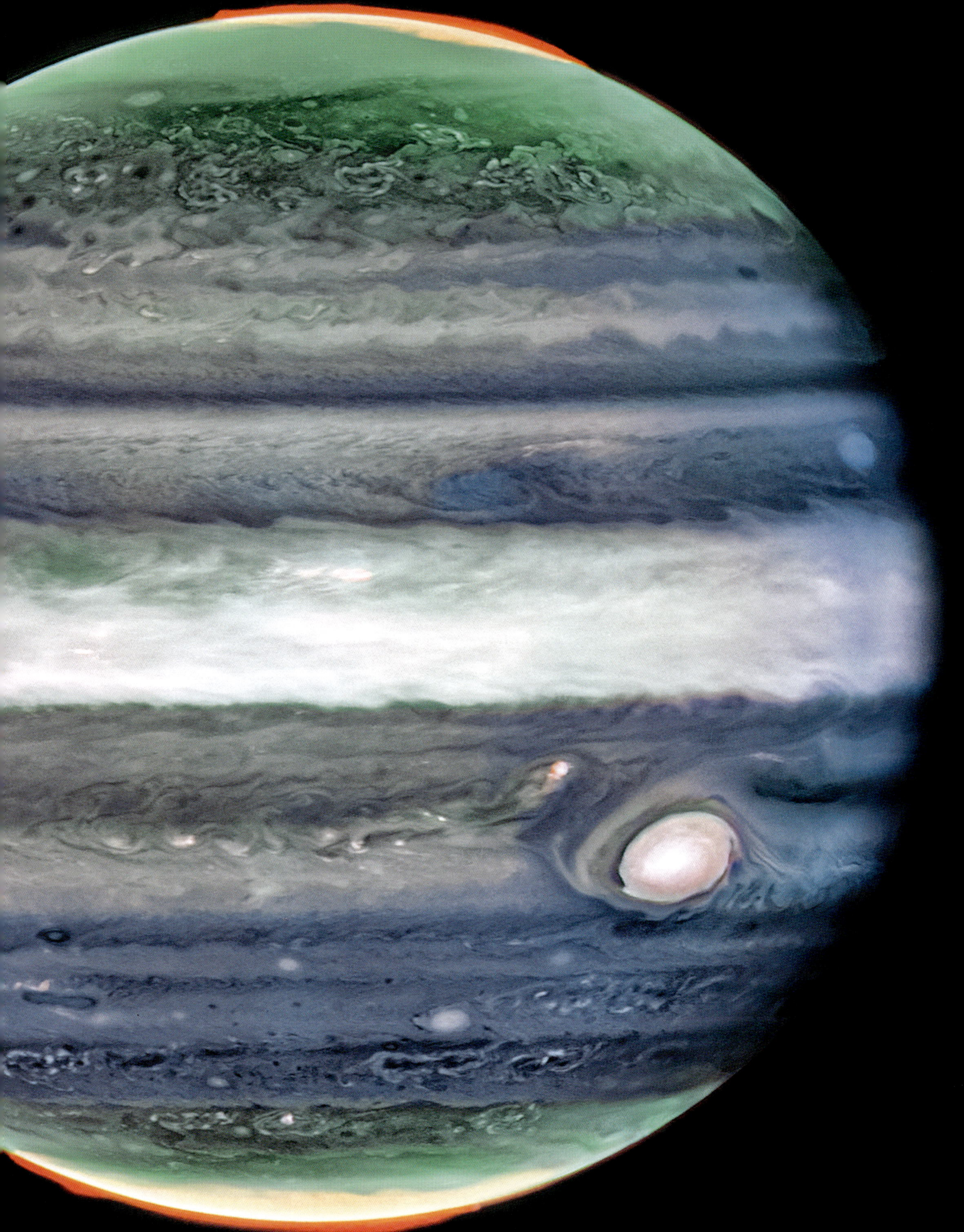

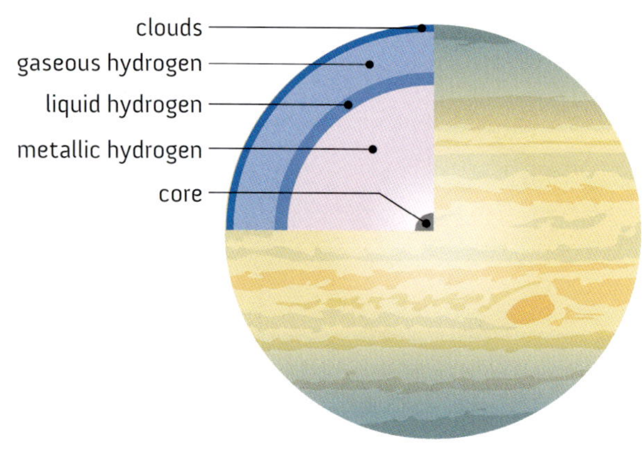

A diagram of Jupiter's layers.

JUPITER IS THE FIFTH PLANET from the sun and the largest planet by a significant margin. For some context: about 1,300 Earths could fit into Jupiter and the planet is around 320 times the mass of Earth. Jupiter is named after the Roman god of the sky and thunder, who was also known as Jove.

Unlike Earth, which has a rocky core, Jupiter is a gas giant. It's thought to have an outer layer of liquid metallic hydrogen, and so has a powerful magnetosphere. Earth also has a magnetosphere, which is the magnetic field that encases our planet, but Jupiter's is an order of magnitude stronger. This magnetic field means that the giant planet boasts magnificent auroras at its poles, which are showcased in red in the main image. They extend many kilometres into its atmosphere, which also reflects their radiation back to Jupiter's surface.

PLANET OF STORMS

Jupiter is a planet of storms, the most famous of which is the Great Red Spot. In the image below, we can see it on the right, below the equator. This giant storm was first observed in 1831 but has probably been raging a lot longer than that. In visible wavelengths, it appears reddish (although no one is sure why). In Webb's image, it is a swirling white knot.

NIRCam images of Jupiter's equatorial jet stream, which spans more than 4,800 kilometres (3,000 miles) wide.

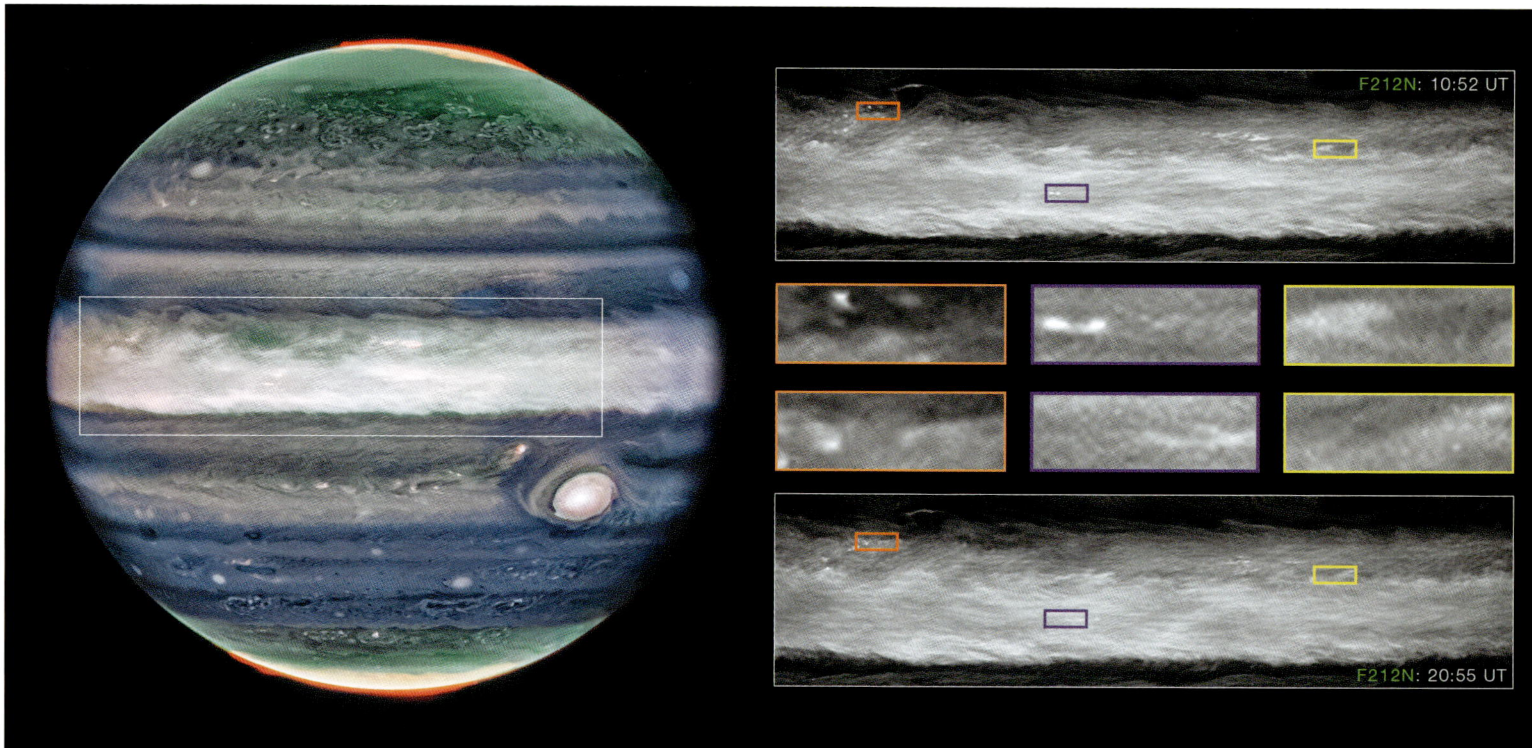

The planet's atmosphere is multilayered, like ours, with different wind speeds at various levels. Using this image, a high-speed jet stream was discovered around Jupiter's equator. It spans about 4,800 kilometres (3,000 miles) and is about twice as fast as the most severe hurricane on Earth.

The bright white spots and stripes in the image accentuate very high-altitude clouds and maybe storms swirling above the surface of the huge planet. One of the ways that we can understand the dynamic and turbulent atmosphere of a planet like Jupiter is by viewing it in several different wavebands. By combining infrared data from Webb with visible data from ground-based telescopes like Gemini and optical space telescopes such as Hubble, we can glean valuable information that is more than the sum of the different data sets.

MANY MOONS OF JUPITER

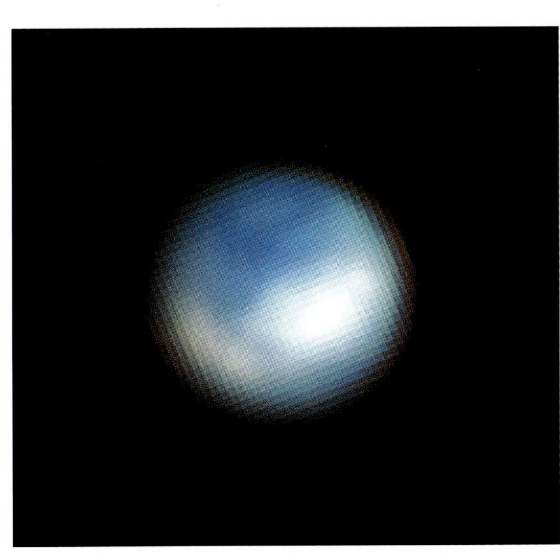

Scientists suspect that Jovian moon Europa (imaged here with NIRCam) has an ocean of water beneath its surface.

Aside from its gargantuan size, Jupiter is best known for its many moons – some of which, such as Europa, offer the tantalizing possibility of water. It also has very faint rings, which are difficult to see. Webb, using the supersensitive NIRCam instrument, has been able to image its rings.

Jupiter has ninety-five moons that have been officially recognized by the International Astronomical Union (IAU). Some of these are small, most likely the remnants of collisions between larger moons, while others are as big as planets. Jupiter's icy moon, Ganymede, is the largest moon in our solar system. It's bigger than the planet Mercury and is the only known moon to have its own magnetic field – a characteristic usually confined to planets like Earth. This magnetic field means that this moon also has auroras. It is thought that Ganymede has an underground saltwater ocean that contains more water than all the water on the surface of our own little blue-green planet.

SATURN

Distance from Earth: about 1.4 billion kilometres (870 million miles)

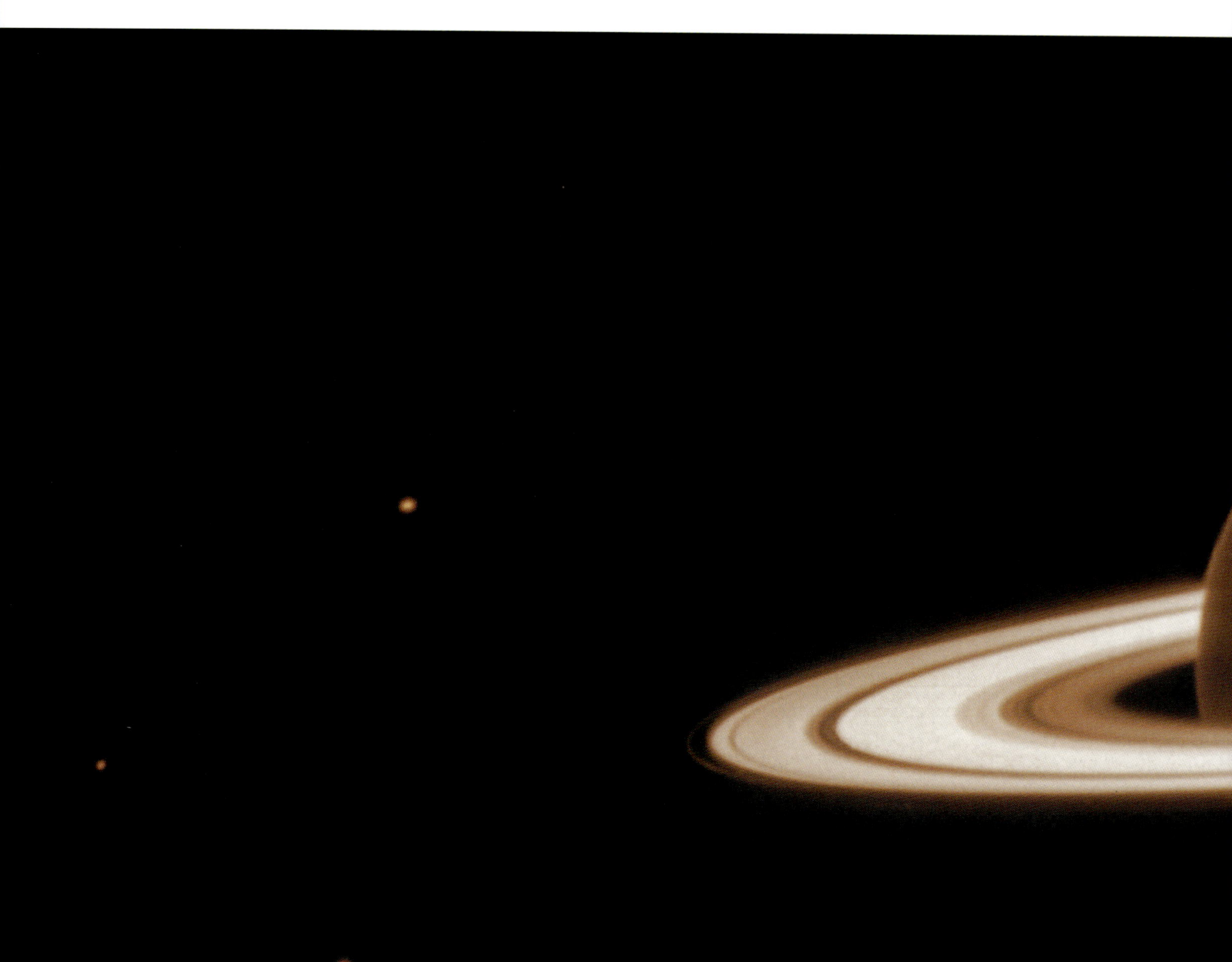

Saturn with three of its moons Dione, Enceladus and Tethys (using data from NIRCam). It is possible to see the giant planet's icy ring in bright detail.

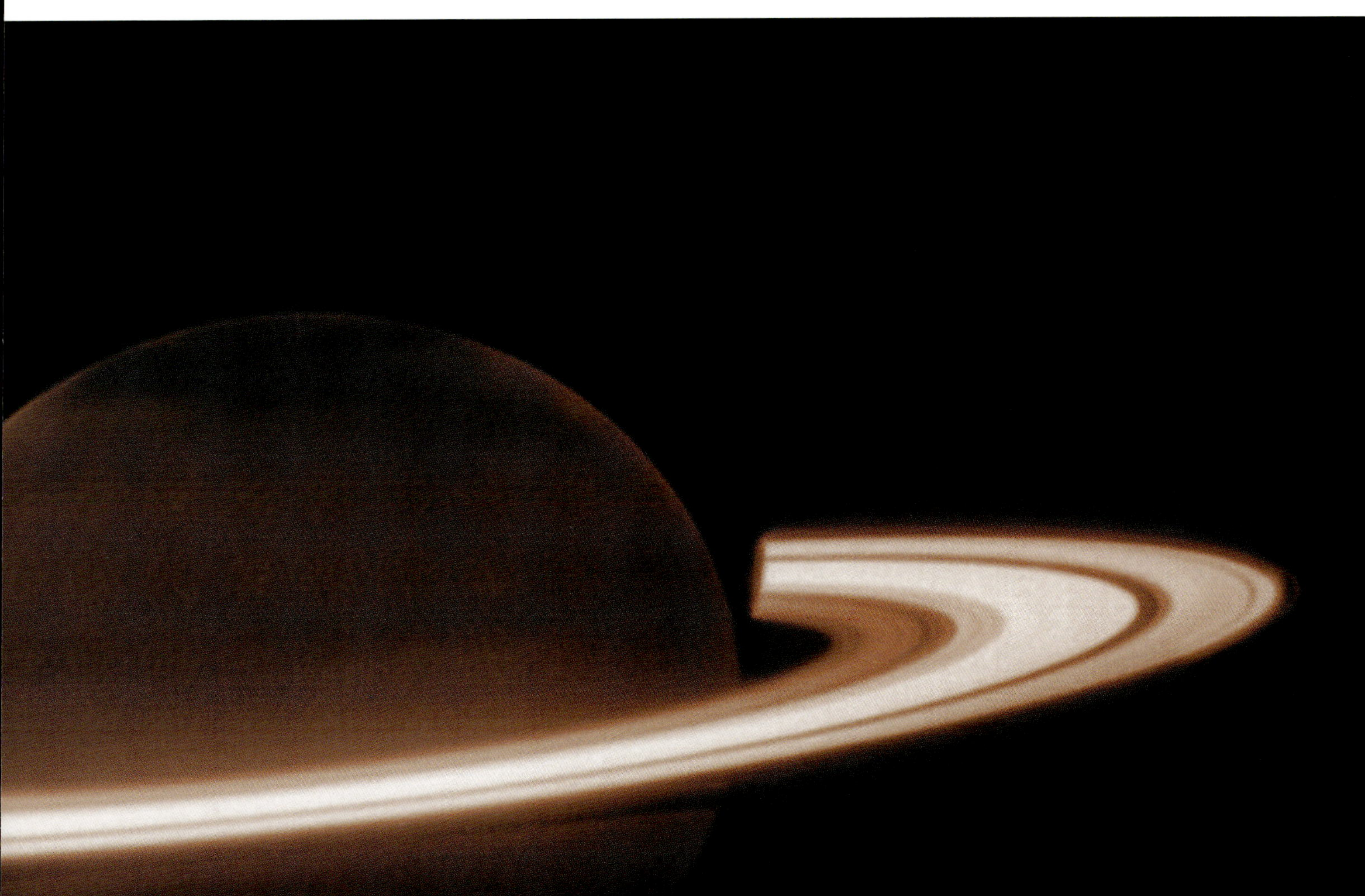

LIKE ITS MUCH LARGER SIBLING Jupiter, Saturn is a gas giant. We suspect that its heart harbours a rocky core, but we can't tell for certain because it is encased in many layers of gas, including hydrogen and helium. Saturn gets its characteristic pale-yellow colour from the ammonia particles in its upper atmosphere.

Saturn is the sixth planet from the sun and is named after a Roman god with rather a large portfolio of responsibilities, which included wealth, agriculture and time. The Romans named Saturday ('Saturn's day') after the planet.

Saturn is the second-largest planet in our solar system and is more than ninety-five times more massive than Earth – but its gaseous layers mean that it is much lighter and is about one eighth of our planet's density. That makes Saturn the only planet in the solar system that is less dense than water. Its rotation also gives it the appearance of a squashed ball, more so than other gas giants in our system. Together, Jupiter and Saturn are responsible for about 92 per cent of the planetary mass in our solar system.

MAGNIFICENT RINGS

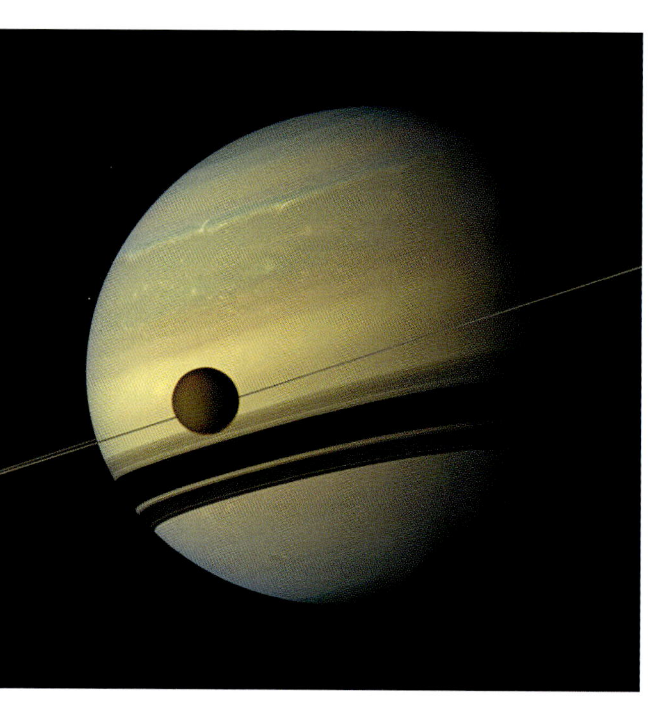

This image of Saturn and one of its moons was created by combining six images captured by NASA's Cassini spacecraft.

Saturn is perhaps best known for its rings. These spectacular disks circle the planet's equator and are mainly made of water ice – although they also have a pinch of other elements, including carbon. The purity of the water composition increases the further away you get from Saturn, so the outermost ring is the most 'pure'.

The rings' constituent pieces range from tiny specks of dust up to objects 10 metres (33 feet) across. At their widest, the rings are about 270,000 kilometres (170,000 miles) in diameter (that is twenty-one Earth diameters), and yet each one is just 20 metres (66 feet) thick – which is about the height of a five-storey building. With Webb's infrared gaze on the previous page, they look particularly magnificent; and Webb's sensitivity shows the delicate gaps between the rings.

Their visual splendour is in part thanks to the fact that Saturn looks rather dull by comparison. This is because its atmosphere absorbs infrared radiation, which is what Webb is designed to detect. Saturn's rings, on the other hand, reflect the sun's infrared light and so they appear luminous against the dark backdrop of space.

MANY MOONS

Webb's image also captures some of Saturn's crowd of moons. More than 140 moons are known to orbit the gas giant. Not all of them have names, but in the image we can see some of the larger (and named) ones: Dione, Enceladus and Tethys.

In 2023, thanks to Webb, astronomers discovered that Enceladus had a giant plume of water vapour erupting from its surface. Plumes had previously been detected emerging from this ocean world, but the newly discovered jet is particularly massive. Exploding out of a geyser-like volcano, the plume extends almost 10,000 kilometres (6,200 miles) upwards – which is about the distance from Los Angeles to Moscow. We think that Enceladus's water plumes are ejecting water into the surrounding ring system, leaving a halo of water in its wake.

Titan is the most impressive of Saturn's many moons. It's big – it is the second-largest moon, after Jupiter's Ganymede, and also bigger than the planet Mercury – but what makes it really special is that it is the only moon known to have a dense atmosphere. Most moons do not maintain an atmosphere because they are less massive than planets and therefore don't have sufficient gravitational force to keep hold of an atmosphere.

One of my most vivid memories of a space mission to date is the European Space Agency's Huygens space probe landing on Titan in January 2005. Once it had landed, we had views of an old riverbed with pebbles shaped by liquid erosion. It was thought that these pebbles were possibly shaped by the movement of liquid methane. Titan was also found to have stable bodies of liquid on its surface, a characteristic that only Earth is known to share. This probe was a true landmark in space exploration when it successfully landed 1.2 billion kilometres (746 million miles) away from Earth, the furthest distance that any probe has landed so far.

In 2022, the first Webb images of Titan sent our astronomical communities into a flurry of excitement and activity. They showed clouds moving above the surface of Titan. Their movement validated computer models about the moon's climate and behaviour. Using Webb's infrared instruments, the team can pierce the atmospheric haze surrounding Titan and probe the composition of the moon's lower atmosphere to understand its inner workings. They hope to discover whether Titan always had an atmosphere, and what could happen to it in the future.

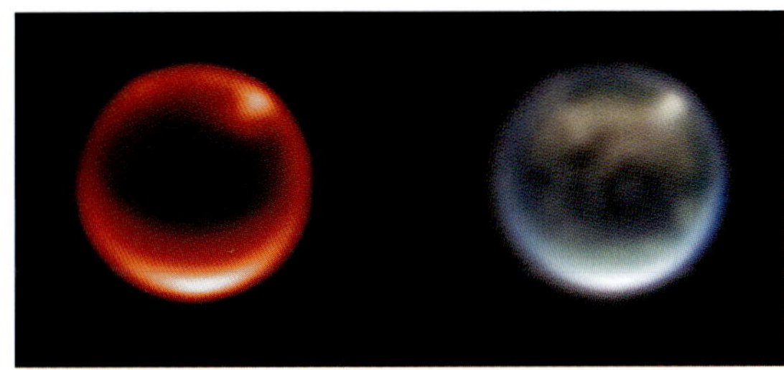

Two images of Saturn's moon Titan (NIRCam). They highlight the atmosphere and clouds swirling around the moon.

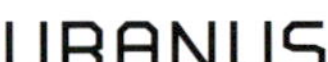

Distance from Earth: 2.6 billion to 3.2 billion kilometres (1.6 billion to 2 billion miles)

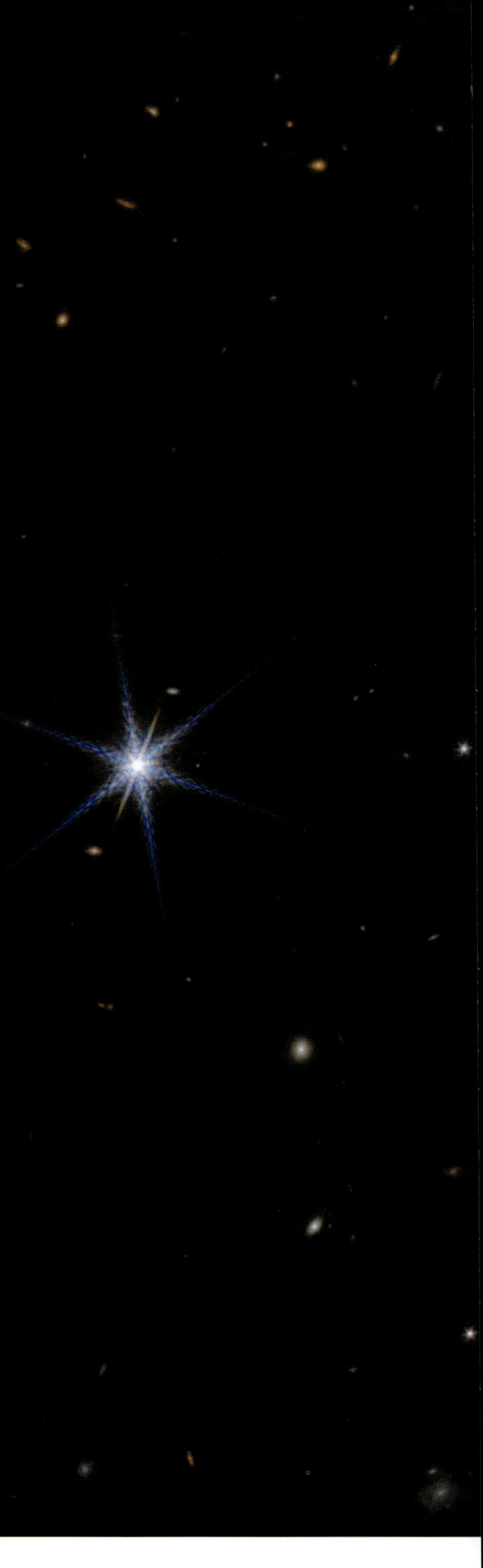

This NIRCam image of Uranus puts the planet's north polar cap on display, along with its glowing outer rings.

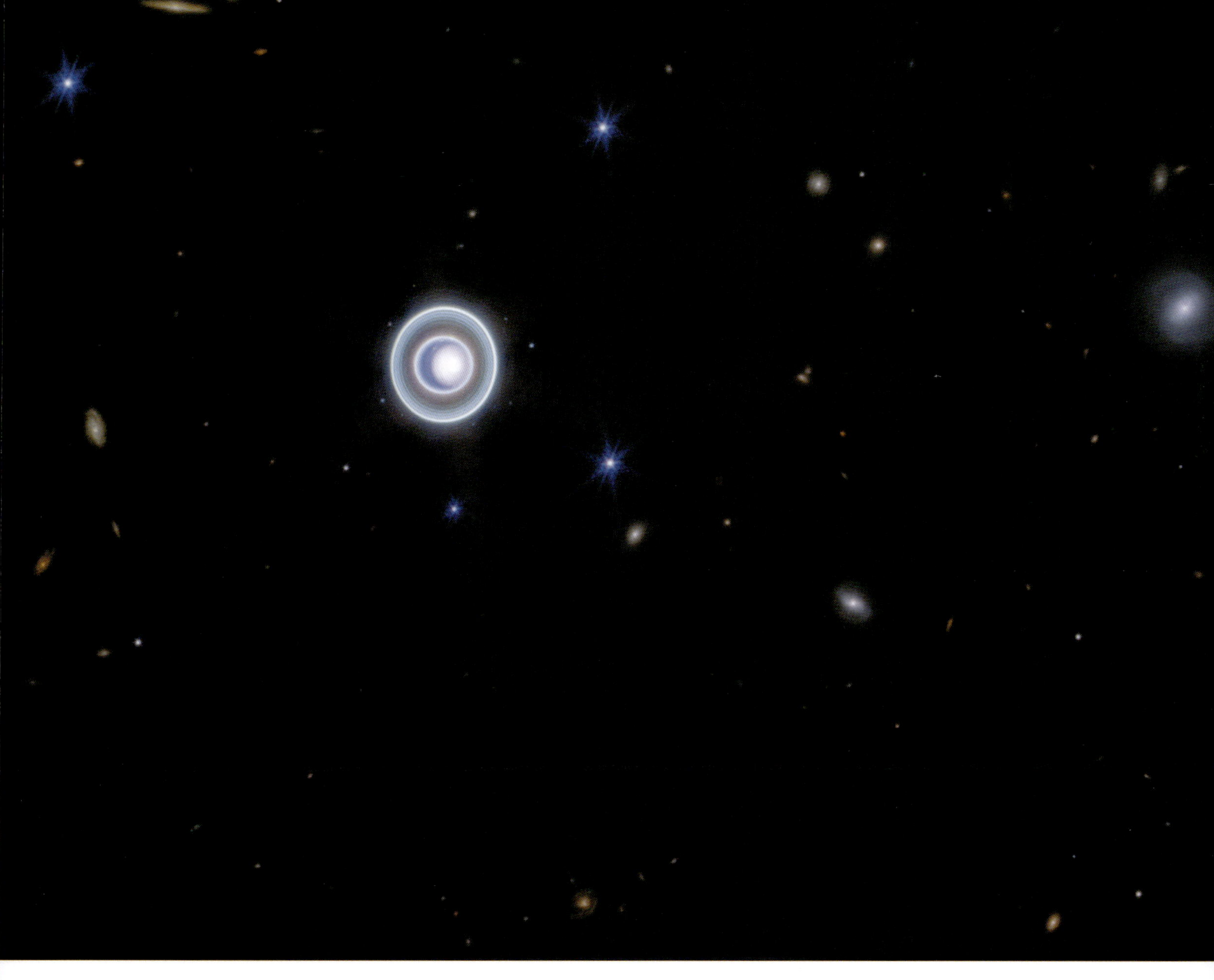

URANUS IS A MASSIVE ICE giant and was the first planet to be discovered with a telescope. It was identified in 1781 by German-British astronomer William Herschel, who initially thought it was a comet or a star.

Named after the ancient Greek sky god, Uranus takes an incredibly long time to orbit the sun – eighty-four years! That means it will be another ten years before Uranus completes its third rotation around the sun since it was discovered all those years ago. Its orbit is something that sets it apart from its fellow planets in the solar system: it appears to rotate on its side, and so looks like a ball rolling around the sun. Practically, this means that each pole gets forty-two years of continuous sun, followed by forty-two years of darkness. It is possible to see Uranus's north pole in the startling Webb image on pages 78–79. The giant white circle on the right half of the planet is its polar cap, dazzling in direct sunlight. The blinding white spots – a large one on the left edge of the planet and another on the top-left of the polar cap – are, in fact, clouds.

Both Uranus and Neptune also have an unusual and irregularly shaped magnetic field. Earth's magnetic field, along with that of most other planets, is typically in alignment with our planet's rotation. But Uranus's odd rotation means that its magnetic field is tipped over: the magnetic axis is tilted nearly 60 degrees from the planet's axis of rotation and is also offset from the centre of the planet by one third of the planet's radius. Auroras have been detected in the atmosphere of Uranus; but these occur not at the poles but rather at odd angles, which could give some insight into how these outer planets generate their magnetic fields.

Uranus appears to orbit around the sun on its side, rolling like a ball, and its equator is nearly at a right angle to its orbit, with a tilt of 97°.

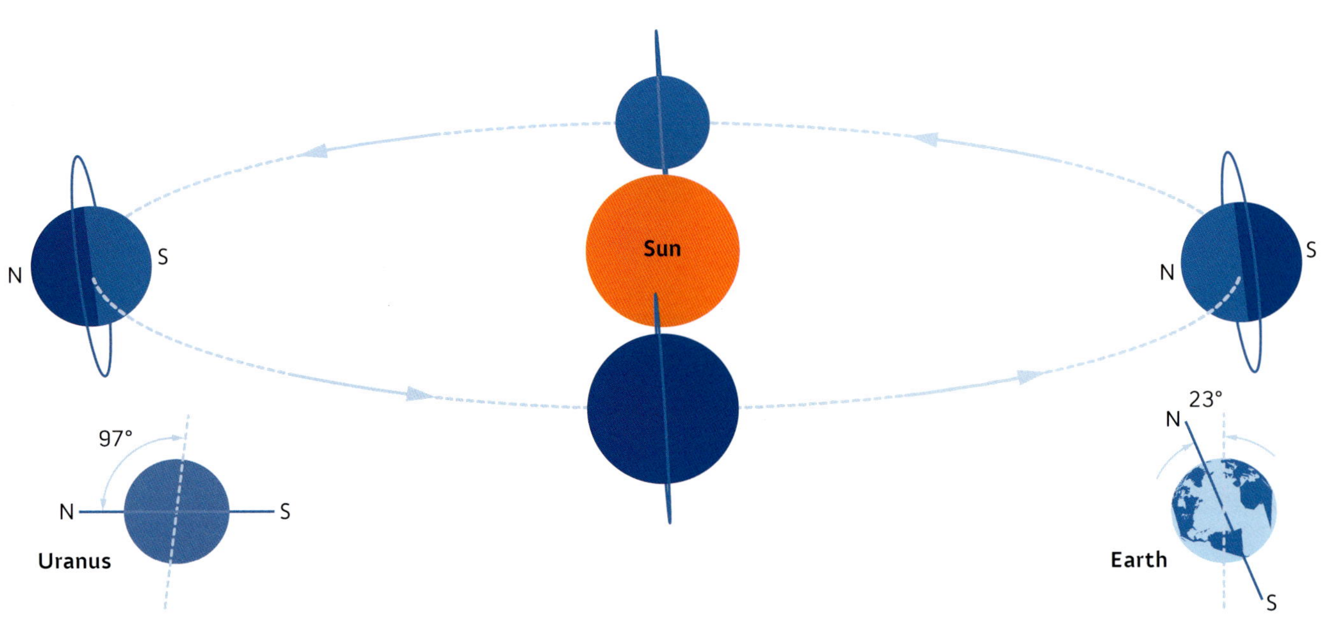

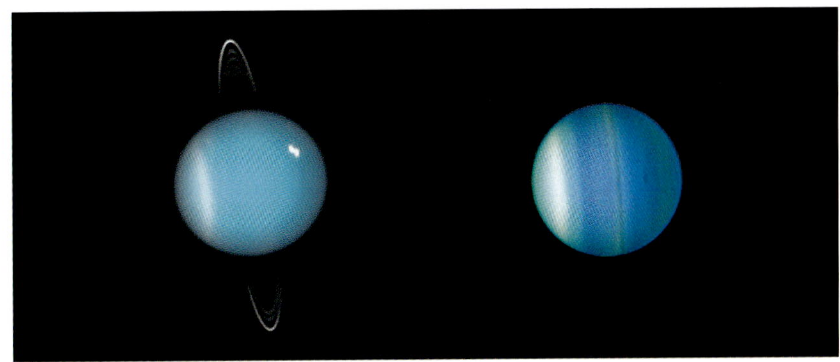

Hubble's views of Uranus.

FROZEN PLANET

Uranus looks like a solid snowball, but its mantle is mostly made up of slushy water, ammonia and methane. Uranus gets its blue-green colour in visible light from that methane gas in the atmosphere. As sunlight passes through the atmosphere and is reflected back out by Uranus's cloud tops, methane gas absorbs the red portion of the light, resulting in its blue-green colour. This ice giant has the coldest recorded temperature of any planet in our solar system, at a rather frigid -195°C (-320°F).

Perhaps the most eye-catching part of the Webb image is Uranus's glorious rings. We can see only eleven of its thirteen rings, but this is the most clearly we've ever been able to see them. Most of Uranus's rings are very narrow, only a few kilometres wide, which is one of the reasons they are so difficult to image.

A wide-field image, taken with NIRCam, managed to capture six of Uranus's elusive moons, with a handful of galaxies photobombing in the background.

Webb has managed to capture some of Uranus' elusive moons, circled in blue (clockwise from bottom left: Titania, Ariel, Puck, Miranda, Umbriel, Oberon).

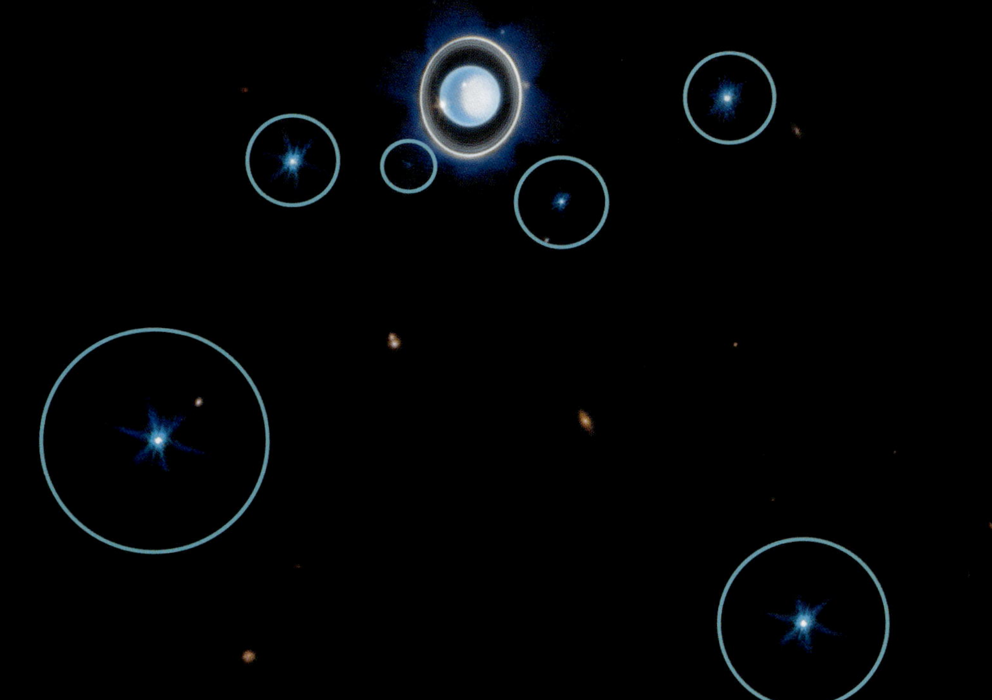

NEPTUNE

Distance from Earth: about 4.5 billion kilometres (2.8 billion miles)

An image of Neptune, captured with Webb's NIRCam instrument. One of its moons, Triton, glows with Webb's distinctive eight-spike diffraction pattern.

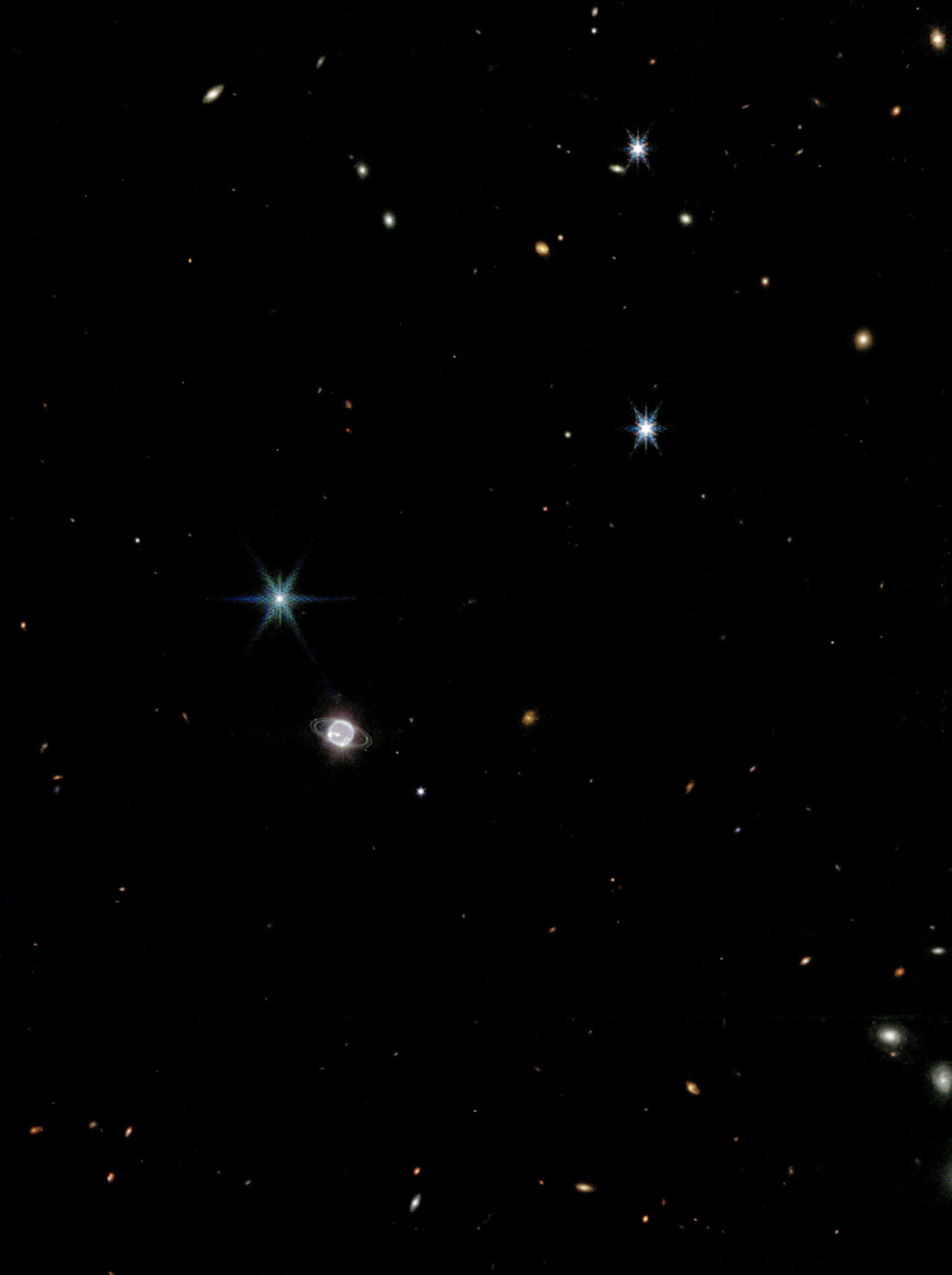

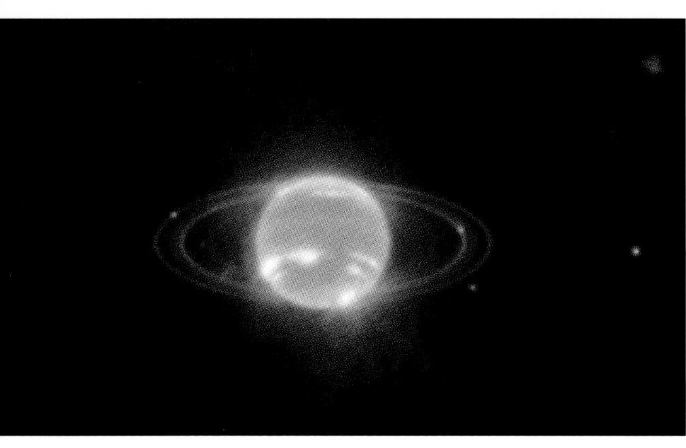

Neptune, with glowing clouds and delicate rings.

BILLIONS OF KILOMETRES FROM EARTH, ferocious winds bombard the surface of Neptune. It has the strongest sustained winds of any planet in the solar system, reaching up to 2,100 kilometres (1,300 miles) per hour. To put this into context, the fastest windspeed recorded on Earth to date is just over 400 kilometres (250 miles) per hour, detected on 10 April 1996 in Barrow Island, Australia, during Severe Tropical Cyclone Olivia.

Neptune is an ice giant and its interior is mostly made up of ice and rock. It has seasonal weather and its atmosphere comprises mainly hydrogen and helium, with a large dose of icy volatiles (atoms and compounds that can easily become vapour) like ammonia and methane. Since Pluto's demotion to 'dwarf planet', Neptune is now the outermost planet in our planetary system.

Neptune has the honour of being the only planet in our solar system discovered thanks to maths. It's not possible to see it with the naked eye from Earth, but astronomers in the nineteenth century noticed that Uranus's orbit was behaving strangely and unexpectedly. French astronomer Alexis Bouvard hypothesized that there was an invisible planet lurking nearby, and the (surprisingly giant) planet was finally observed in 1846. It is seventeen times the mass of Earth and orbits the sun every 165 years.

PLANET OF SEA GODS

Neptune is named after the Roman god of the seas and freshwater because of its apparent blue colour when viewed through an optical telescope. Most languages use some variant of 'Neptune' when talking about the ice giant. In the Māori language, for example, it is called Tangaroa, after the Māori god of the sea.

Because NIRCam observes objects in the near-infrared range of 0.6 to 5 microns, Neptune does not seem blue to Webb. In fact, the methane gas absorbs so much red and infrared light that the planet appears dark at these near-infrared wavelengths, except where high-altitude clouds are present. These methane ice clouds appear as dazzling streaks and patches that reflect the infrared portion of the sunlight before it is absorbed by methane gas.

We can also see how these clouds congregate at the poles. Although the north pole faces away from Webb, it has a luminous halo that is intriguing to astrophysicists because they don't know what causes it. They have also found that Neptune's south pole is continuously hidden by a cap of high-altitude clouds, which appear as a blinding white spot in the image.

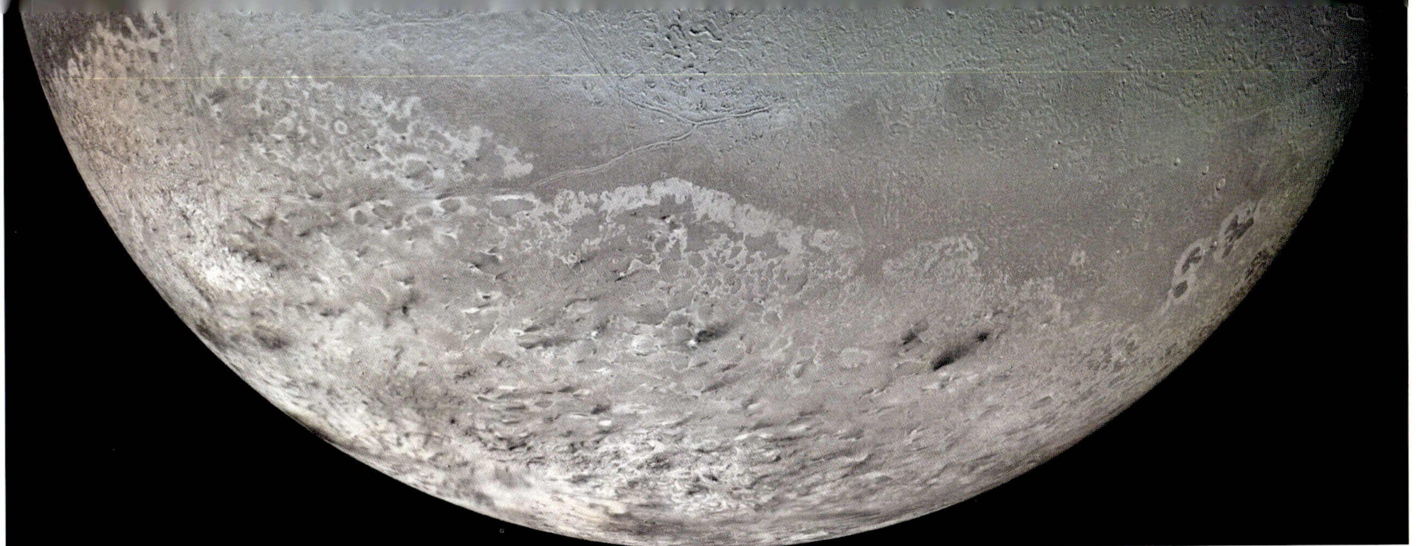

Triton's cratered surface, imaged with Voyager 2, was carved by icy lava flows.

CAPTURED MOON

Neptune has fourteen moons but only one – Triton – is worthy of special mention. Triton is Neptune's largest moon by a big margin. It comprises more than 99.5 per cent of the mass that orbits the ice giant and is also remarkably cold at -235°C (-391°F).

Unlike other large moons, Triton has a retrograde orbit, travelling in the opposite direction to Neptune's rotation. This has led astronomers to hypothesize that it was once a dwarf planet in the outer solar system.

In the Webb image, Triton glows like a beacon at the top left of Neptune. It is so bright that it has Webb's characteristic diffraction spikes. Triton glows much brighter than Neptune because it reflects more of the sunlight that hits it and it doesn't have Neptune's infrared-capturing, methane-rich atmosphere.

An image of Neptune, snapped by Voyager 2.

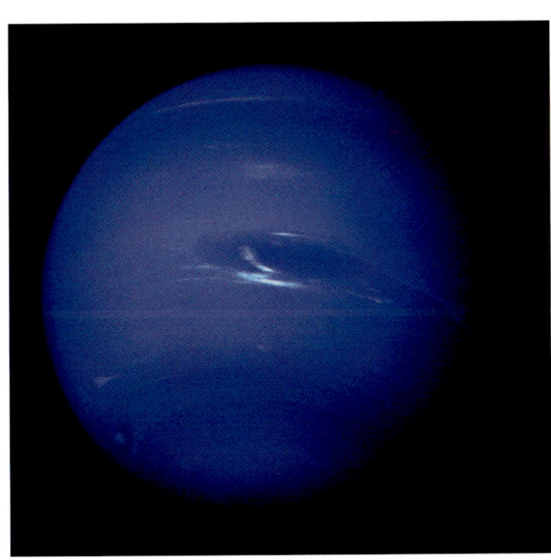

NEPTUNE'S RINGS

Much of what we know about Neptune is thanks to NASA's Voyager 2 space probe. Launched in 1977, it aimed to study the outer planets of our solar system. It's the only spacecraft to have visited the ice giants. Voyager 2 reached Neptune in 1989, twelve years after it left Earth, and discovered the planet's rings. Thirty years later, Webb has now offered a new perspective on these thin icy bands.

Neptune has five main rings, but much of the ring system is dusty and quite unstable. Some of the planet's moons orbit within the ring system and can be seen in Webb's image. Scientists suspect that the rings are fairly new and are most likely the result of a moon collision. Webb offers an unprecedented view of the difficult-to-see inner rings.

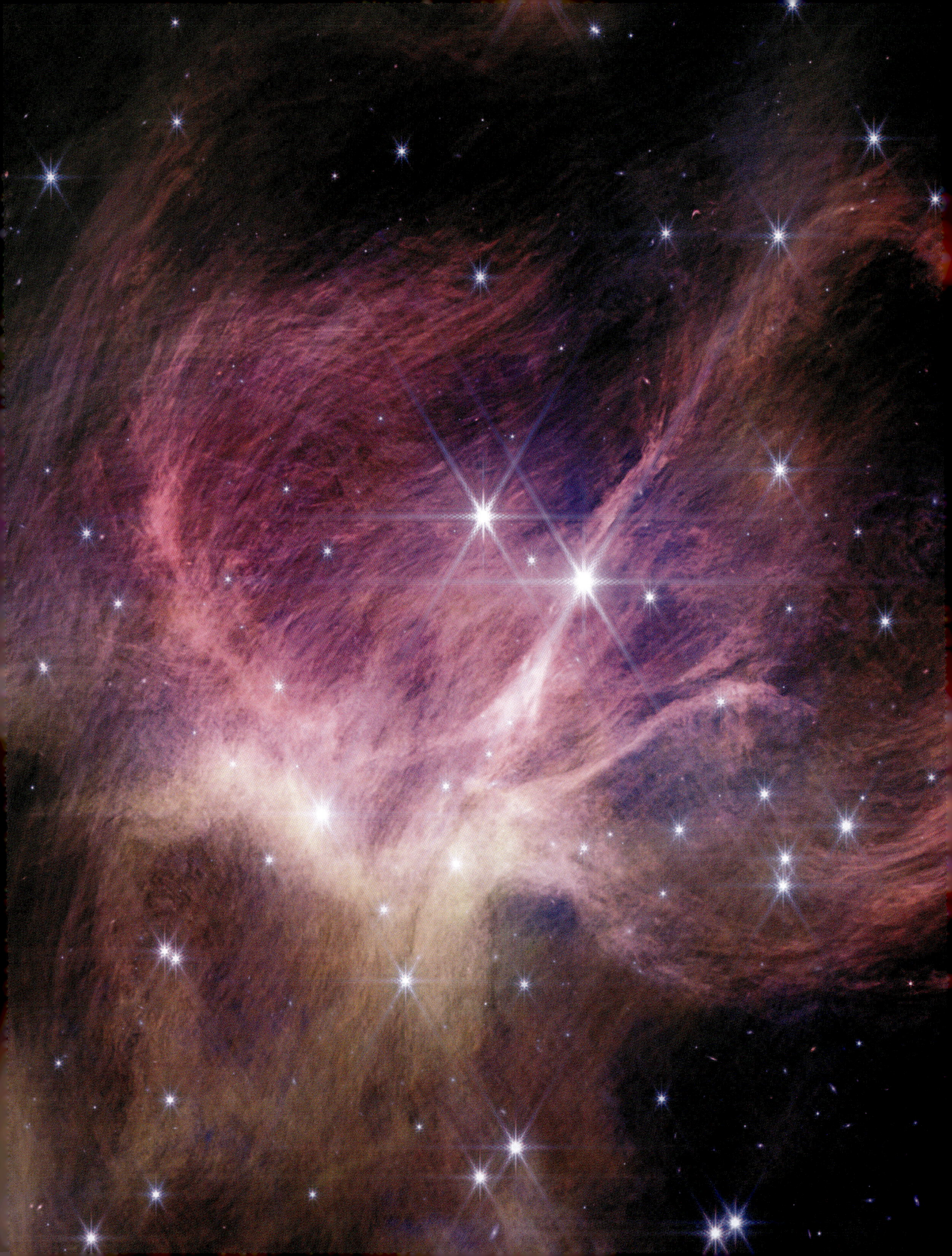

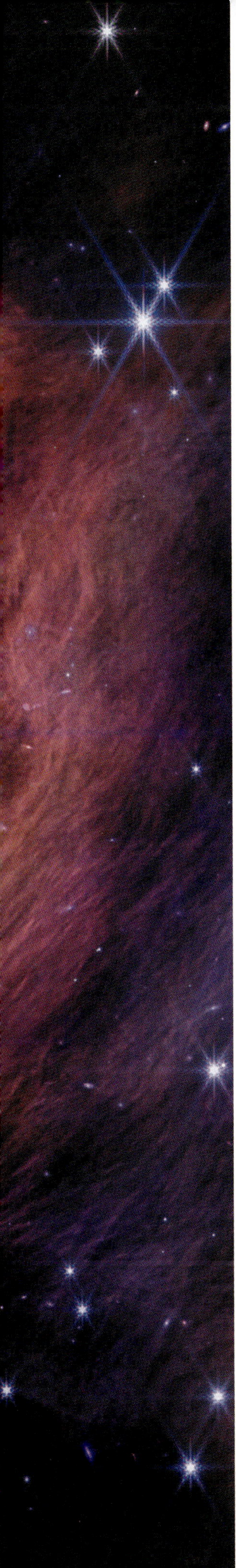

EXOPLANETS

Exoplanets are planet-sized objects that are found beyond our solar system, although an exact definition of an exoplanet is still up for debate. They can orbit stars like the planets of our solar system, or they can be 'rogue' planets that are not tethered to a specific star. As with all planets, they must be big enough to have a nearly round shape and have 'cleared' the neighbourhood around their own orbit.

Exoplanets are hard to detect. Just like the planets within our solar system, they do not generate significant amounts of electromagnetic energy as stars do. We can see the planets of our solar system only because they reflect light from our sun. But exoplanets orbit stars trillions of kilometres away from us, and any starlight that they reflect is too dim for us to detect on Earth. Exoplanets are also hard to detect as most of these planets sit next to bright stars which will swamp the dim light that they reflect. So, we mainly detect exoplanets the using the transit method (see page 28).

Up until now, we have the Kepler Space Telescope to thank for much of our understanding of exoplanets. There are currently more than 5,500 confirmed exoplanets, and their variety is just mind-boggling. There are planets made of diamond, while others are giant balls of water, and some (such as KELT-9b) are hotter than some stars. And – excitingly – there are rocky exoplanets similar to Earth that could possibly even sustain life.

Webb will be able to provide us with more data on exoplanets than we could have imagined. Webb will not only help us to discover more exoplanets, but its onboard instrumentation will also allow us to analyse their atmospheres and see whether there are molecules that could indicate biological life.

IC 348 is a young star-forming region in the constellation Perseus.

LHS 475 B

Location: orbiting star LHS 475
Constellation: Octans
Distance from Earth: 40.7 light-years

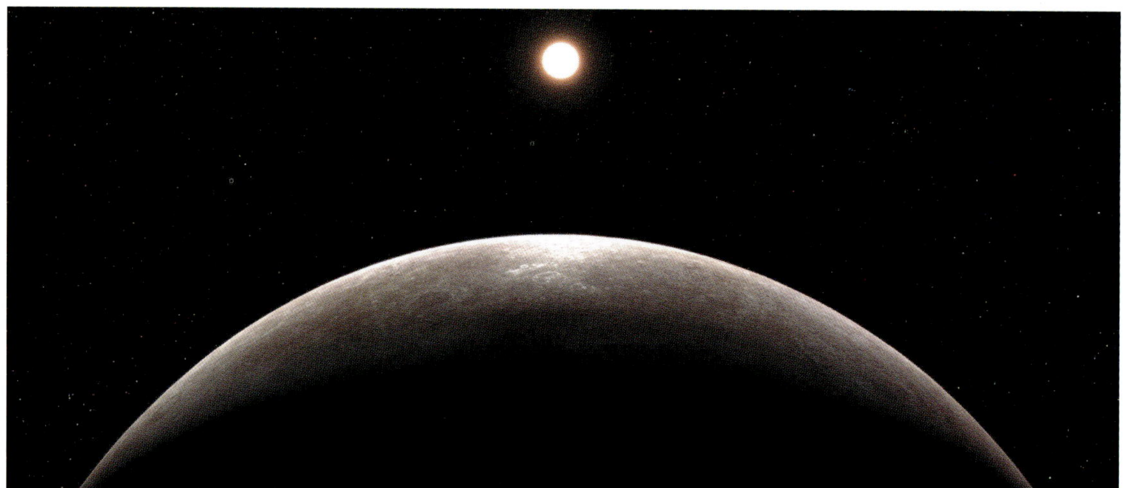

An artist's illustration of exoplanet LHS 475 b and its star.

LHS 475 B WAS WEBB'S first confirmed exoplanet. Using the NIRSpec instrument, it detected the planet using the transit method. But Webb can go one step further. It can measure the exoplanet's transmission spectrum by comparing what happens to the star's light as it passes through the exoplanet's atmosphere. In the case of LHS 475 b, astronomers could not detect any element or molecule, such as hydrogen or methane. This led them to deduce that the rocky exoplanet does not have an atmosphere.

In the diagram below, the flat yellow line represents the transmission spectrum of an exoplanet with no atmosphere, while the purple line shows what it would look like if it had a methane-rich atmosphere, and the green line models a carbon-heavy atmosphere. The grey data points of LHS 475 b shows that it aligns most closely with the featureless model.

The data from LHS 475 b points to an exoplanet without an atmosphere.

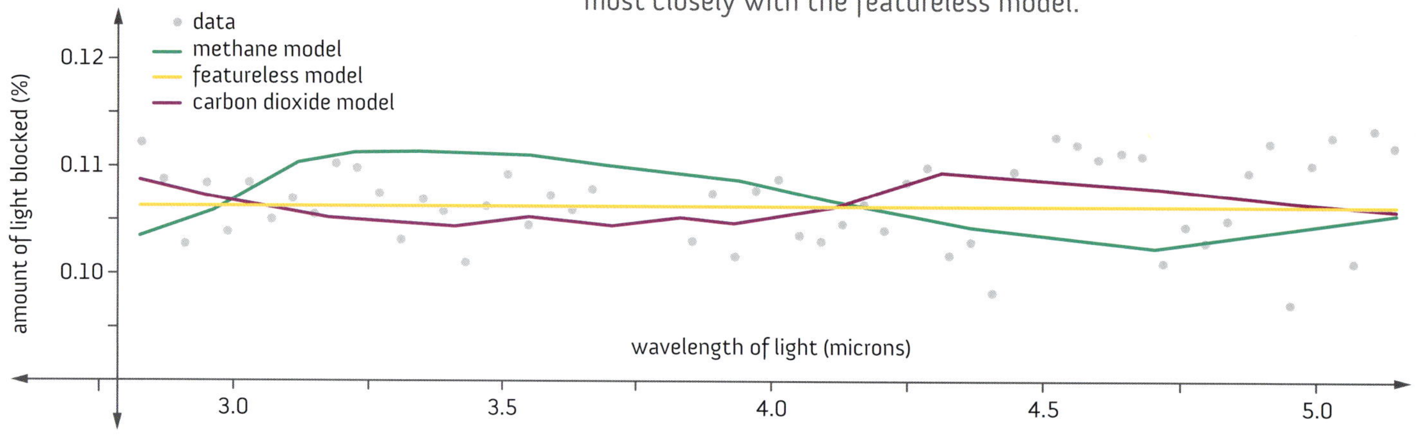

WASP-80 B ALSO KNOWN AS WADIRUM

Location: orbiting star WASP-80
Constellation: Aquila
Distance from Earth: 162 light-years

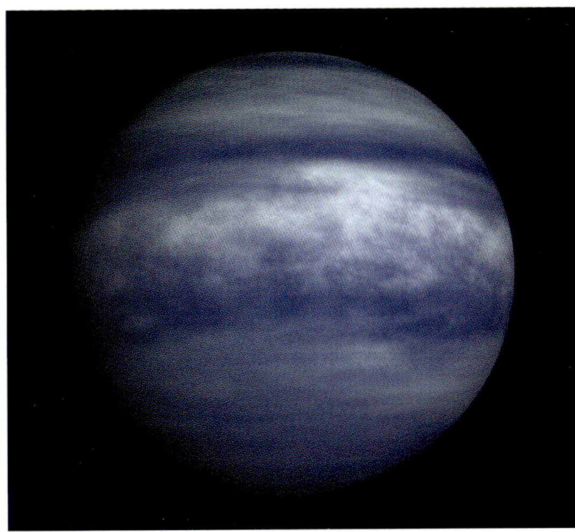

An artist's impression of exoplanet WASP-80 b.

WE'VE KNOWN ABOUT WASP-80 B since 2013. A project called IAU100 NameExoWorlds was held in 2019 to celebrate the 100th anniversary of the founding of the IAU. Through this project, the WASP-80 parent star was renamed Petra, after the Jordanian World Heritage Site, and the exoplanet WASP-80 b was officially named Wadirum, after the famous Wadi Rum region in Jordan. This ancient site contains over 20,000 petroglyphs and inscriptions tracing human history back over 12,000 years. Even now, some nomadic Bedouin people live here along one of the migration routes modern people took out of Africa, offering a living depiction of our human origins. The area is also known as the Valley of the Moon.

WASP-80 b's close proximity to its star means that it is very difficult to image, but Webb's sensitive spectrographs, in tandem with atmospheric transit analysis techniques, managed to yield new insights into this giant exoplanet. We now know that its atmosphere contains methane and water vapour.

Astronomers are particularly interested in water vapour and methane because both are associated with carbon-based life, which is what we have on Earth. Future analysis of the atmosphere, with MIRI looking at different wavelengths and monitoring other carbon-based molecules, could reveal more about this fascinating exoplanet. The presence of water on a planet's surface and the presence of methane in its atmosphere can be indicators of the possibility of life, but these molecules can potentially originate from non-biological activity so their presence can't tell us for certain that a planet may host life.

The discovery of methane and water vapour in the atmosphere of Wadirum (WASP-80 b) will also enable us to compare exoplanets outside of our solar system with planets closer to home, giving us a better understanding of how our planet evolved.

HIP 65426 B

ALSO KNOWN AS NAJSAKOPAJK

Location: orbiting star HIP 65426
Constellation: Centaurus
Distance from Earth: 385 light-years

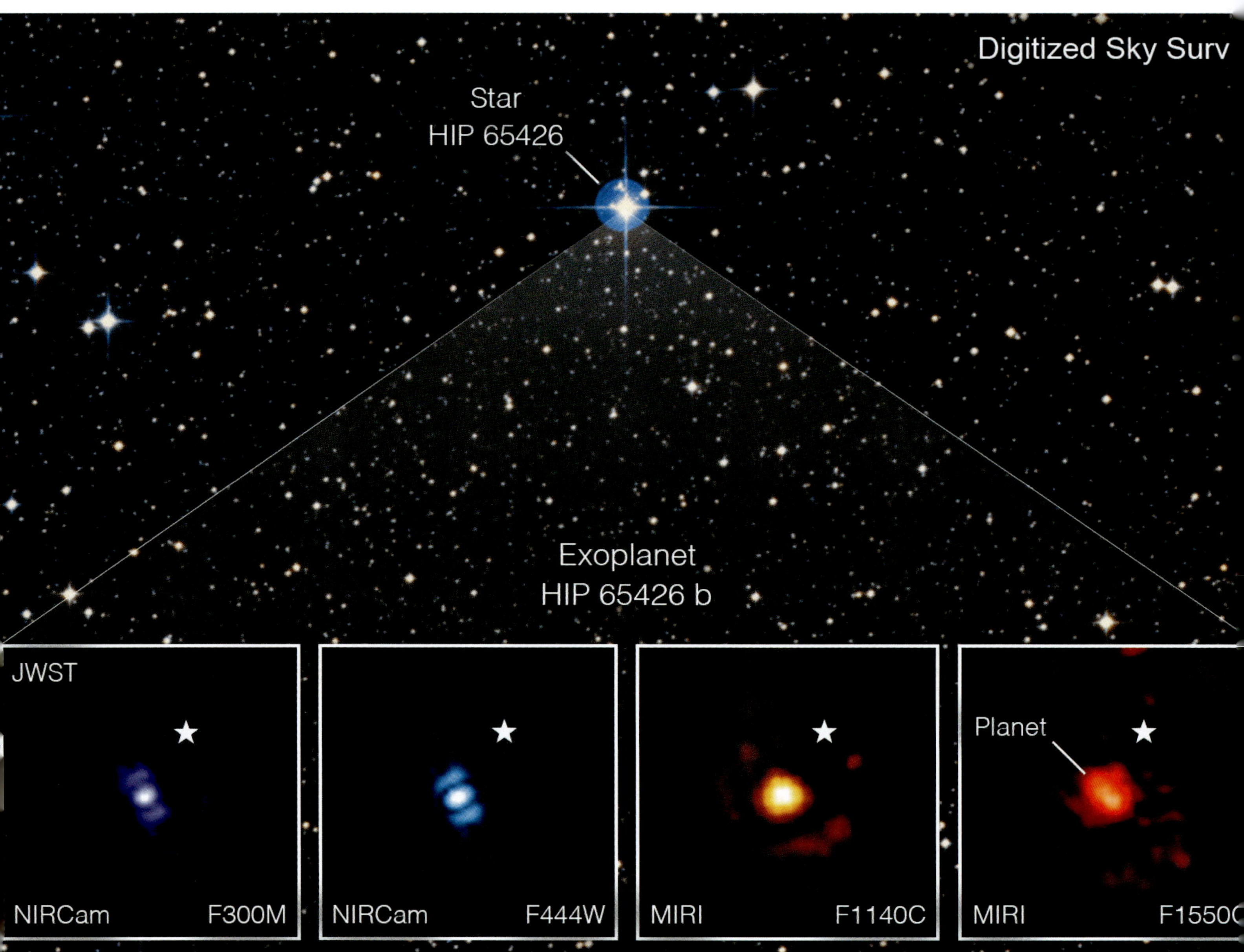

The exoplanet HIP 65426 b in different bands of infrared light, captured with the NIRCam and MIRI instruments.

FOR A PLANET, HIP 65426 b is unfathomably large. It is known as a 'super-Jupiter exoplanet' because it is between six and twelve times the mass of Jupiter. It's the first exoplanet that Webb observed directly, probably chosen due to its large size. In NameExoWorlds 2022, it was given the official name Najsakopajk, which means 'Mother Earth' in the indigenous Mexican language Zoque, and its star was officially named Matza, meaning 'star'.

Najsakopajk is of great interest to planetary scientists because it doesn't fit into our current understanding of planet formation. It sits very far from its parent star, Matza, at just under 100 astronomical units (AU) away. An astronomical unit is a distance measurement defined as the average distance between planet Earth and our sun – approximately 150 million kilometres (just over 90 million miles). In our solar system, Neptune is the furthest planet from the sun, sitting at a distance of 30 AU.

The great distance between Najsakopajk and its parent star leads to a number of questions about this exoplanet. Did it form at this distance? If it did, how big was the accretion disk around the star? If it formed closer and was projected outwards, what mechanism enabled its departure? Another alternative is that it was a rogue planet wandering interstellar space and got caught in Matza's gravitational pull.

This distance was useful when it came to imaging the planet. It meant that Webb was able to capture the exoplanet's dim light without it being overwhelmed by the light of its parent star. Najsakopajk is more than 10,000 times fainter than its host star in the near-infrared, and a few thousand times fainter in the mid-infrared.

To image the planet, both NIRCam and MIRI use coronagraphs. A coronagraph effectively blocks out a star's light so that the instrument can snap a picture of the planet. By using both the MIRI and NIRCam instruments, Webb was able to image HIP 65426 b at a variety of wavelengths to reveal different characteristics of the exoplanet. The acquisition of this image was a great showcase of the trailblazing techniques that Webb can deploy and we will no doubt be seeing these used to image other exoplanets in the future.

IC 348

Constellation: Perseus
Distance from Earth: 1,000 light-years

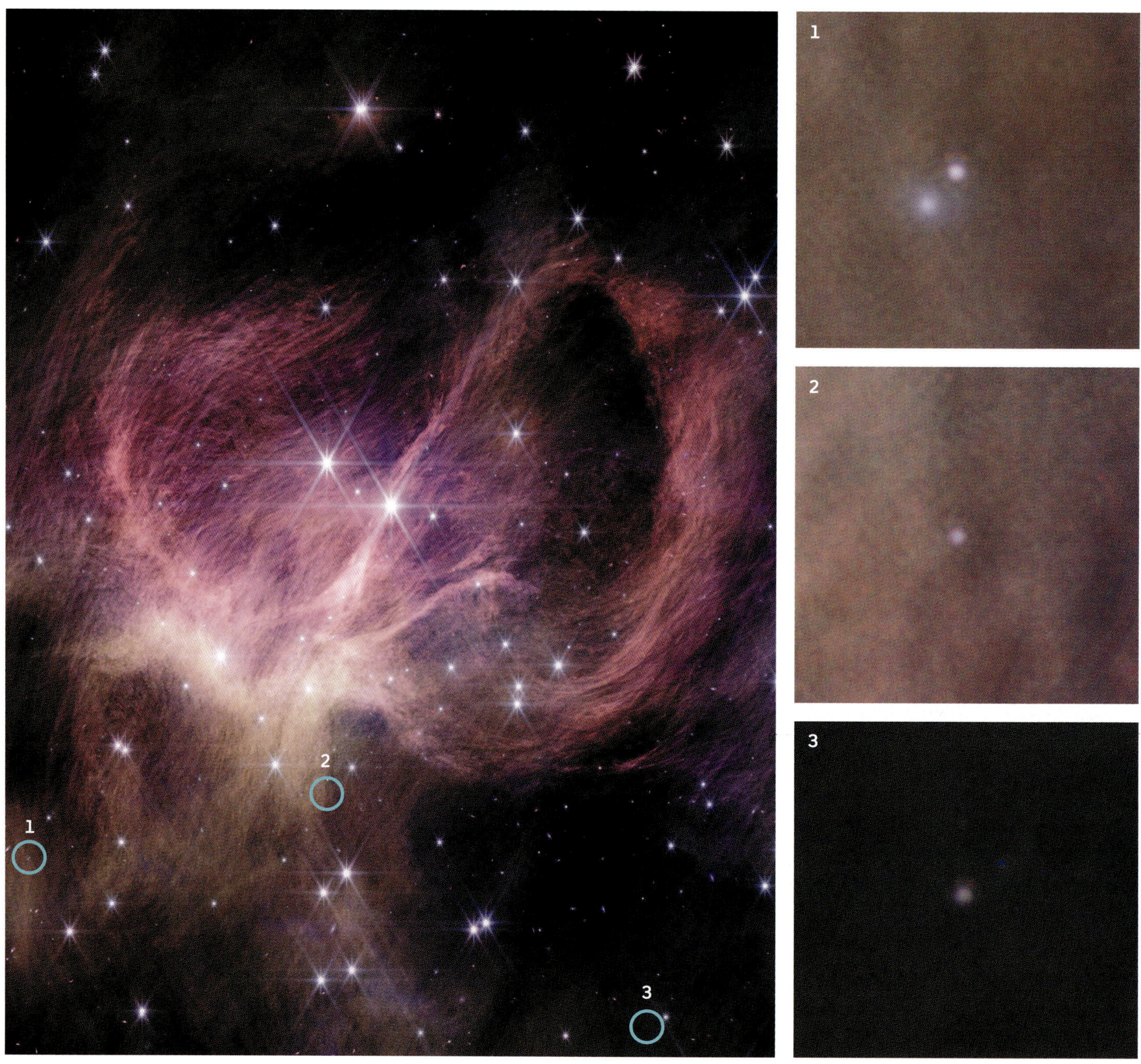

Using Webb data, scientists found at least three brown dwarfs in star cluster IC 348. They are circled in the image.

WHEN IT COMES TO CELESTIAL objects, there is a big difference between the size of the largest planet and the size of the smallest star. The middle ground is inhabited by objects called brown dwarfs. The existence of brown dwarfs was predicted in the 1960s, but it took thirty years of looking before one was finally discovered. They're called brown dwarfs because they occupy the ambiguous position between bright white dwarfs and non-luminous planets. A white dwarf is what stars like our sun will become once they have burned up all their fuel. They were originally named black dwarfs, but that title was already being used to describe a white dwarf that had cooled to the point where it was no longer emitting radiation.

Using Chandra data, scientists suspect that within galaxy NGC 6388, a white dwarf star tore apart a planet that came too close.

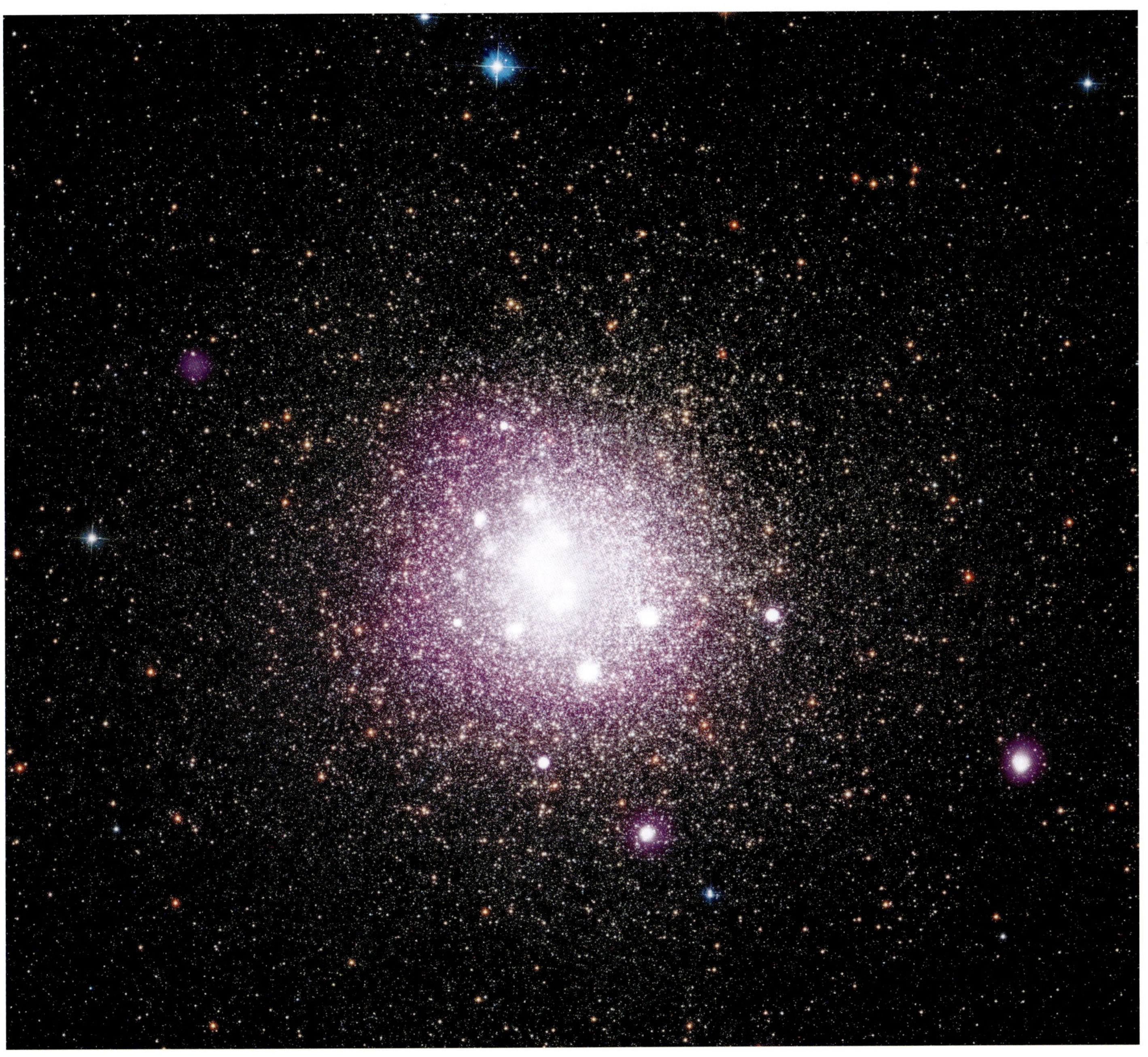

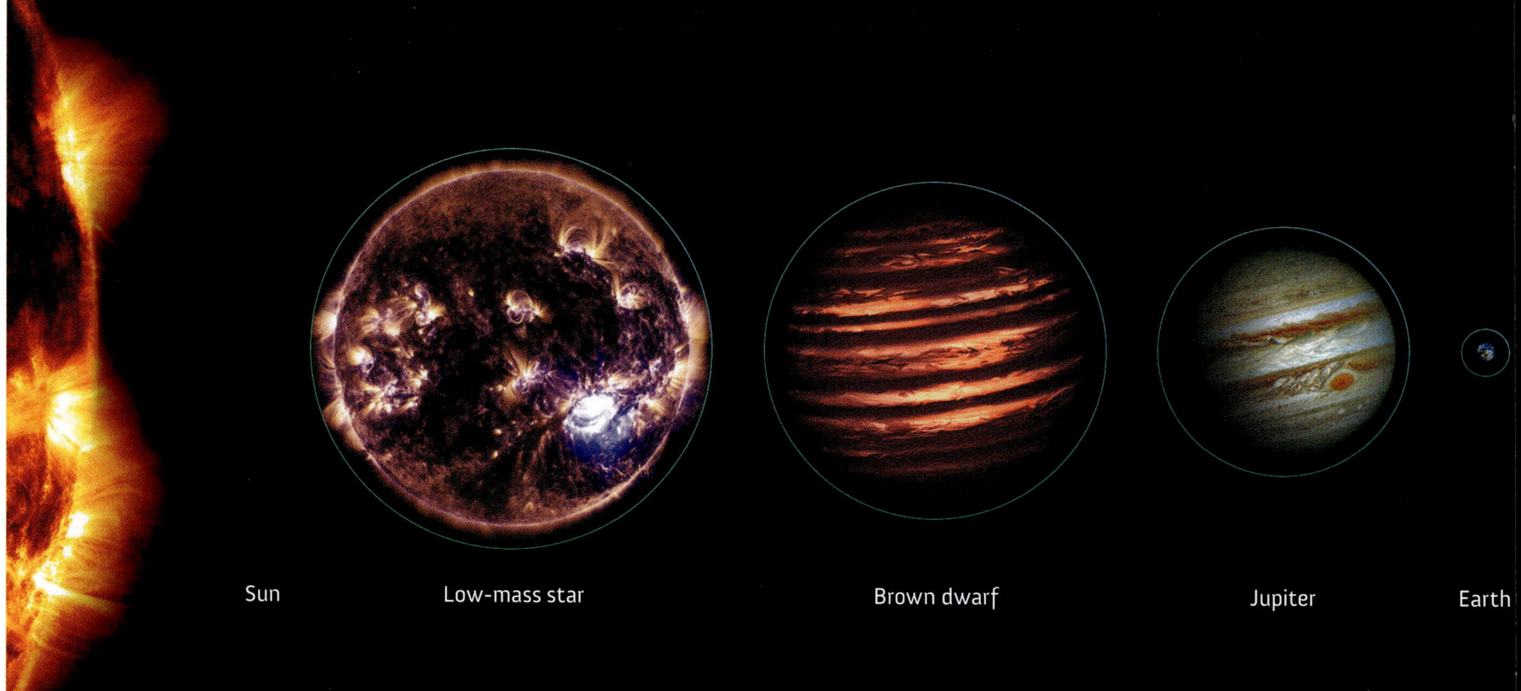

Brown dwarfs inhabit the large gap in size between small stars and large planets.

Brown dwarfs are the ultimate 'in between' object. They usually range between fifteen and eighty times the mass of Jupiter, making them much bigger than a normal planet but smaller than stars. Their size means that they aren't big enough to fuse hydrogen in their cores but are still big enough to emit some light. The name actually has nothing to do with their colour – their colour depends on their temperature. The particularly hot ones can be red or orange, while their cooler siblings may even appear dark.

A SWIRLING NEST OF STARS

To hunt for brown dwarfs, a team of astronomers looked into the heart of IC 348, a young star-forming region that is only 5 million years old (see page 92). For some context, 5 million years ago, humans and chimps were in the process of splitting from a shared common ancestor. While 5 million years is a long time in human terms, it is a cosmic blink of an eye. With the region being so relatively young, it would mean that brown dwarfs found in this volume are likely to be still emitting radiation and so Webb may be able to detect them.

The image on page 92 contains the red wisps of interstellar clouds and the distinctive eight spikes of incredibly bright stars. This tapestry of interwoven material is called a reflection nebula, which – as the name suggests – reflects the light of the nebula's stars. Our sun produces a stream of charged particles that stream out to the outer reaches of the solar system. In a similar way, charged particles or stellar winds from the stars in this nebula could be responsible for hollowing out the caverns in these clouds.

A COTERIE OF BROWN DWARFS

In the image on page 92, what interests brown-dwarf hunters are the faint points in the background. First, astronomers looked at IC 348 with NIRCam to identify possible brown dwarfs. The problem was that it is difficult to tell which small red pinpricks were brown dwarfs and which were distant background galaxies. So, using NIRSpec's microshutter array (which can block out the bright light of nearby stars), the astronomers could narrow down their list of targets.

They found three potential candidates; but surprisingly, these are smaller than we expect brown dwarfs to be. They range from three to eight times the mass of Jupiter. If confirmed, the smallest will be the tiniest brown dwarf found to date. The challenge then will be to try to understand how such a small brown dwarf could form.

One possibility that's been considered is that they are not brown dwarfs but instead giant rogue planets roaming interstellar space without being tethered to a parent star. However, this has been dismissed, as the stars in IC 348 tend to be of low mass. Small stars of this type are unlikely to have generated such large planets.

Adding to the mystery, two of the discovered brown dwarfs have also been found to contain hydrocarbons – molecules that contain both carbon and hydrogen atoms. The same infrared signature has been detected on Saturn and its moon, Titan, and in the gas between stars. This is the first time it has been detected on a body outside our solar system. These molecules are of interest as it is thought that they may be important in the seeding of life.

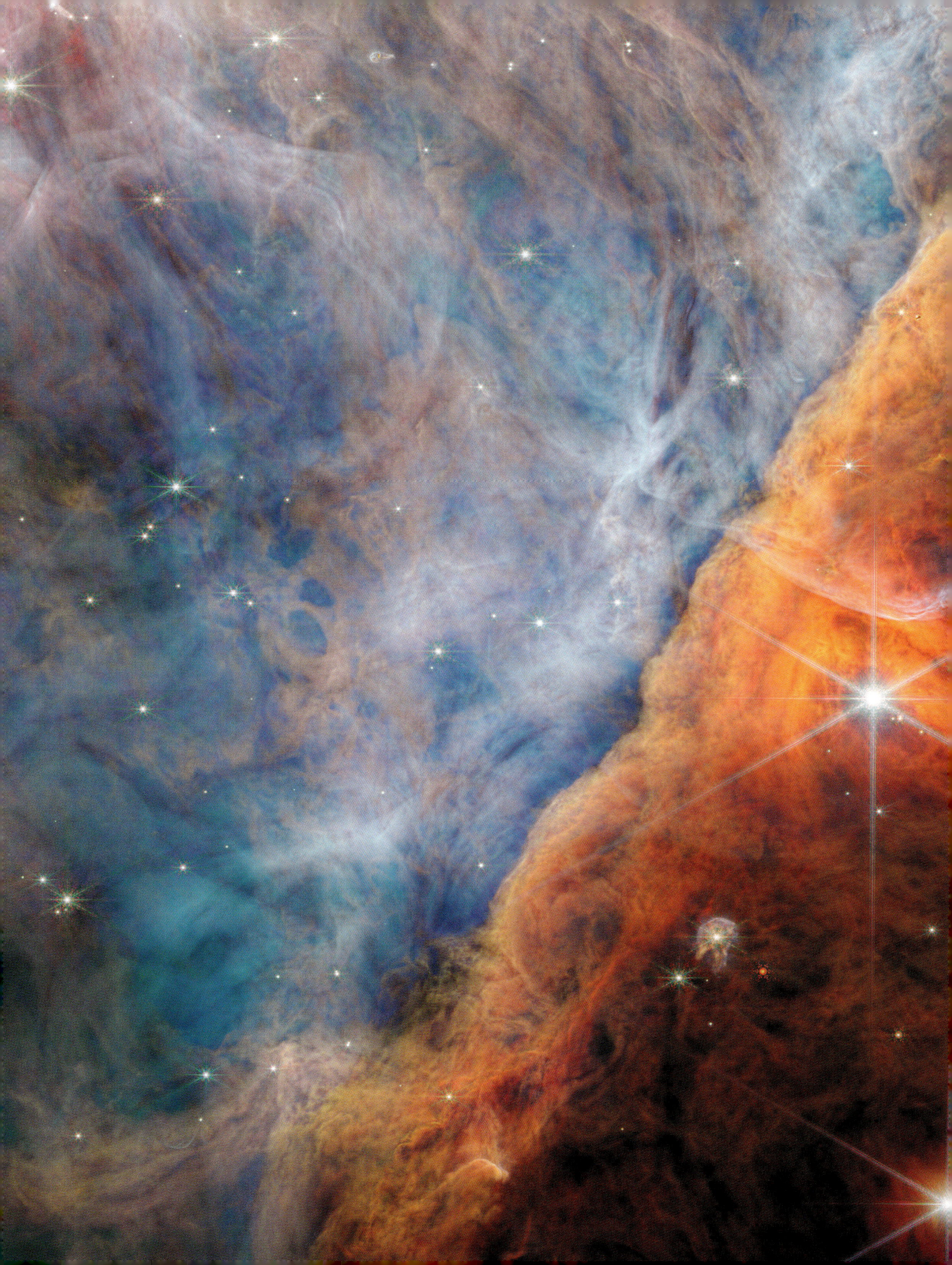

NEBULAE

Nebulae are giant swirls of interstellar gas and dust. Some nebulae form out of the remnants of supernovae – massive stars that exploded at the end of their lifecycle when they run out of fuel for the fusion process. Often, nebulae are inconceivably large, with even light taking millions of years to travel from one side of a large nebula to the other. They are mostly made from hydrogen and helium, the simplest atoms.

In this cosmic soup, amazing things can happen – such as the birth of new stars. Some types of nebulae are called 'stellar nurseries' because the eddies of gas and dust begin to coalesce and their gravity starts to attract more matter. Eventually these clumps collapse in on themselves and stars are born.

There are many types of nebulae, including H II regions (which are large, diffuse nebulae made up of positively charged hydrogen atoms), planetary nebulae (which are gaseous shells around a dying star – our sun will form one of these in a few billion years when it runs out of fuel for fusion), supernova remnant (the exploding remnants of a high-mass star), and dark nebulae (an interstellar cloud that is so dense that visible light cannot penetrate it).

The tricky thing about some nebulae is that it is difficult to figure out what is going on inside them. The density of particles can mean that light cannot escape the swirling dust and gas, so we cannot see into them. That is where Webb's infrared capabilities really come into their own. The telescope has been designed to detect the infrared radiation that can escape these interstellar clouds and reveal the phenomena behind their veil.

A NIRCam image of a part of the Orion Nebula known as the Orion Bar.

CARINA NEBULA

Location: in the Carina–Sagittarius Arm of the Milky Way
Constellation: Carina
Distance from Earth: 7,500 to 8,500 light-years

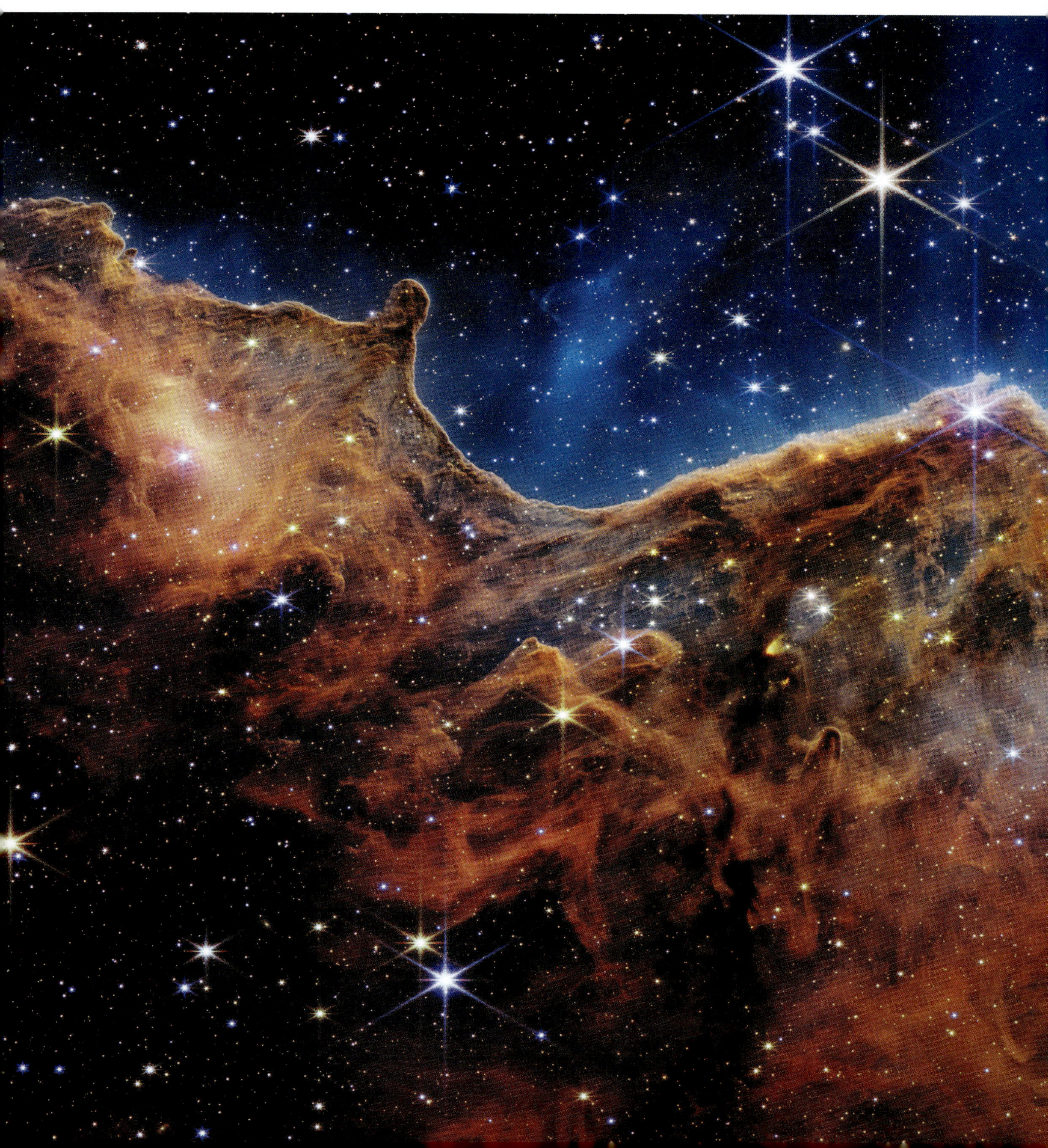

Known as the 'Cosmic Cliffs', this is the edge of star-forming region NGC 3324 in the Carina Nebula, imaged using NIRCam data.

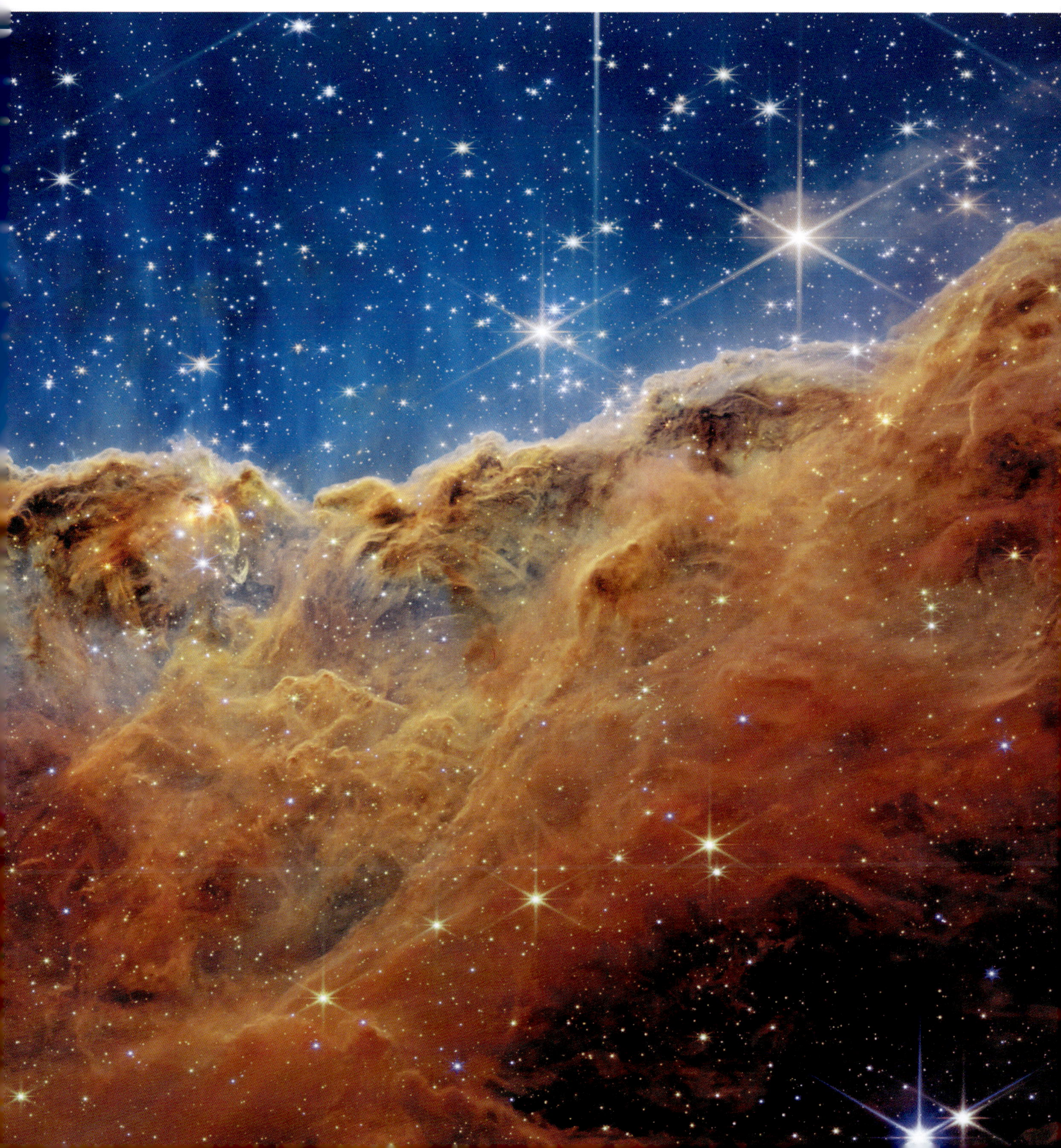

THE CARINA NEBULA IS ONE of the largest diffuse nebulae that we can see from Earth. It's as bright as the Orion Nebula and four times as large. It was discovered by French astronomer Nicolas-Louis de Lacaille in 1752 from the Cape of Good Hope. Carina means 'keel' or the bottom of a ship. Many ancient civilizations saw Carina as part of a larger constellation, the Argo Navis, which resembled a ship. Later astronomers decided that such a constellation was too large and so broke it up into smaller components.

COSMIC CLIFFS

In the image on the previous page, Webb has captured the early star-forming region in NGC 3324, known as the Cosmic Cliffs, against the backdrop of space. It is in the north-west corner of the

An image of the Cosmic Cliffs created through a combination of NIRCam and MIRI data.

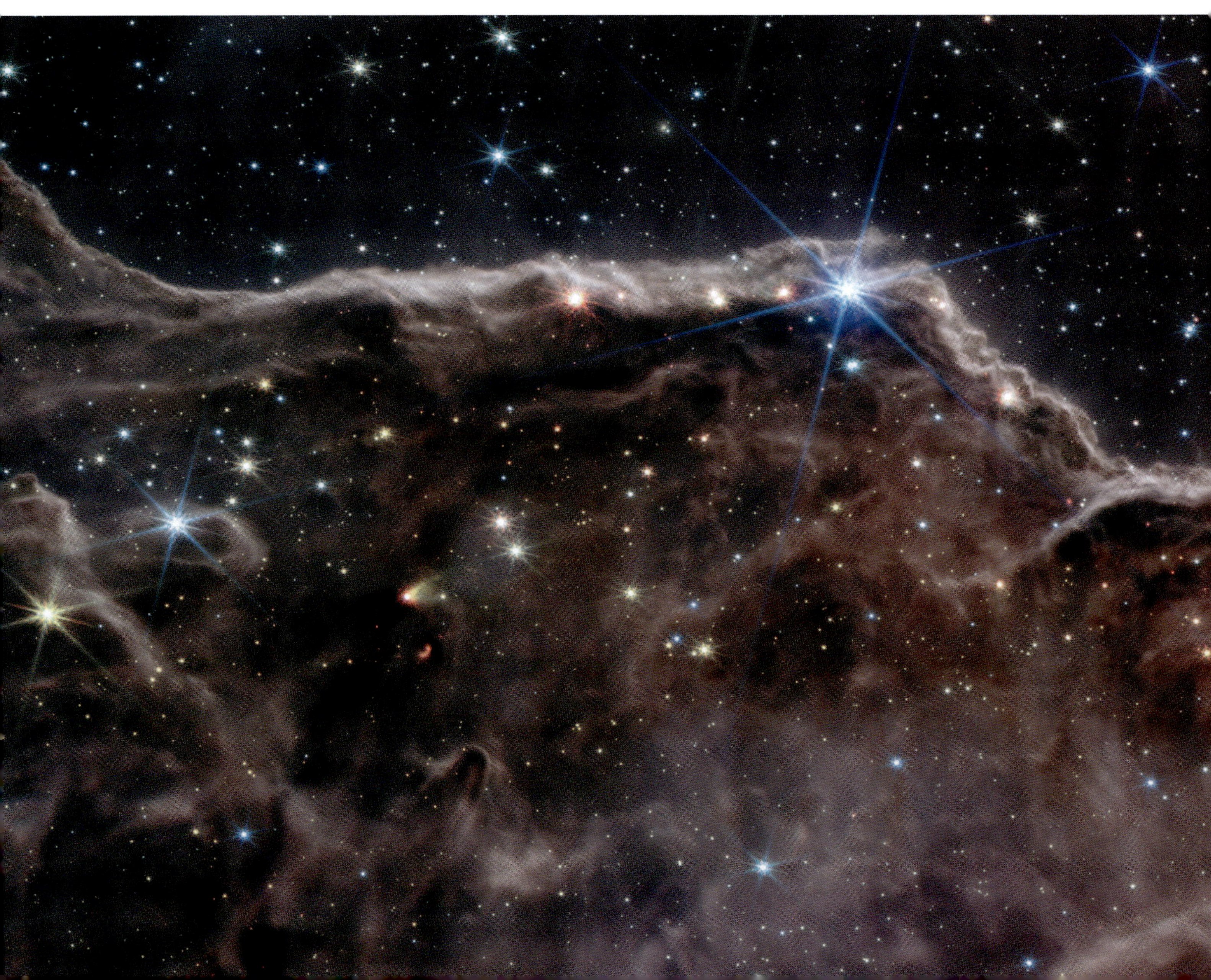

Carina nebula. The red, brown and yellow half of the image, which resembles mountains, is actually a gaseous cavity within NGC 3324. It is surrounded by extremely massive, very hot stars, and their radiation is slowly eroding the wall of the nebula.

Webb's sensitivity has uncovered hundreds of previously hidden stars in this region. Although Hubble has looked at this section of space before, it wasn't able to see these stars. The largest stars appear closer and brighter than the smaller, fainter ones. In the top-right corner, you can see the light spokes of a very large, very bright star that is just out of view.

The intense radiation makes the 'mountains' look as though they're steaming, as gas and dust flow away from the nebula. Golden light streams from dust-enveloped young stars.

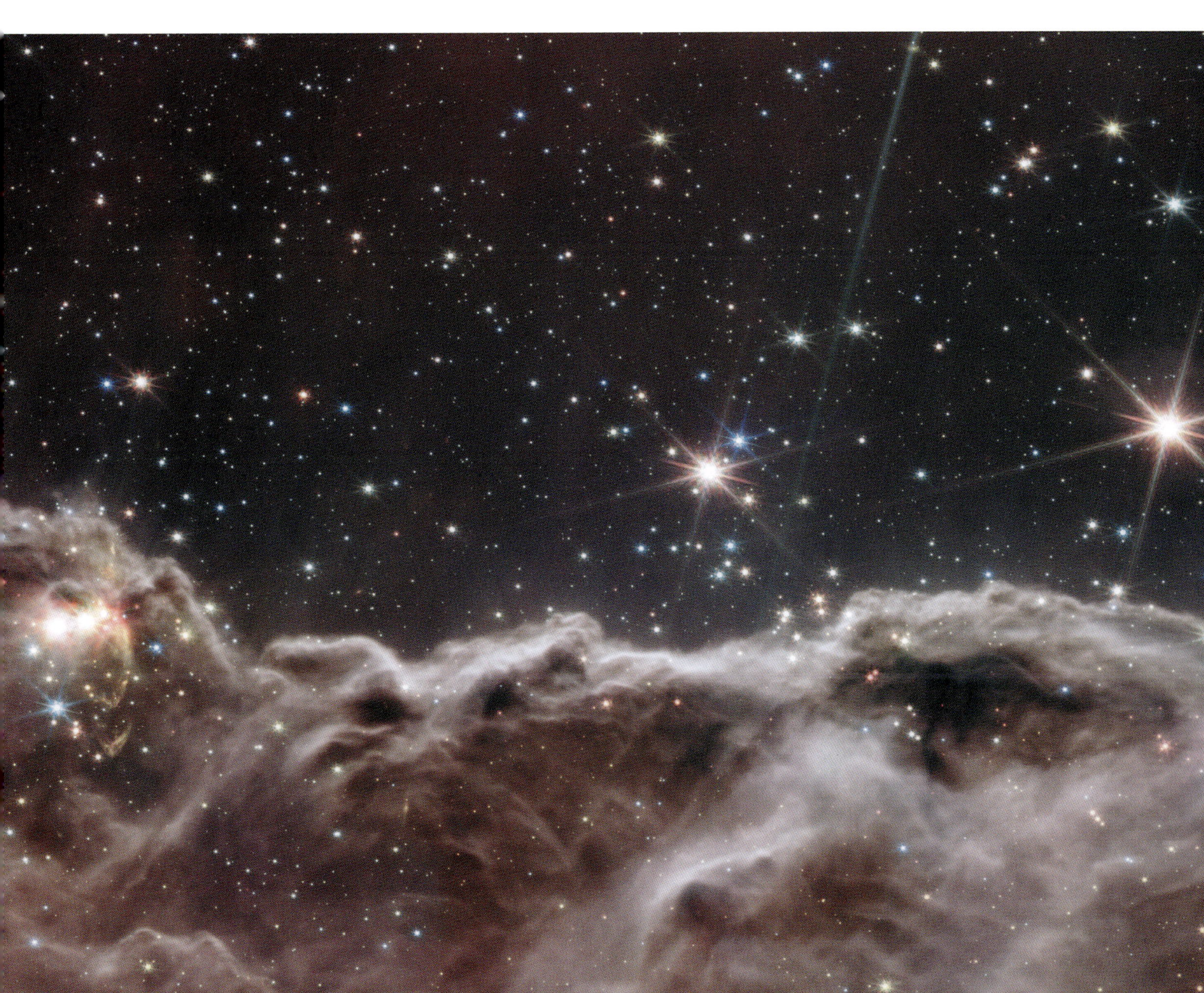

TARANTULA NEBULA

ALSO KNOWN AS 30 DORADUS

Location: in the Large Magellanic Cloud (LMC), a satellite galaxy of the Milky Way
Constellation: it straddles Mensa and Dorado
Distance from Earth: 159,800 light-years

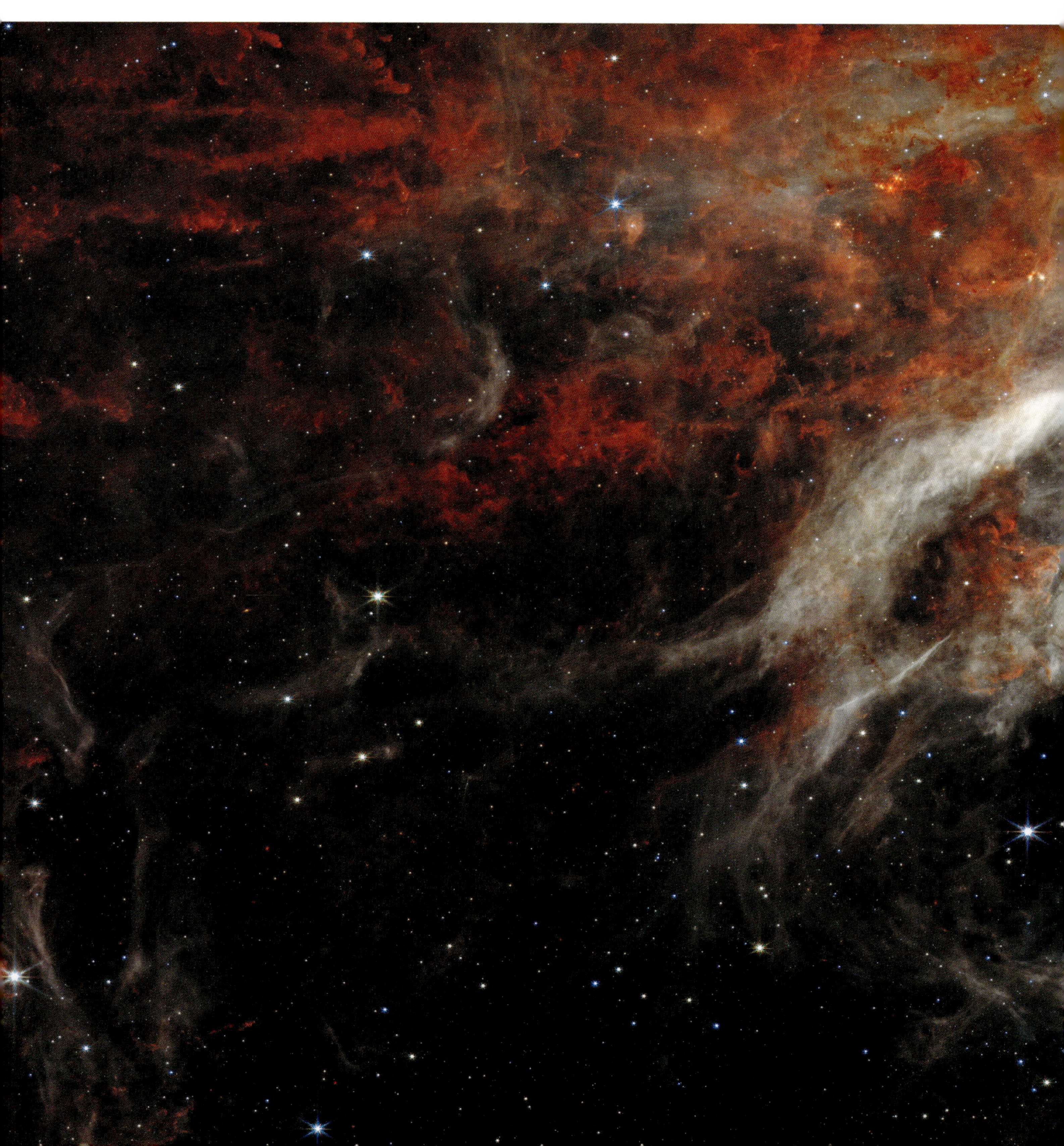

This image of the Tarantula Nebula's star-forming region, captured with NIRCam, spans 340 light-years.

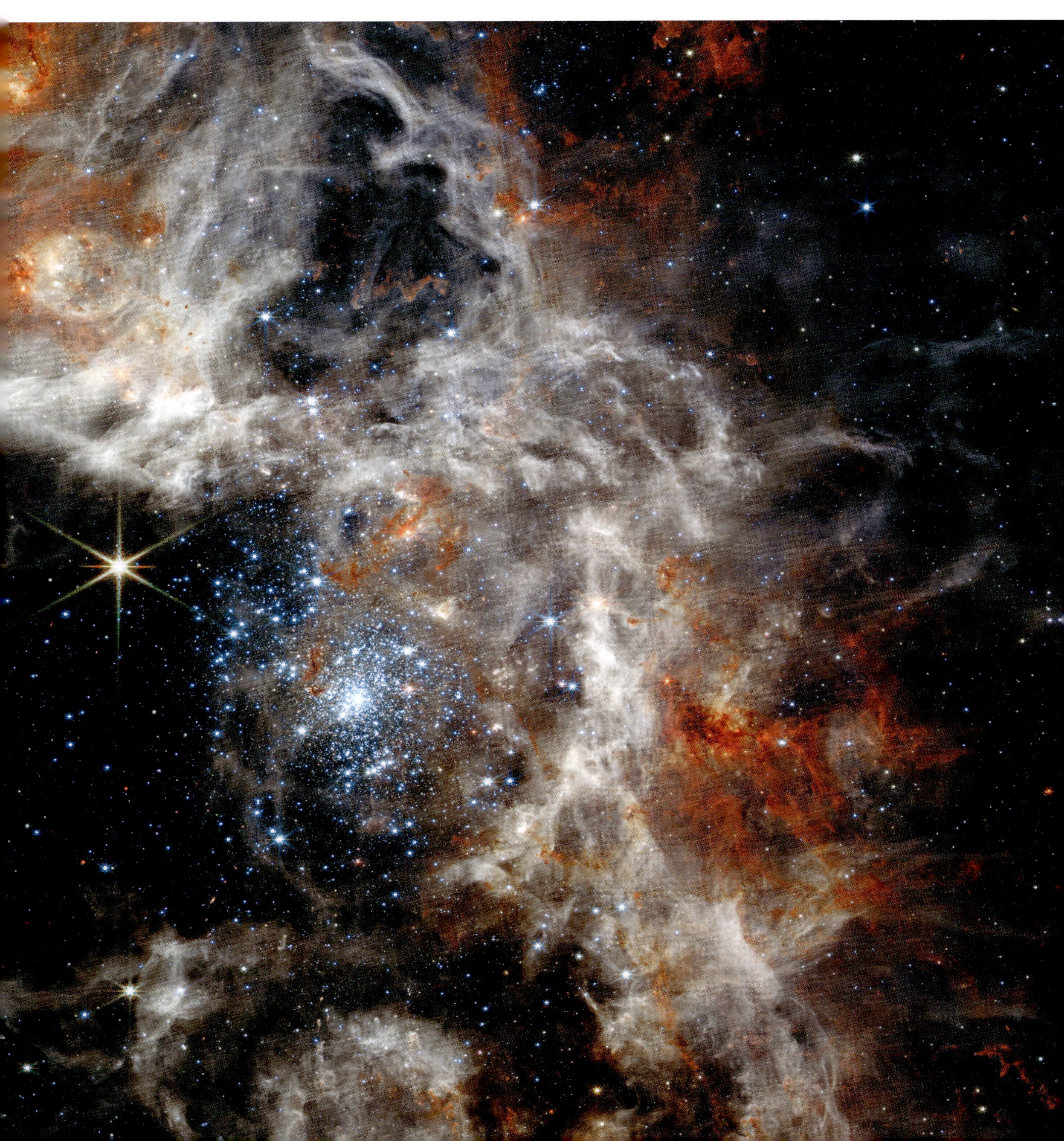

THE TARANTULA NEBULA IS VISIBLE with the naked eye, and can be spotted as a large milky patch. It is a humongous 160,000 light-years across. It was originally named 30 Doradus, but its complicated internal structures and criss-crossing dusty filaments led people to begin calling it the Tarantula Nebula.

The nebula is a H II region, which means that the hydrogen contained within it is not neutral but is ionized (positively charged). The nebula is home to some of the hottest and biggest stars we know. In Webb's images of this nebula, we can see the sensitivities and strengths of the telescope's four scientific instruments.

THE MYSTERIOUS LIVES OF STARS

When we overlay Webb's glorious infrared detection with Chandra's X-ray data, we see a version of the universe that we couldn't have imagined. In this image (bottom right), the infrared signals are red, orange, green and light blue, but with X-ray data included (coloured dark blue and purple). The active whorl of stars in the cavity are visible in startling clarity.

The royal-blue and purple swathes of the image show gases that stellar shockwaves have heated to unfathomably high temperatures. The infrared observations highlight the protostars swaddled in the nebula and surrounded by clouds of cooler gas that these stars will consume as they grow.

LEFT: In MIRI's image (left), the psychedelic hydrocarbons make the nebula appear more ghostly.

RIGHT: A composite image of the Tarantula Nebula using data from Chandra and Webb.

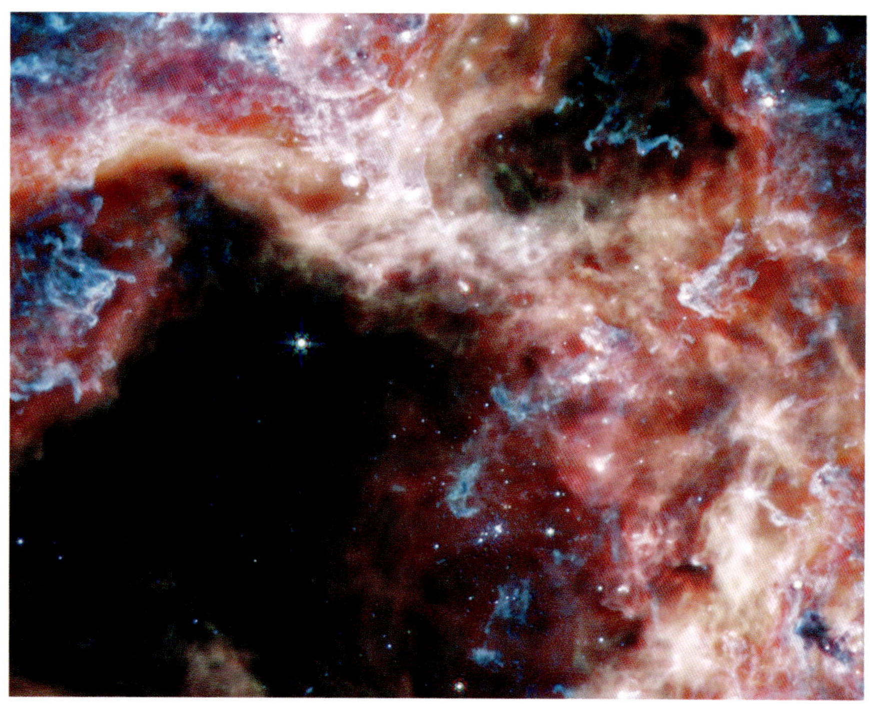

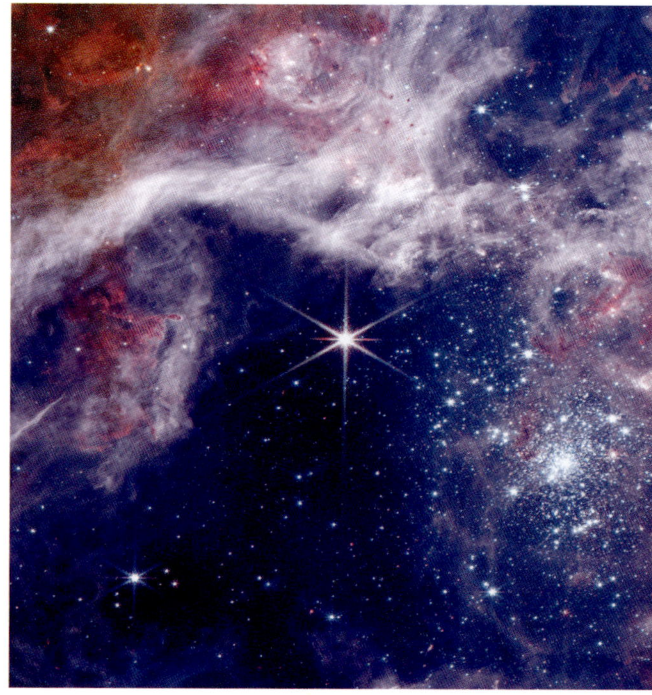

Scientists have found that the Tarantula Nebula's chemical composition differs from that of most nebulae in our galaxy, suggesting that it is like the conditions prevalent when the Milky Way was much younger. In practical terms, this means that the Tarantula Nebula allows us the opportunity to learn how stars formed in our galaxy long ago – which is why the region is of such great interest.

A SWIRL OF ENERGETIC YOUNG STARS

In the main image, NIRCam was able to detect radiation that passed through the dust clouds to reveal tens of thousands of previously unseen stars. A cavity dominates the image, and inside it there is a swirl of bright-blue stars that are young, huge and active. Interspersed between them are red dots, which are stars still enveloped in the nebula's dust.

A large, older star shines prominently out of the cavity in bright gold. Just above it, within the dust cloud, there is an almost round bubble with a golden halo. This is the beginning of a new cavity, its edges gouged out of the dust by nearby young stars. Away from the cavity, drifts of red clouds hang in the nebula. These comprise cooler gas that is full of complex hydrocarbons, which will ultimately form the core of new stars.

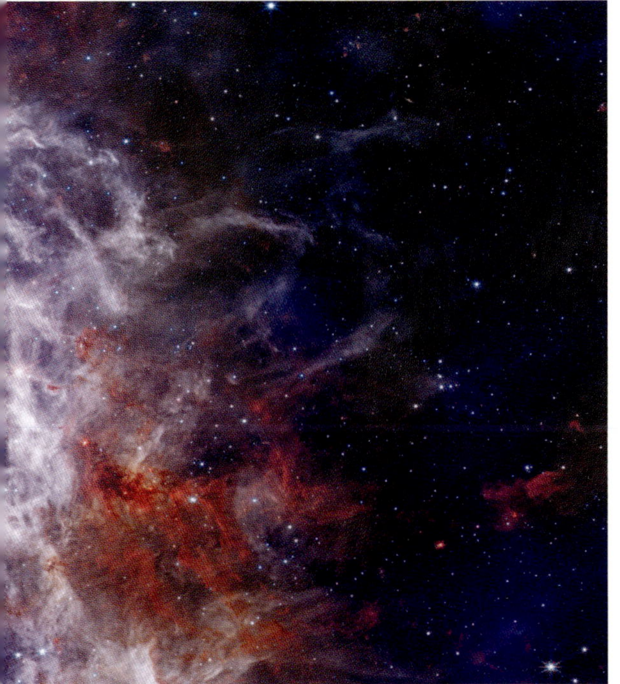

PSYCHEDELIC HYDROCARBONS

In the bottom left image, MIRI offers a very different and more psychedelic interpretation of the Tarantula Nebula. With its longer wavelength detection, the hot bright stars fade into obscurity while the rich hydrocarbons become more pronounced. Coloured here in eye-catching cyan and purple, these molecules outline the dust clouds.

Longer wavelengths can escape from deeper within the nebula, showing the phenomena occurring inside its cloudy shroud. So, in this MIRI image, we can see new protostars being born. But there are things that even Webb cannot see, such as the dark regions of dust from which not even the infrared signals can escape.

CRAB NEBULA

ALSO KNOWN AS M1, NGC 1952, TAURUS A

Location: in the Perseus Arm of the Milky Way
Constellation: Taurus
Distance from Earth: 6,500 light-years

By combining MIRI and NIRCam data, scientists have created a new perspective on the Crab Nebula, which is actually the remnants of a supernova.

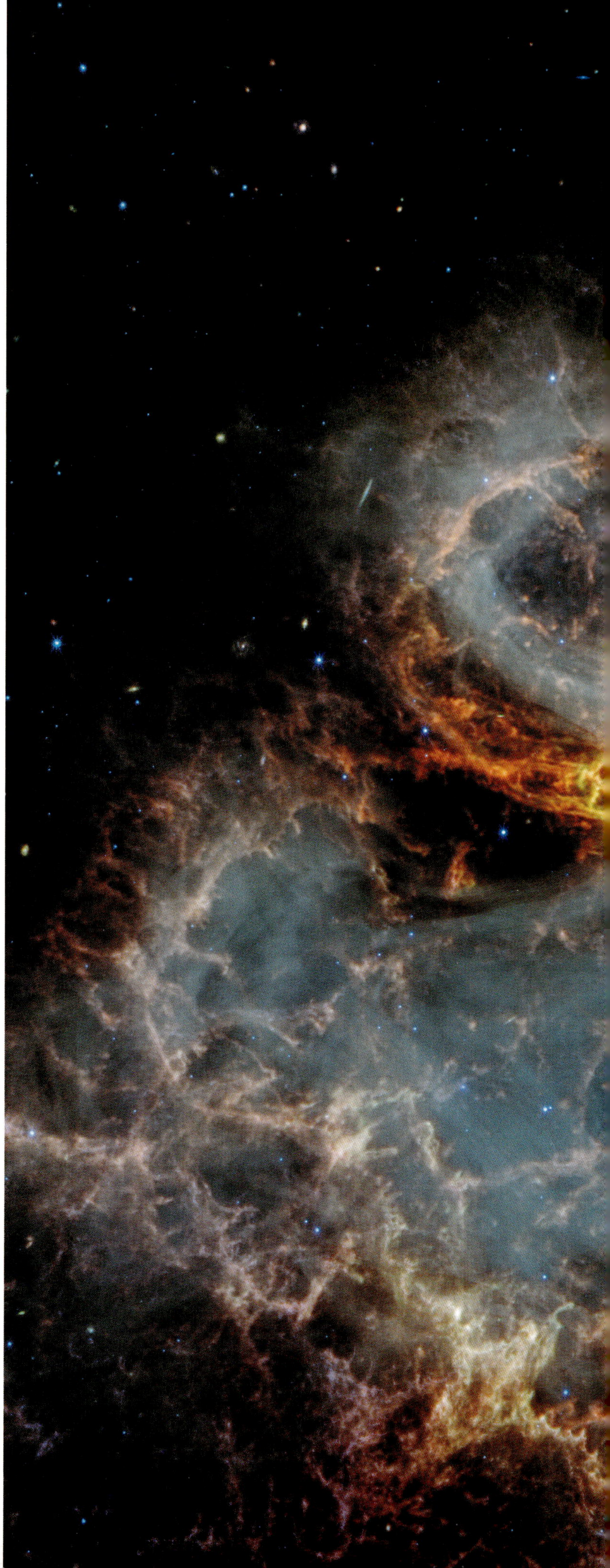

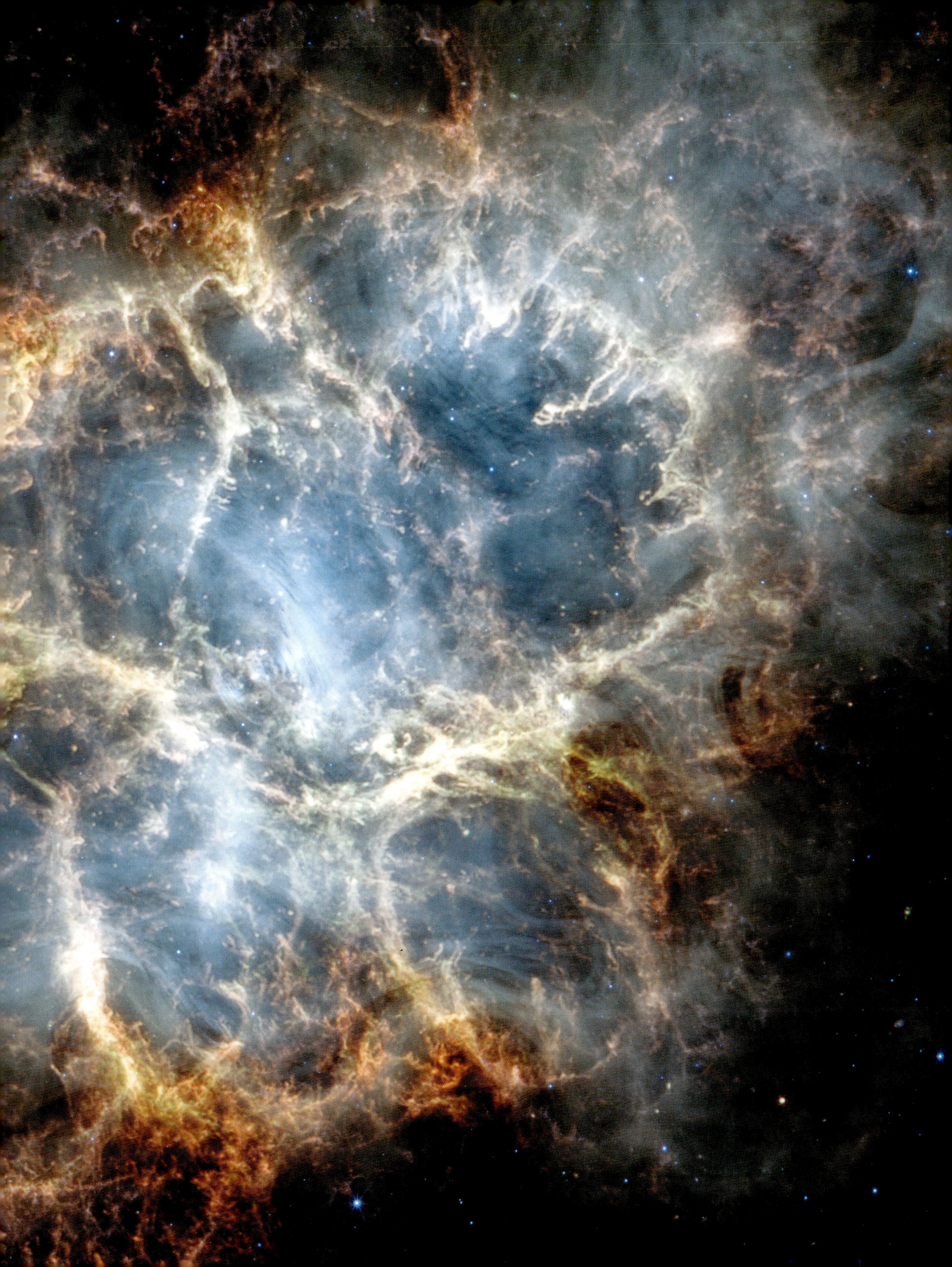

IN 1054, CHINESE ASTRONOMERS OBSERVED a 'guest star' in the sky. It was so bright that witnesses could see it in the daytime for nearly a month. It was actually a supernova explosion thousands of light-years away. Seven hundred years later, English astronomer John Bevis identified it as the Crab Nebula. It is the first identified astronomical object that corresponded with a historically observed supernova explosion. Today, it is not visible to the naked eye, but can sometimes be seen with good binoculars.

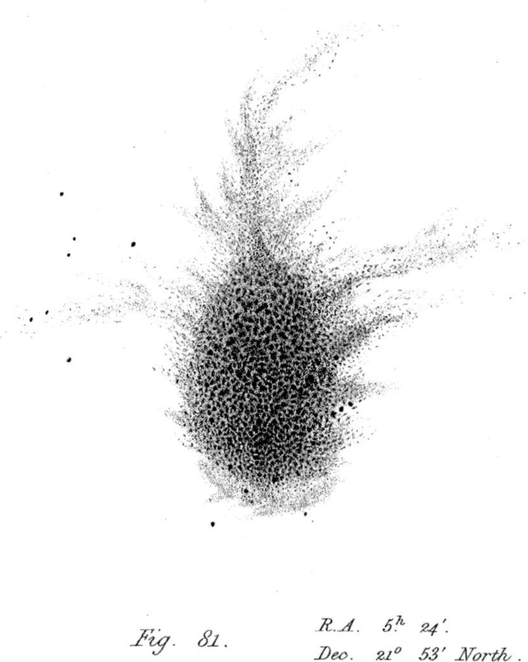

British astronomer William Parsons' drawing of the Crab Nebula.

When British astronomer William Parsons observed it in the 1800s, he drew it with arms like a crab. Later observations were unable to replicate these arms, and the nebula doesn't actually look much like a crab, but the name stuck.

At its heart, the Crab Nebula is a pulsar, a special type of fast-rotating neutron star with a strong magnetic field that funnels particles along their magnetic poles. This results in the poles emitting a beam of electromagnetic waves that travel far out into space. Each time this beam sweeps past Earth, we can detect it as pulses of radiation. Hence they get their name 'pulsar'. Neutron stars are thought to be the remnants of supernovae after gravitational collapse has occurred. They are usually the size of a large town, around 30 kilometres (19 miles) across, but are also among the densest things in the universe – only black holes are more dense.

Pulsars can give us insight into the strange objects that are neutron stars. These stars rotate with incredible regularity, so we have been able to observe their pulses as very accurate natural clocks. If any deviations occur from their regular natural frequency, it is likely that this is caused by internal mechanisms taking place or interference with the waves as they pass through space. So any change in the frequency of their pulses allows us to study some of their inner workings or analyse objects that sit between us and the pulsar. For example, radio astronomers used radio waves emitted by the Crab Nebula's pulsar to map the sun's corona. As the pulses passed through the sun's outer atmosphere, distortion in the atmosphere disrupted the frequency of the pulsar's pulses.

The Crab Nebula is also known as a 'plerion', which is a term to describe a pulsar wind nebula. This special class of nebula is powered by the winds from the neutron star that lies within it. Neutron stars have a powerful magnetic field, and as these stars spin very rapidly so does the magnetic field. This spinning magnetic field accelerates charge particles that encounter it, creating a wind. When these particles interact with the surrounding medium, radiation is generated. The Crab Nebula is one of the best examples of this phenomena.

THE CRAB IN INFRARED

By combining data from NIRCam and MIRI, Webb has given us an unprecedented view of the Crab Nebula in the infrared (see previous page). While its general shape mimics what we see with Hubble (below), the images from Webb tell us about its inner workings and the elements and molecules that make up this grand, iconic structure.

Pulsars warp space and matter around them. Their very strong magnetic field is a great source of synchrotron radiation, which is the emissions produced when charged particles move through a magnetic field at high speeds. In the Webb image of the Crab Nebula, synchrotron radiation is white and looks like smoke billowing out of the nebula. By tracing the smoke back to its source, you can see the white pulsar at the nebula's heart. The smoky wisps also curve along the pulsar's magnetic fields, which mould and manipulate the structure of the nebula.

Charged sulphur, coloured red-orange, encases the structure, like vines growing up a building, while blue iron dots the image. Unlike its predecessors, Webb has managed to capture grains of dust that billow within the nebula's cage, seen in yellow-white and green. These new insights will help us gain a better understanding of what kind of star the Crab Nebula's pulsar once was and what will happen to its remnants in the future.

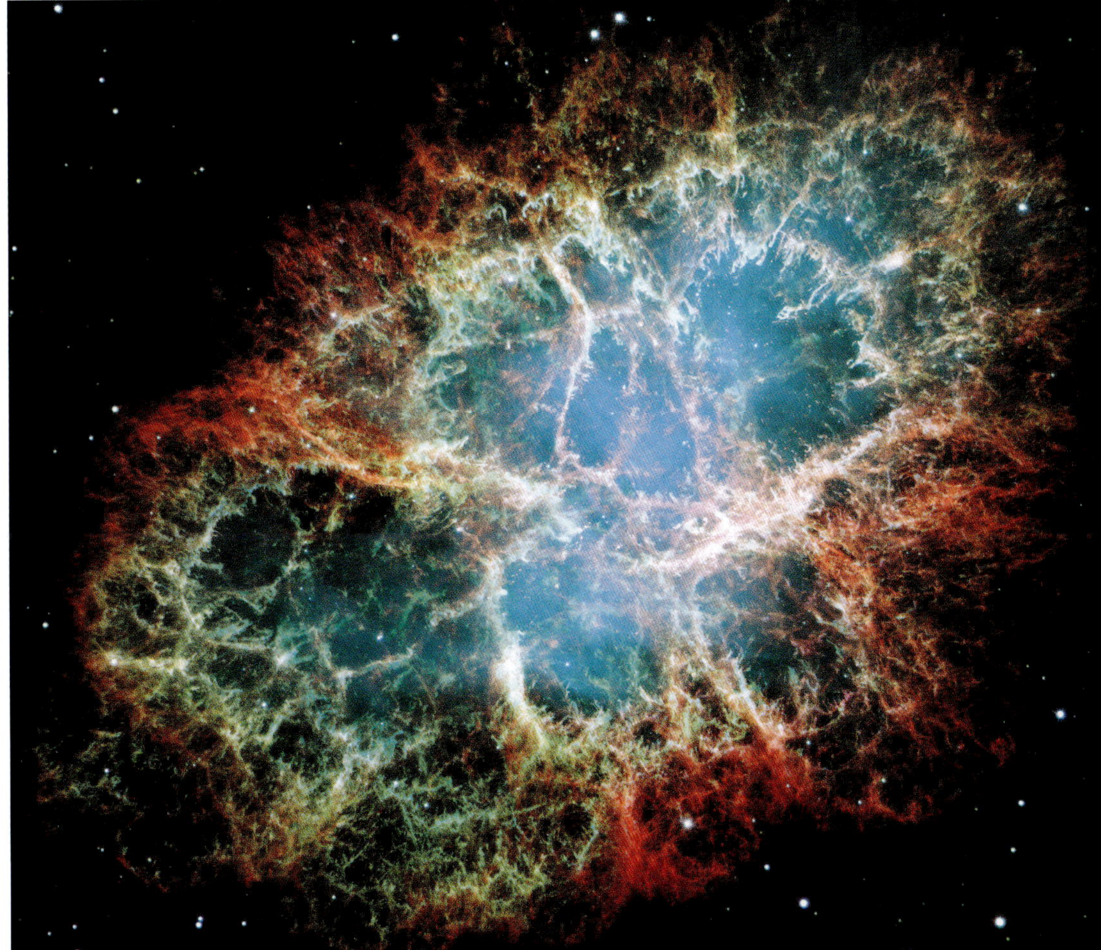

A view of the Crab Nebula from the Hubble Space Telescope.

THE PILLARS OF CREATION

Location: in the Eagle or Star Queen Nebula
Constellation: Serpens
Distance from Earth: 6,500 to 7,000 light-years

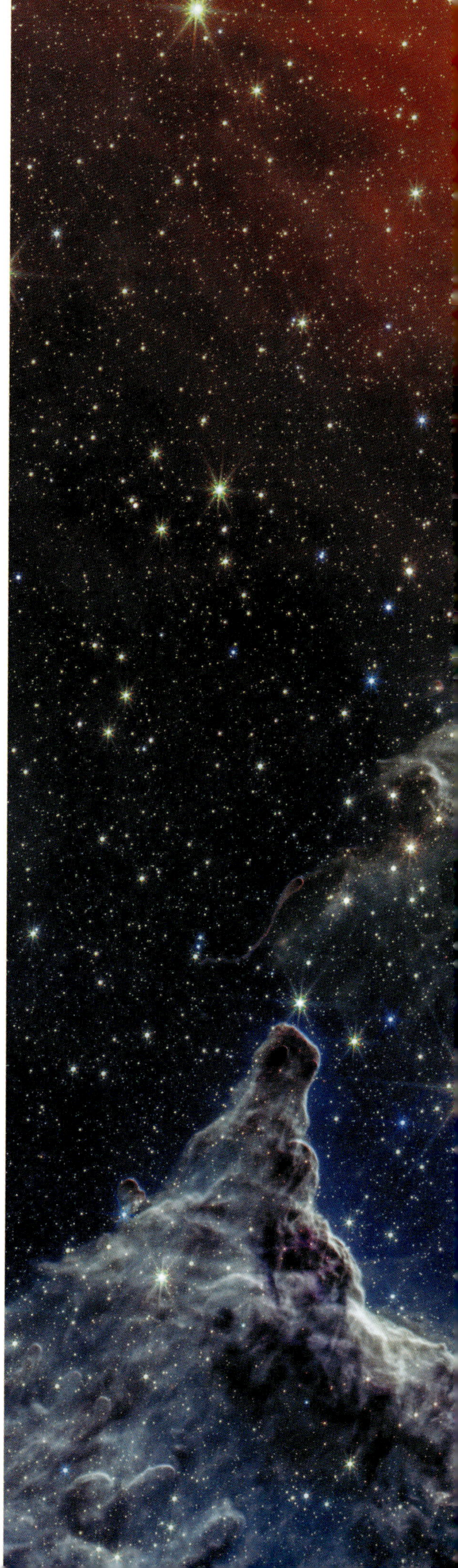

This image of the Pillars of Creation in the Eagle Nebula, made up of MIRI an NIRCam data, shows some amazing new detail of the famous star-forming region.

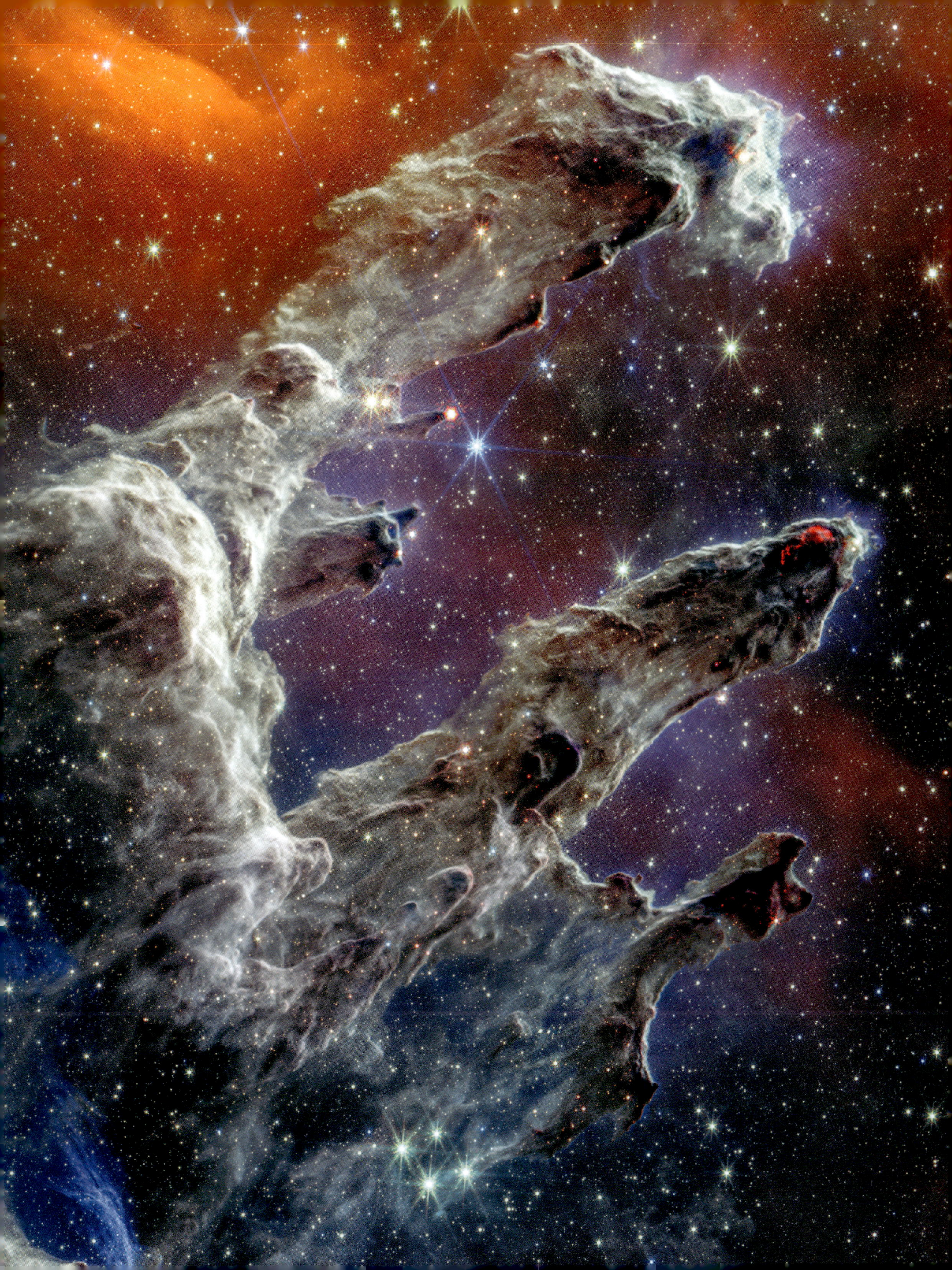

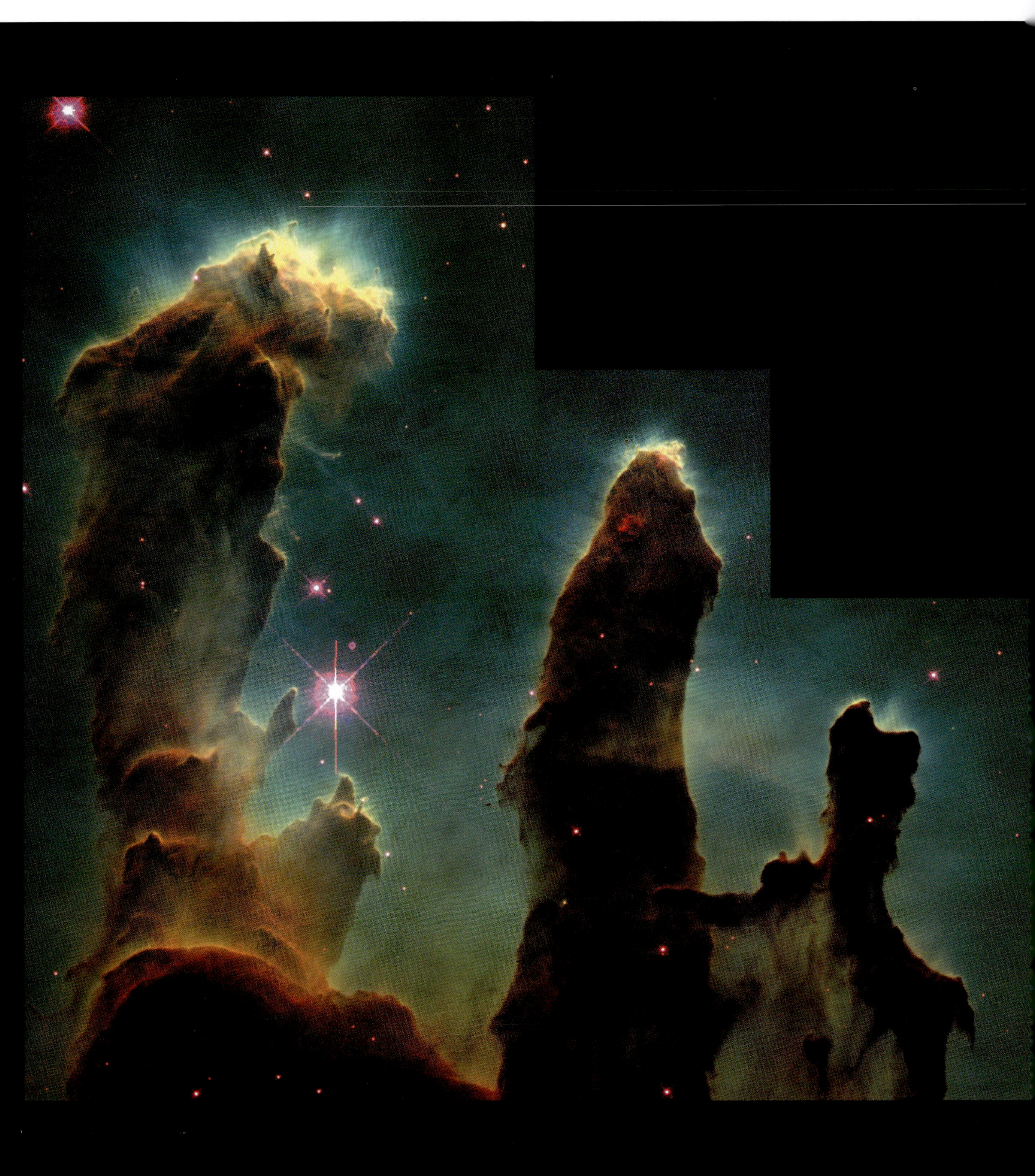

LEFT: This picture of the pillars was taken on 1 April 1995 with the Hubble Space Telescope Wide Field and Planetary Camera 2. The missing blacked-out part of the photo is because the camera for that quadrant had a magnified view, meaning that when the image was scaled down to fit with the rest, the gap appeared.

THE PILLARS OF CREATION WERE first brought to the attention of the general public when they were photographed by the Hubble Space Telescope in 1995. But we believe that an image of this region was first captured on photographic plate some seventy-five years earlier, in 1920, by John Charles Duncan using the 1.5-metre (5-feet) Mount Wilson Telescope, the largest telescope in the world at that time (see black-and-white image below).

To give you some idea of the scale of what we're looking at, the longest pillar on the left-hand side of the image opposite projects some four light-years into space and the small projections on the end are about the size of our whole solar system.

LIFE AT DIFFERENT WAVELENGTHS

The name Pillars of Creation was coined because star formation happens in nebulae like this. The incredible new Webb images of the pillars show the four gargantuan columns of dust and gas jutting out into the nebula where they reside. By investigating this area using different parts of the electromagnetic spectrum in our analysis, we can develop a better understanding of how star formation occurs within these stellar nurseries.

BELOW: American astronomer John C. Duncan is thought to have been the first to capture the great pillars on film. He took many pictures of the nebulae and galaxies from the Mount Wilson Observatory in Los Angeles, California, which he first visited in 1920.

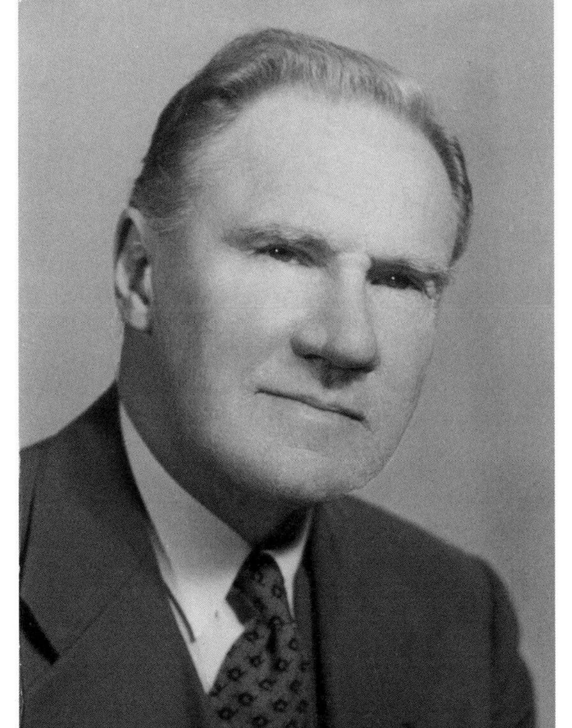

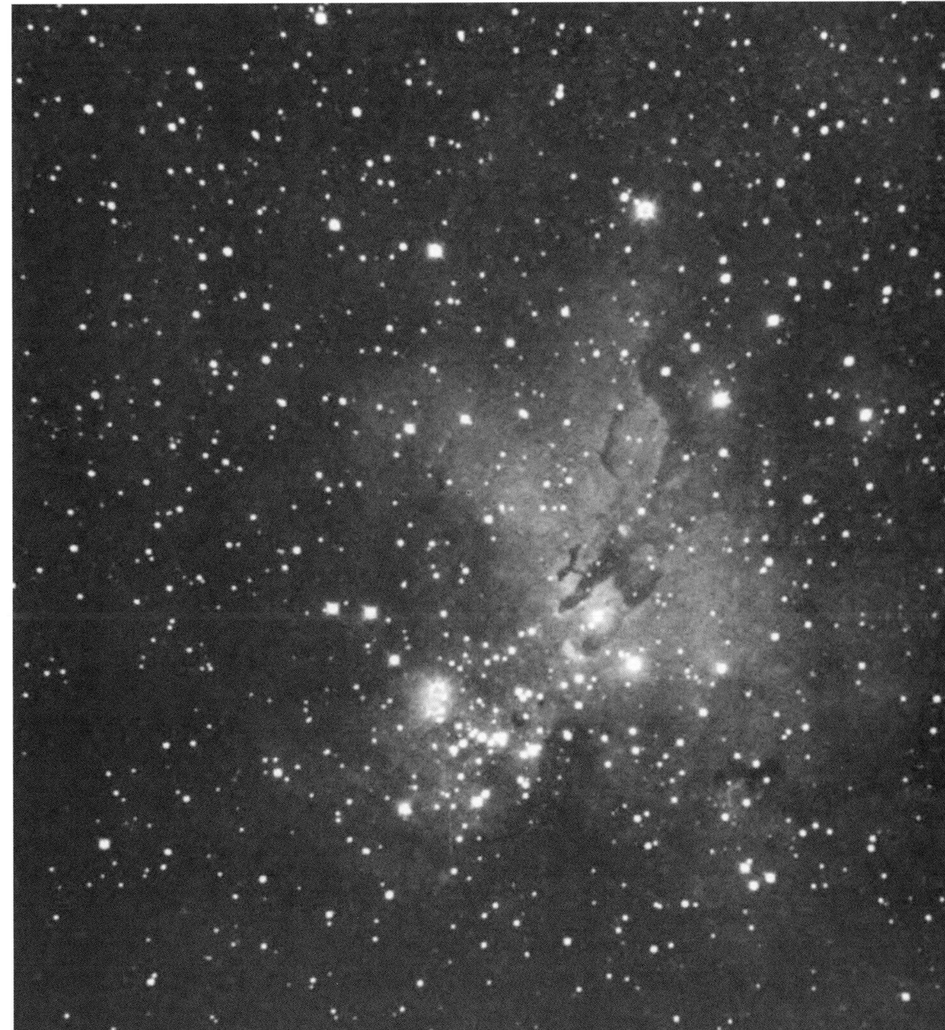

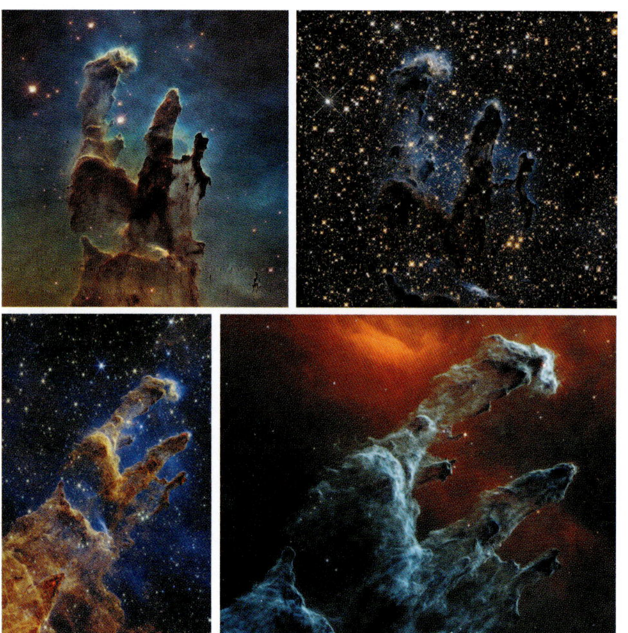

The top left picture is a visible image taken by Hubble. Top right is also by Hubble but using its near-infrared detector. The two Webb images are taken at wavelengths of the NIRCam imager (bottom left) and the MIRI imager (bottom right).

The four images on the left show us the different wavelengths of light emitted by the pillars and give us some insight into their inner workings. As we explored earlier in the book, infrared waves have longer wavelengths than visible waves, allowing them to pass through dense regions of dust and gas with much less scattering and absorption.

Each image shows distinct aspects of the star-formation process. The two Hubble images collecting visible and very-near-infrared light show how the different frequencies reveal different details. Compared to the visible image, the very-near-infrared image allows many more stars to be seen and reveals details of what lies behind the dust clouds in the pillar. But some of the most staggering detail can be seen in the Webb near-infrared image shown in close-up on the opposite page.

THE BIRTH OF STARS

At the ends of the pillars, bright-red patches can be seen. This marks areas where new stars are forming and shooting out supersonic jets of gas that interact with the gas and dust of the nebula, sometimes creating bow shocks that can be seen as wavy lines in the surrounding material. These new stars begin to form when large 'knots' of gas and dust collapse under their own gravitational attraction. Inside dense concentrations of gas and dust known as molecular clouds, molecules (mostly carbon monoxide and hydrogen) bind together at low temperatures to form clumps. When these clumps reach a certain density, stars form as the clouds collapse under their own weight.

Although this image allows us to peer through some of the dust that is more prominent in the visible image from Hubble, dense clouds of interstellar medium mean that what lies behind the pillars is still partly obscured – so we are unable to observe any far-flung galaxies in the background of this image.

We have been taking images of the pillars for many years now, but it's thought that in as little as 500 years' time we will no longer be able to see this epic structure in space. Some observations indicate that a supernova has happened in this region, sending a shockwave of matter out into the nebula and destroying the mighty pillars. However, while this distribution is thought to have occurred already – probably some 6,000 years ago – we will not see the full effects of the devastation for another 500 to 1,000 years because the pillars sit so far away from us.

RIGHT: NIRCam's perspective on the famous Pillars of Creation in the Eagle Nebula.

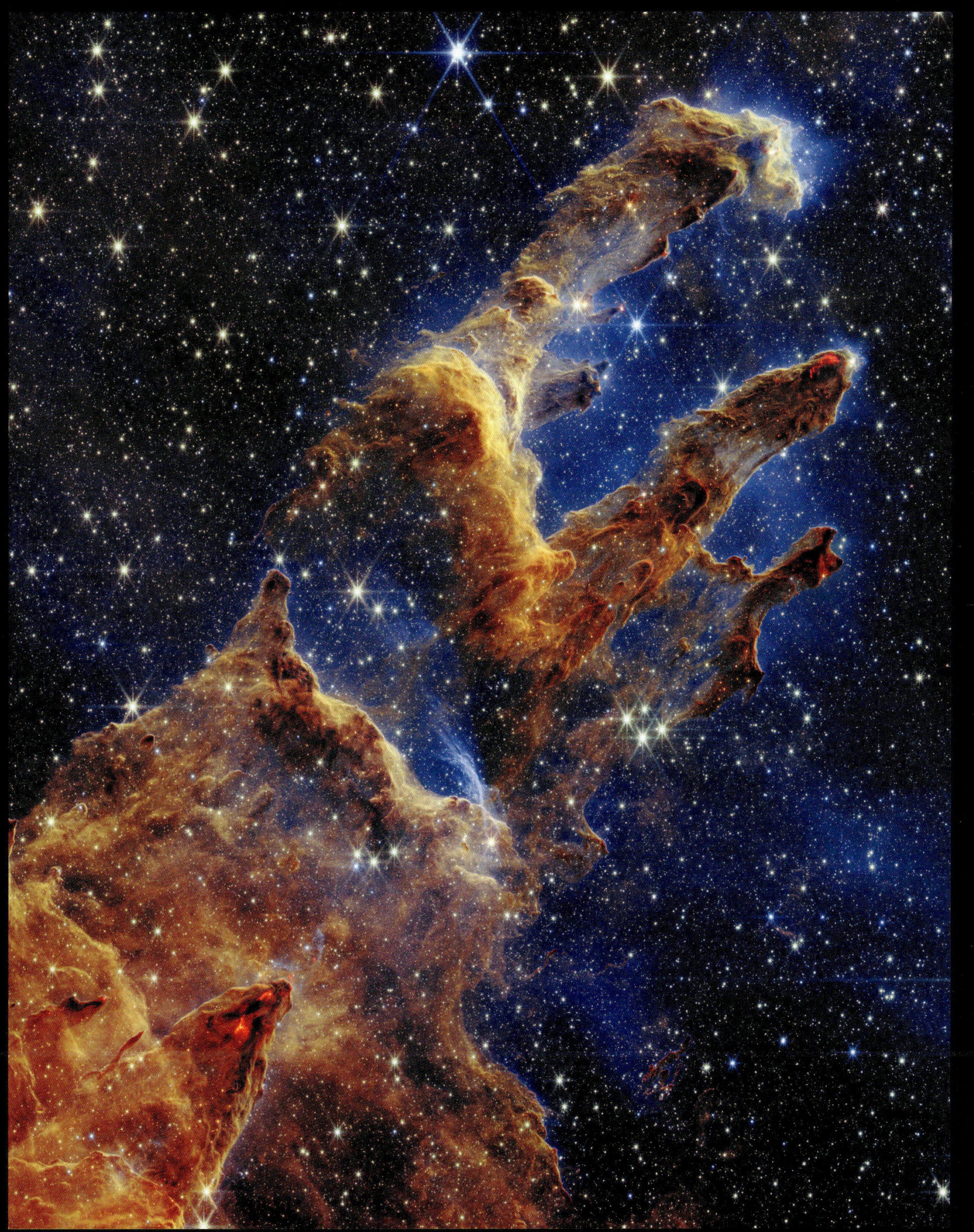

ORION NEBULA
ALSO KNOWN AS MESSIER 42 AND NGC 1976

Location: in the Milky Way, south of Orion's Belt
Constellation: Orion
Distance from Earth: 1,344 light-years

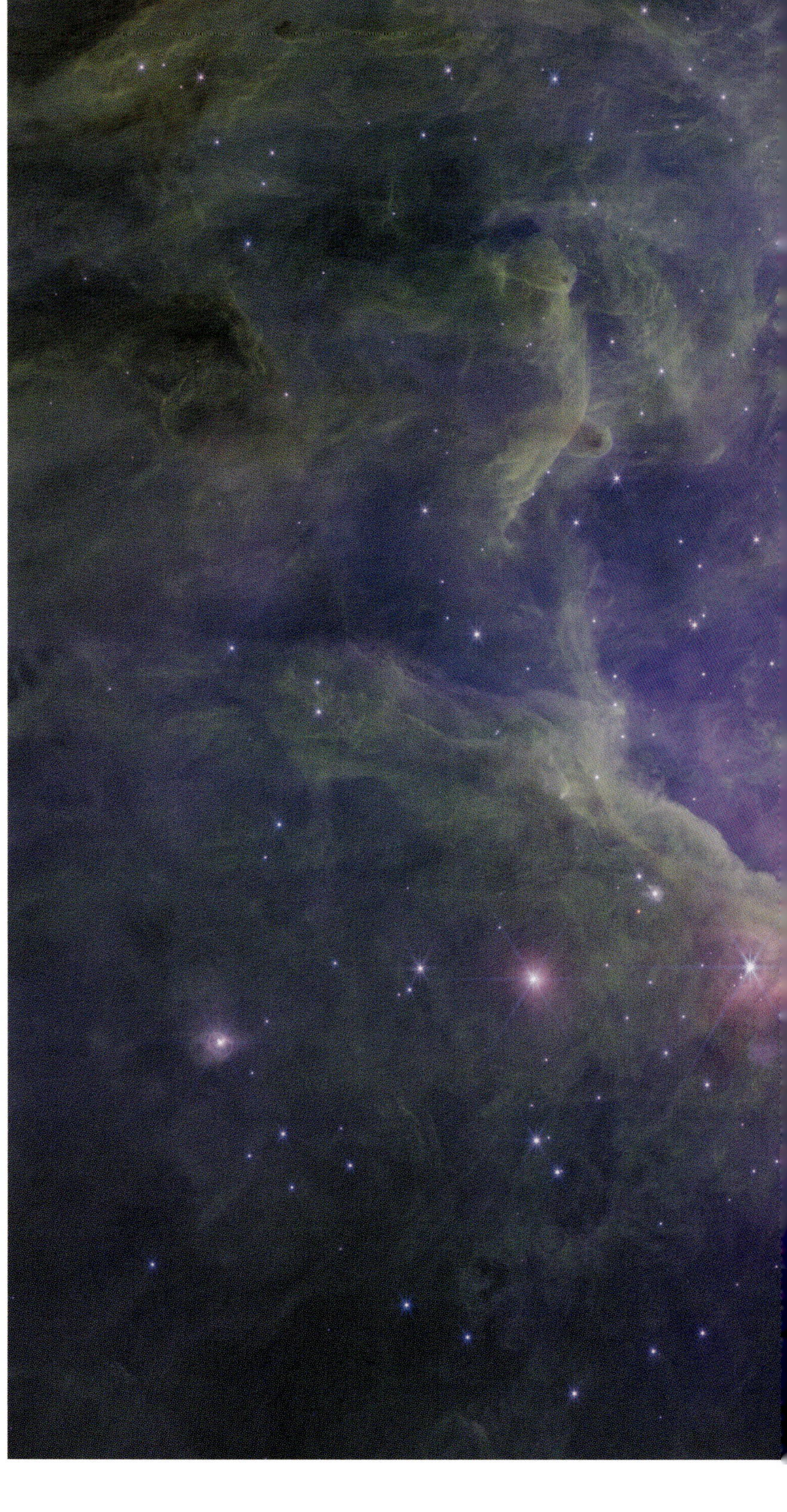

The Orion Nebula observed with NIRCam's long-wavelength channel.

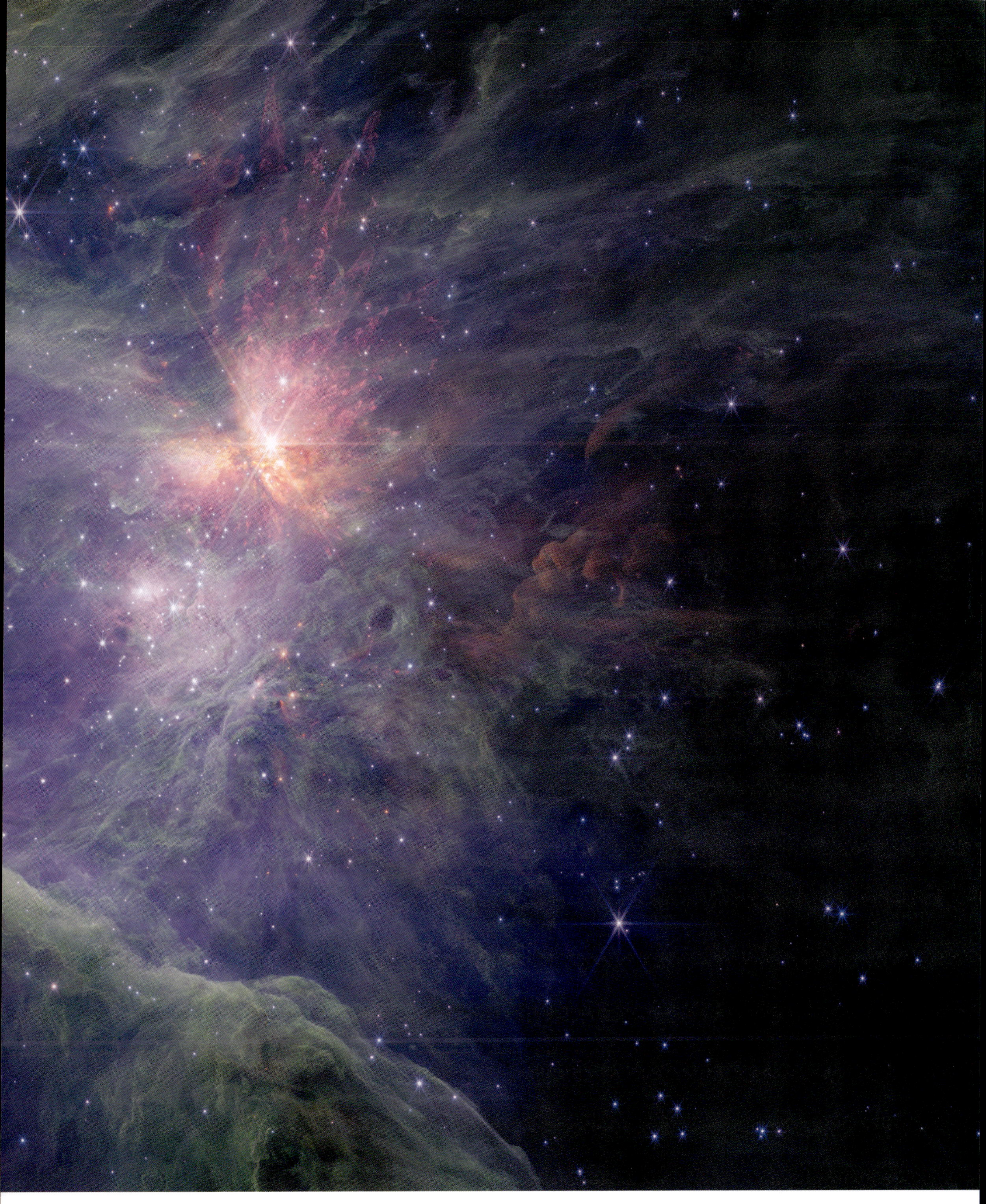

ALTHOUGH THE FIRST 'OFFICIAL' DISCOVERY of the Orion Nebula was in the seventeenth century, there is some evidence that suggests the ancient Mayans described it in their creation myth. Today, it is one of the best-studied nebulae, full of exciting physical phenomena. And we can all enjoy this nebula, as it is visible to the naked eye – it's the middle star in Orion's sword.

The Orion Nebula is a stellar nursery that's about 2,000 times the mass of our sun and is incredibly bright. It's actually part of a much larger nebula, known as the Orion molecular cloud complex, which extends throughout the Orion constellation. Stars are forming throughout the complex, but tend to concentrate in the Orion Nebula.

New stars are being born in its dusty recesses. But Webb's different instruments give us a new perspective on this well-studied object. It is easy to see the brightest stars in this image because they have eight prongs – Webb's famous diffraction spikes.

TRAPEZIUM CLUSTER AND WEBS OF GAS

Within the heart of the Orion Nebula are many thousands of young stars. Perhaps the most famous collection, the Trapezium Cluster, shines out of this Webb image (left), which is made up of short-wavelength infrared signals. Underneath the red volcanic explosion of infrared emissions and slightly to the left, there is a knot of remarkably bright stars. The cluster was first recorded in the seventeenth century, by Galileo Galilei, along with its five brightest stars, whose mass is the equivalent of fifteen to thirty of our suns. Unlike other telescopes, Webb's infrared vision can view the Trapezium's surrounding dust cloud and reveal these stars.

The Trapezium Cluster seen with NIRCam.

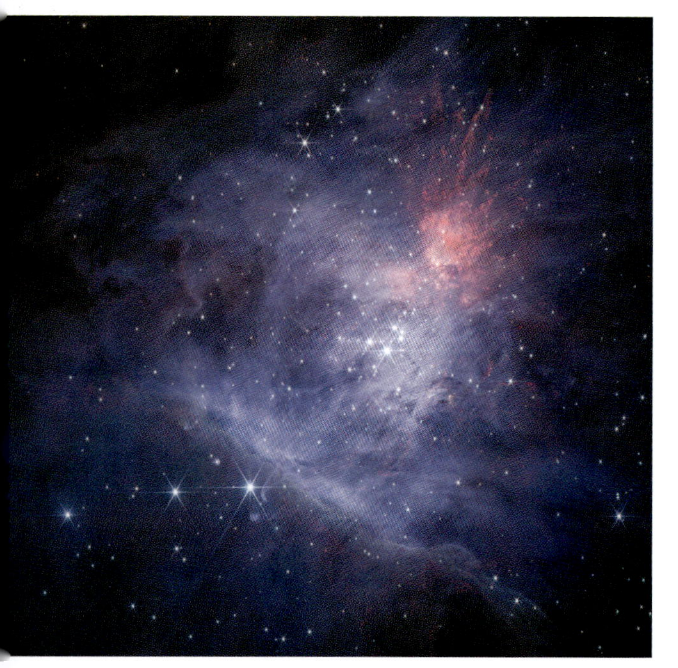

This star-forming region is about 1 million years old, which is relatively young when it comes to stars. And there is a great deal of variety: some of the stars are forty times the size of our sun, while others are a fraction of its mass.

When looking at longer wavelengths, Webb reveals a different facet of the Orion Nebula and its Trapezium Cluster. In the main Webb image, this cluster is less prominent and instead the nebula is a cobwebbed vista of gas and dust. The purple wisps are ionized gas, while the reds, browns and greens are billowing clouds of dust and molecular gas, including polycyclic aromatic hydrocarbons (PAHs) – which are multiple rings of carbon atoms linked together. The giant stars in the centre of the image are consuming the gas and dust, carving out a cavity.

A NEW TYPE OF ORGANIC MOLECULE

Stars change the space around them. The Webb images below home in on the area where the radiation from the Trapezium Cluster interacts with the molecular clouds in the nebula. The stars' ultraviolet light alters the environment and planet-forming areas. The largest image draws on NIRCam data, while the image on the top right is from MIRI. The bottom-right image is a combination of the two instruments and focuses on a relatively small part of space – a protoplanetary disk named d203-506.

Its star is a small red dwarf, and in this disk Webb's data was used to discover a carbon molecule that had never been found in space before. Methyl cation (CH_3^+) promotes the formation of more complex carbon molecules, which are the basis of all known life.

Fierce UV radiation is generally thought to destroy organic molecules like these, but there is now a school of thought that it may actually be necessary to provide the energy for the creation of CH_3^+ and form the basis of life-creating molecules.

This collection of images shows Orion's Bar, part of the Orion Nebula.

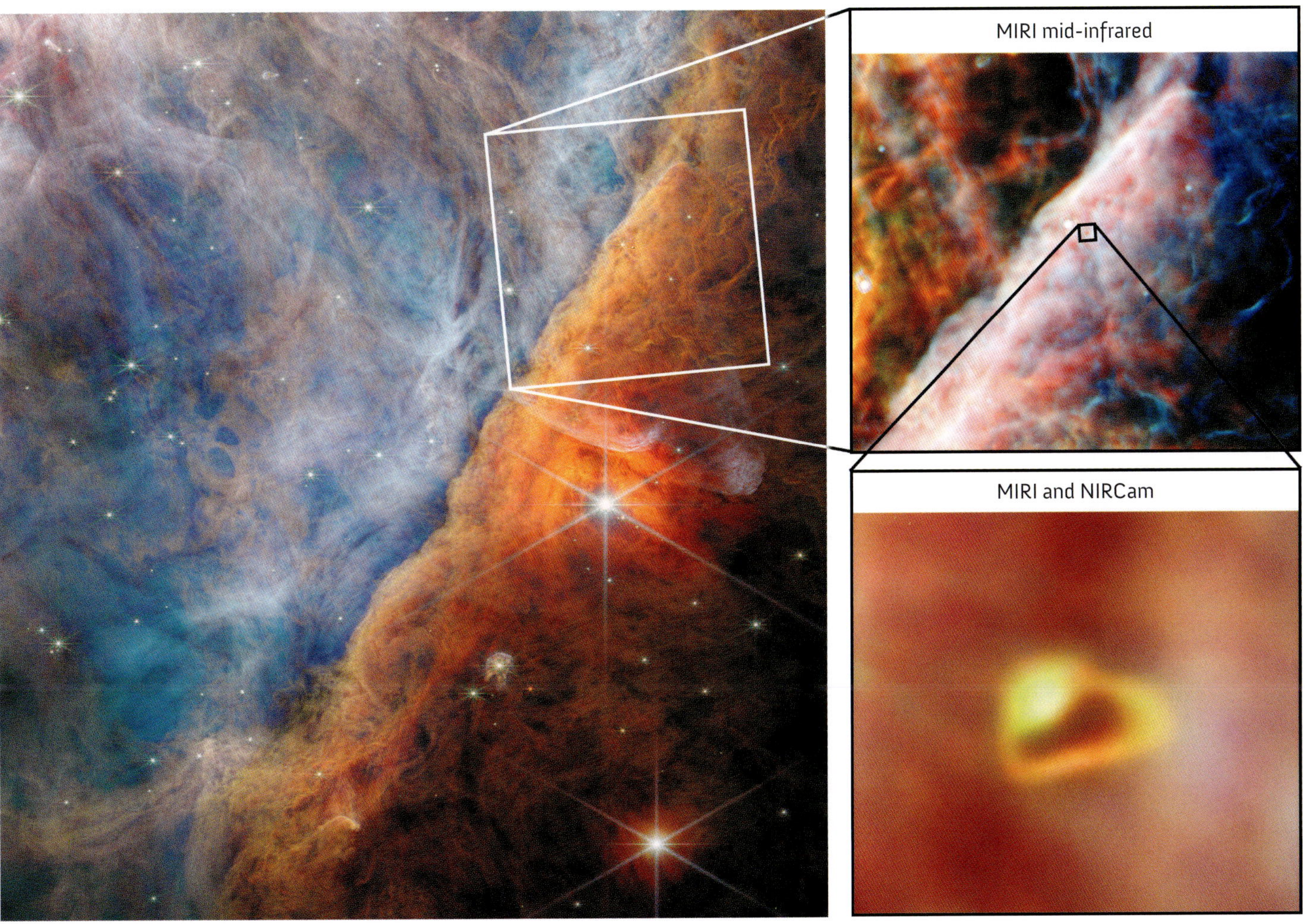

RHO OPHIUCHI CLOUD COMPLEX

Constellation: Ophiuchus
Distance from Earth: 390 light-years

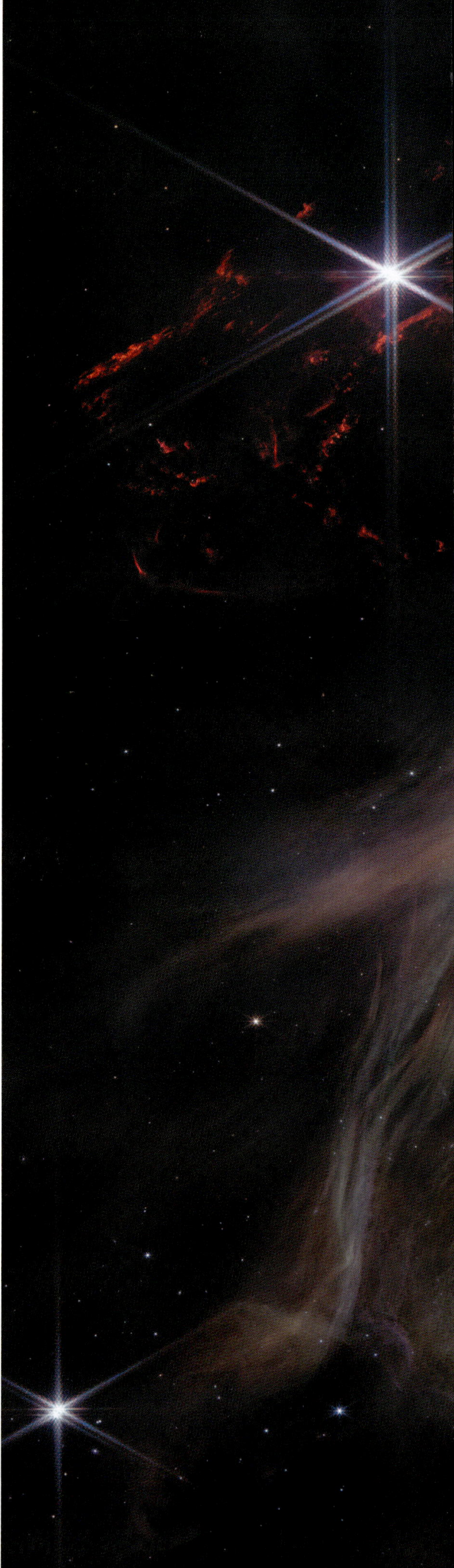

NASA released this startling image to celebrate Webb's first anniversary. It shows the Rho Ophiuchi cloud complex, the closest star-forming region to Earth, in unprecedented detail.

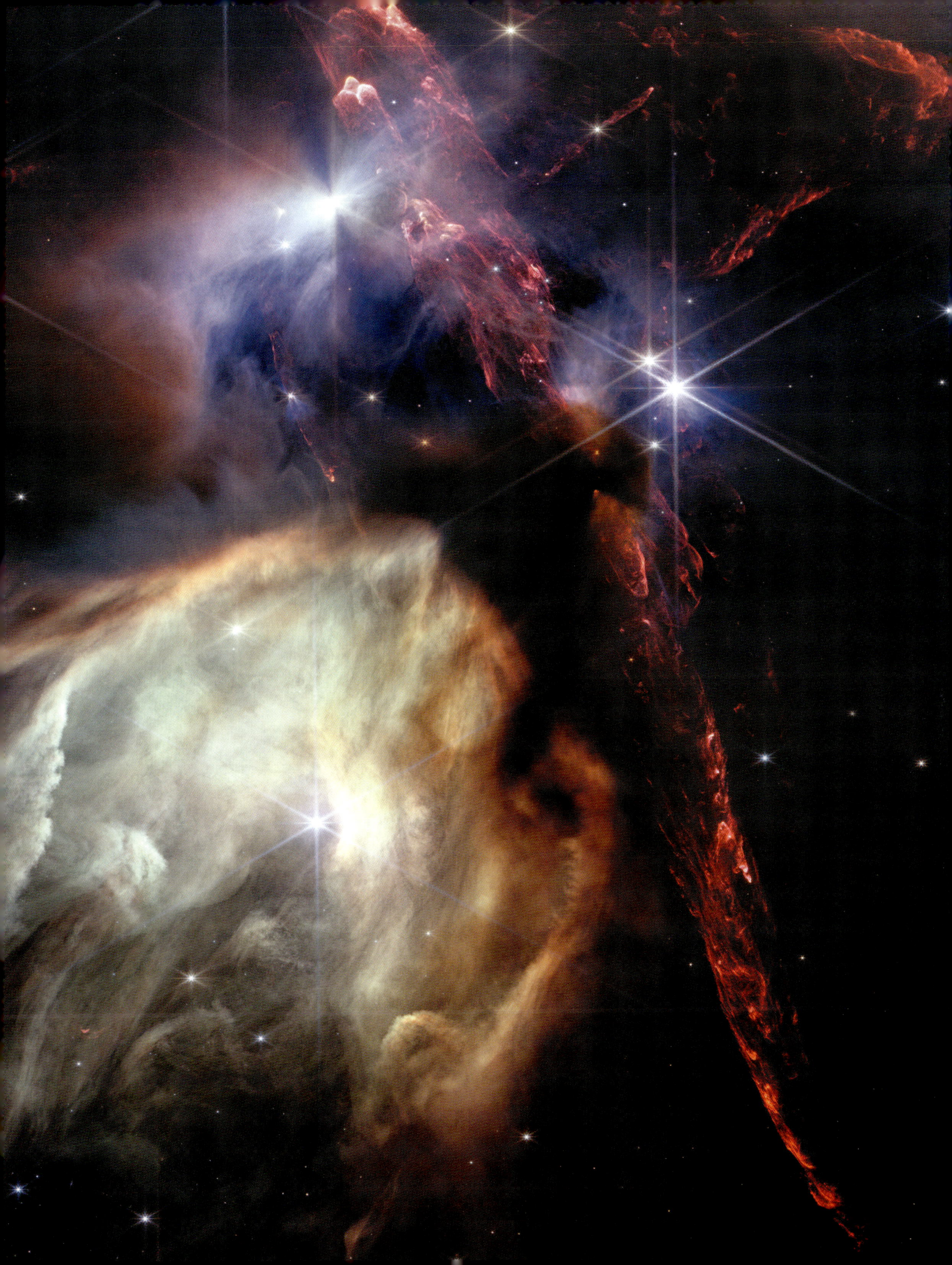

ABOUT 390 LIGHT-YEARS AWAY, IN the Rho Ophiuchi cloud complex, new stars are being born. The Rho Ophiuchi cloud complex is part of the Ophiuchus constellation and is our closest stellar nursery, containing many phenomena that fascinate astronomers. If viewed in only visible light, the scene would appear black. Webb, with its infrared gaze, allows us to see through some of the dust that blinds many other telescopes and into the heart of this pulsing, exciting complex.

Many of the bright sparks of light are fledgling stars, some of which are a similar size to our sun. Size matters when it comes to stars. Along with a star's brightness, size tells astronomers how a star was born, how long it may shine and even how it could die. In this image, Webb has revealed about fifty young sun-like stars. Once upon a time, our own sun would have resembled some of these young, hot stars. By studying these young stars, we can re-evaluate and refine our existing models of how stars such as ours originate and evolve.

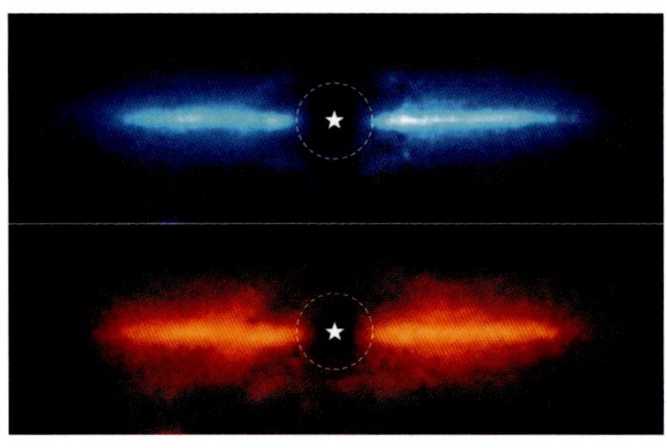

ABOVE: NIRCam images of the protoplanetary disks surrounding the star AU Mic. The coronagraph on the NIRCam allows us to study disks like this in great detail.

BELOW: Gas falls towards the core of a Herbig–Haro object in a rotating disk, while magnetic fields eject plasma out in polar jets.

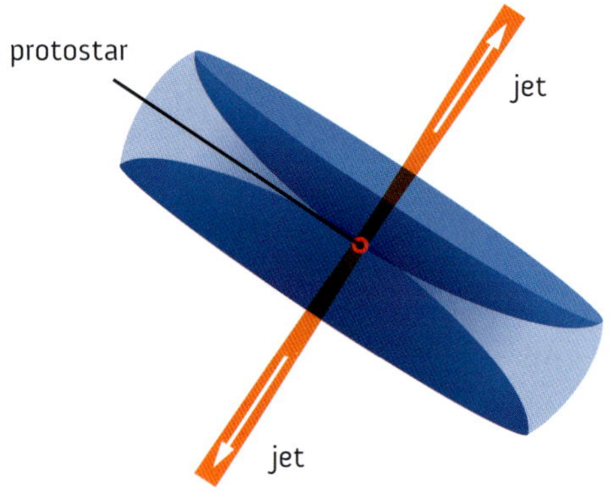

PLANETS IN THE MAKING

People who've studied the image on the previous page have commented that a few of the young stars appear to have distinctive shadows around them that suggest they are surrounded by the building blocks of new planets. Stars form out of swirling clouds of gas and dust, and once they've developed some of that matter orbits in a disk around them. These rotating disks are known as protoplanetary disks because they contain the material that could, one day, form planets.

Some young stars have jets of gas erupting from them. These jets occur when the star first bursts out of the swaddling dust and gas that cocoons it. In the image, Webb has captured this interstellar gas (molecular hydrogen), which has been coloured red and dominates much of the picture. In the top-right corner, there is a particularly impressive example of a new star with two strong jets of molecular hydrogen exploding from it. It almost looks as though the star is impaled on a red stick.

NEW LIFE

A new star named S1 dominates Webb's image of the Rho Ophiuchi cloud complex. Like a pearl in an oyster shell, it is surrounded by a glowing cave of dust (see image opposite). S1 is unlike other stars in the image, as it is significantly larger than our sun.

The light-coloured gas surrounding S1 is different to the red of molecular hydrogen. S1's gas consists of PAHs, which are carbon-based molecules that are commonly found in space. We believe that these molecules are crucial for star and planet formation, but while they are widespread in space, there are still many unanswered questions about how these hydrocarbons form, evolve and react with other particles in space.

But even with its unique gaze, parts of the Rho Ophiuchi cloud complex remain dark to Webb. In these shadows, new protostars are coalescing and solidifying – and in many, many years they will emerge as new hot stars.

S1 shines through the dust.

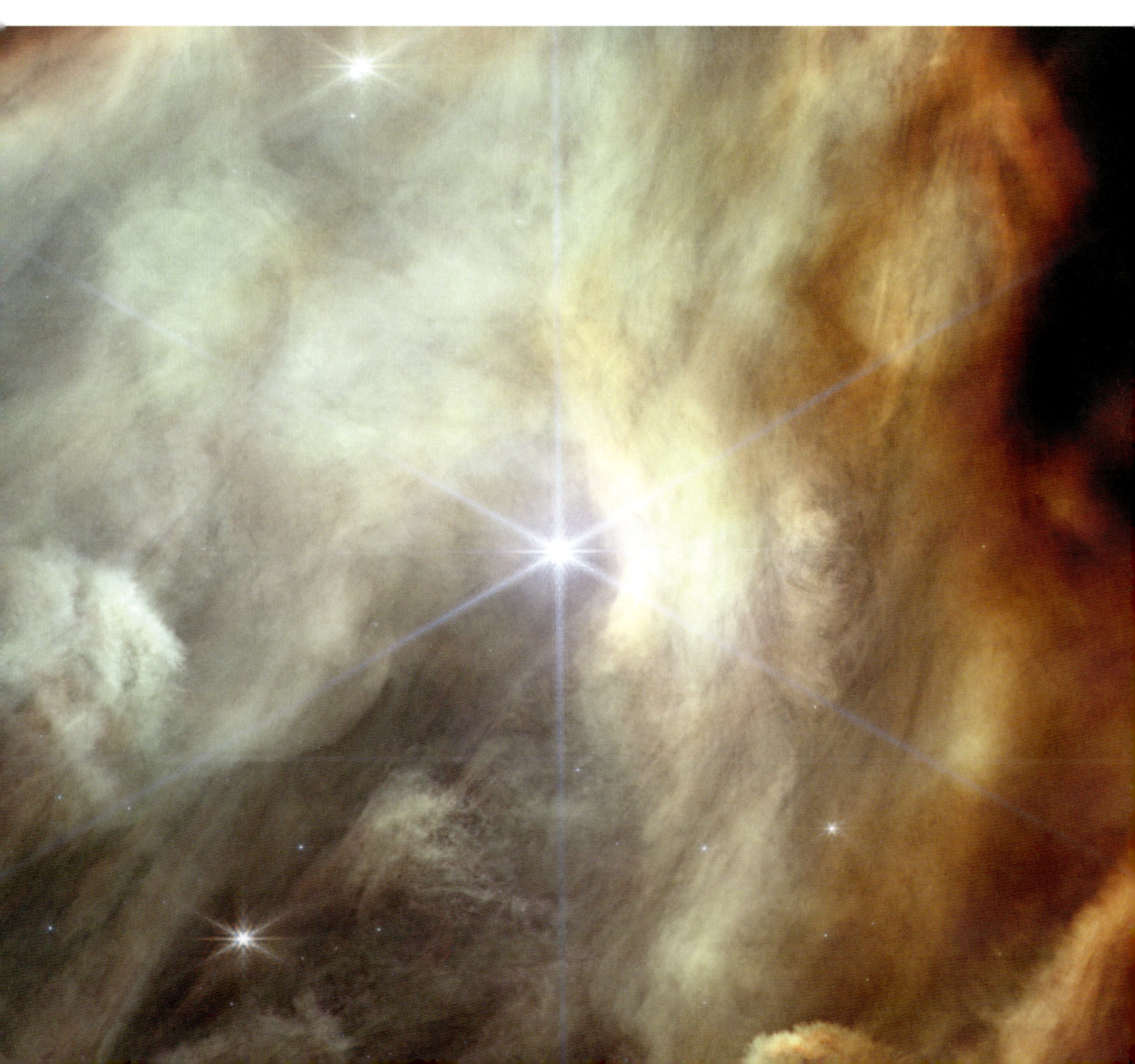

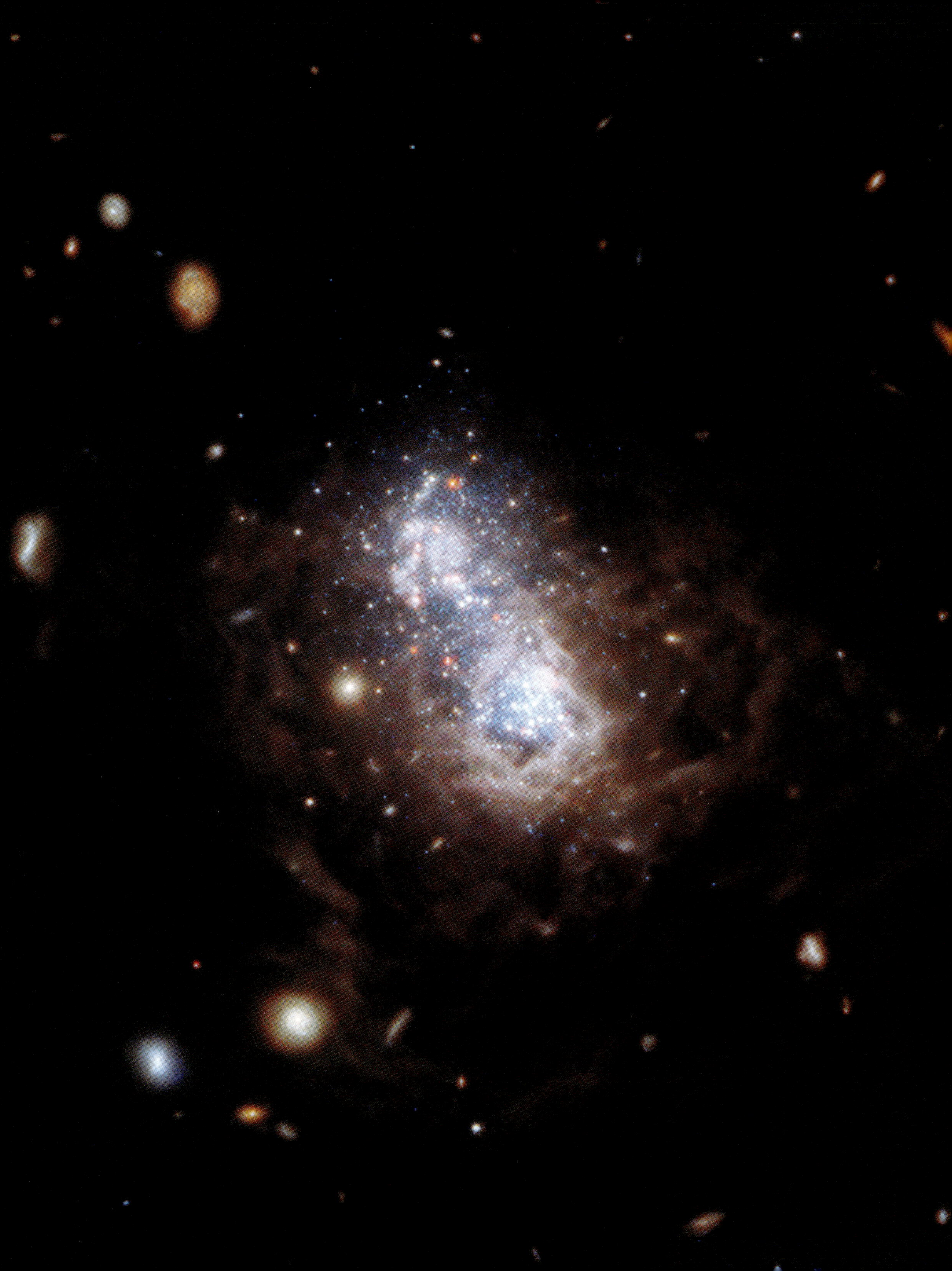

STARS

Our universe contains many stars: giant spheres of plasma (superheated gas) held together by their own gravity. A typical galaxy contains about 100 billion stars, and from Hubble's Deep Field (HDF) imaging we have learned that there are approximately 200 billion galaxies in the known universe. So, using a rough back-of-an-envelope calculation, we can estimate that there are about 200 billion trillion stars in the universe. To put this into context, 1 billion seconds is just over 31.7 years and a trillion seconds is 31,710 years – so 200 billion trillion seconds would be ... a very long time.

Stars vary greatly in mass and size, and these are some of the factors that will ultimately chart the trajectory of the rest of its life. The process that makes a star a star is a continual reaction called fusion that happens at its core, involving turning simpler elements with lower atomic numbers (like hydrogen and helium) into more complex elements with higher atomic numbers (such as carbon, sulphur and iron). This process is governed by the famous equation $E=mc^2$. Here E = energy, m = mass and c = the speed of light in a vacuum, which is a whopping 300,000,000 metres (984,000,000 feet) per second.

This equation shows us that for the tiny amount of mass lost in the fusion process of making new, more complex elements, huge amounts of energy are created and released in the form of electromagnetic radiation that travels out into space. This is why we see stars shine. The fusion process exerts an outwards push from the centre of the star, which is balanced by the inward force of gravity.

A star the size of our sun takes around 50 million years to mature and will inhabit this mature phase for about 8 to 10 billion years. Much larger stars, called hypergiants, emit much more energy but burn out much sooner. Their lifetimes are in the region of a few million years.

This Webb image of the galaxy I Zwicky 18 (I Zw 18), which was discovered by Swiss astronomer Fritz Zwicky in the 1930s, shows the galaxy's star-forming regions in unprecedented detail.

CHAMAELEON I DARK CLOUD

Constellation: Chamaeleon
Distance from Earth: 630 light-years

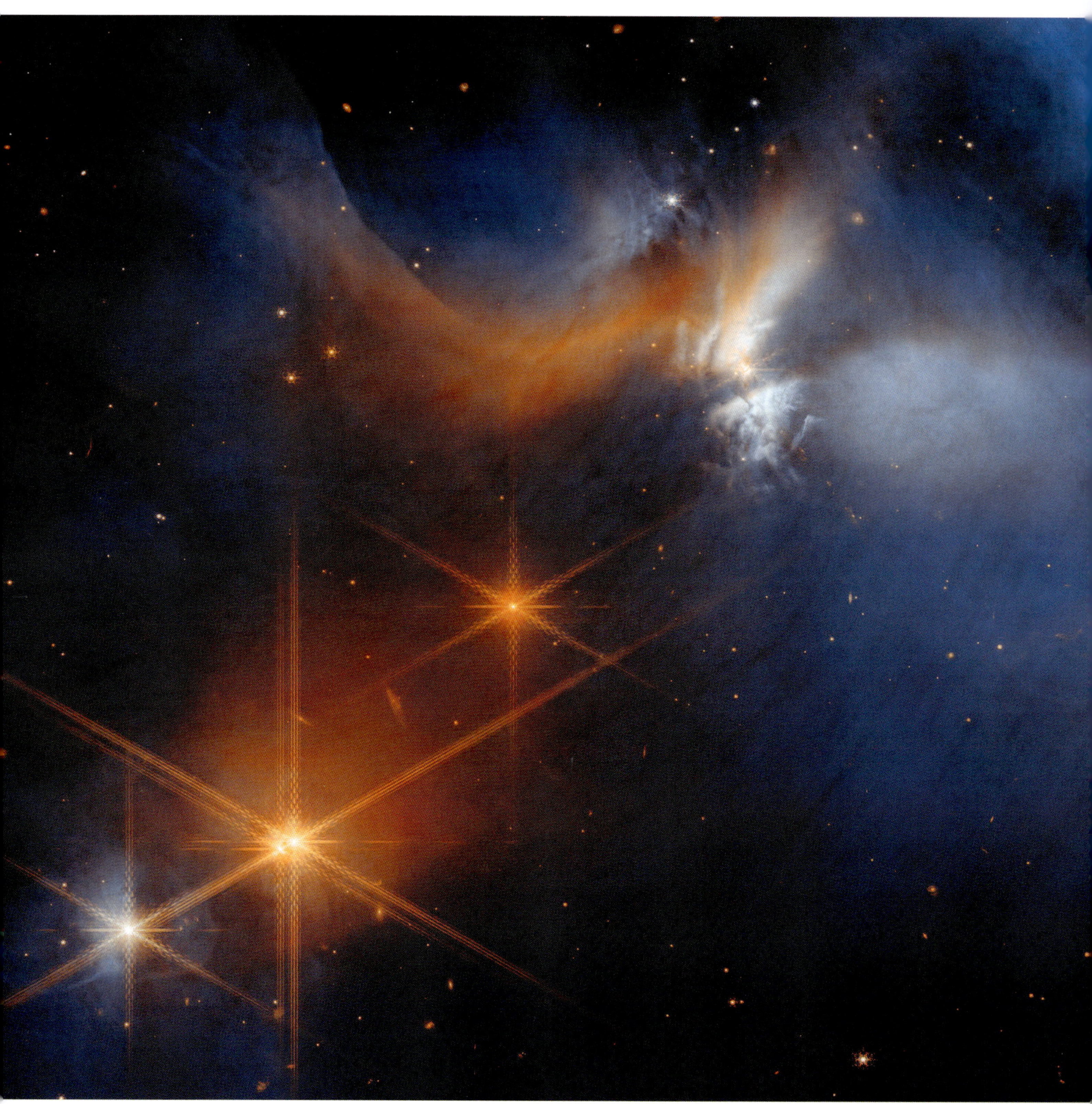

A NIRCam glimpse into the frozen curtains of the Chamaeleon I dark molecular cloud, which resides 630 light-years away.

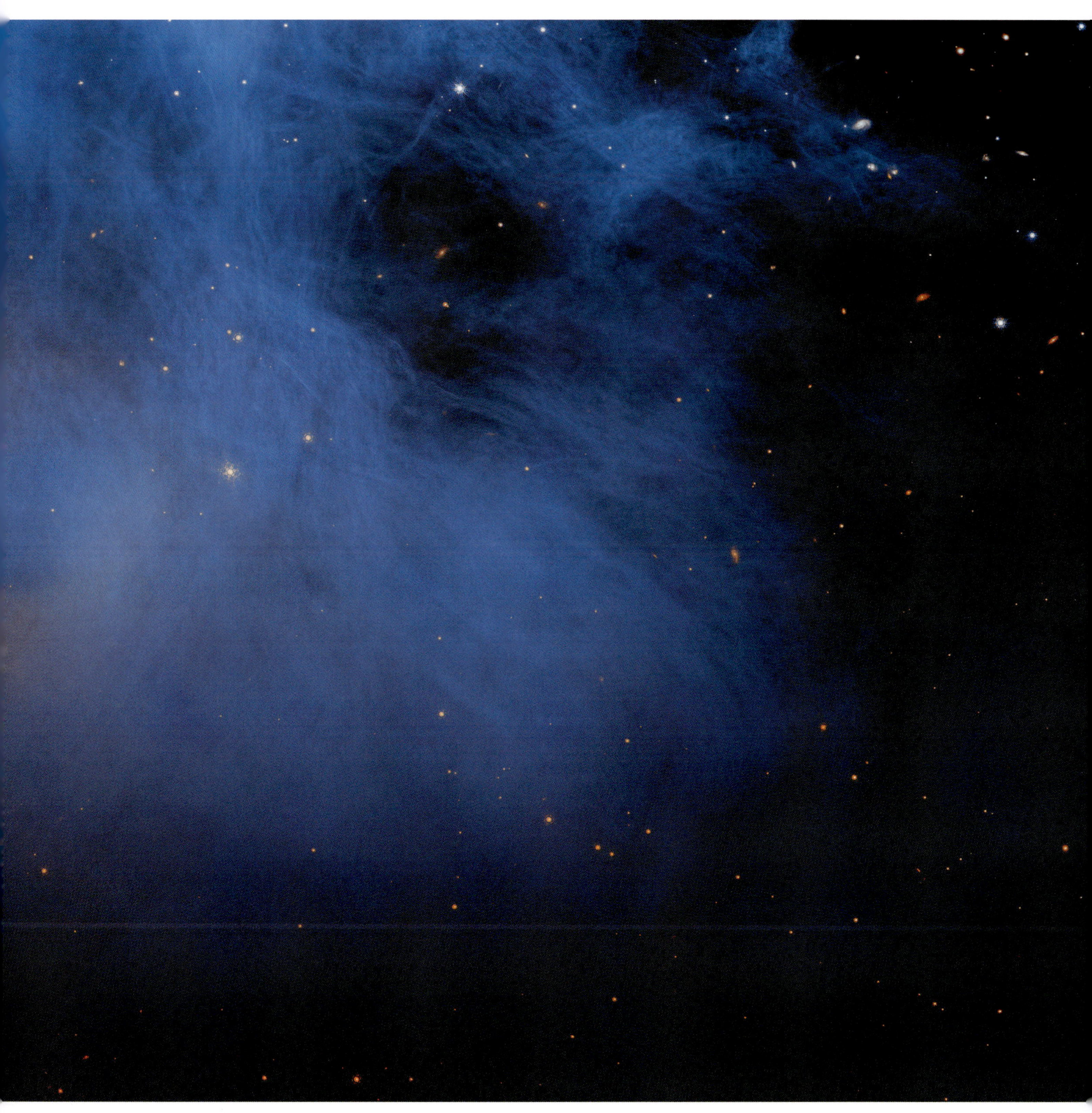

RIGHT: A Hubble view of star-forming regions in the Chamaeleon I dark molecular cloud. It is a 315-million-pixel composite image, comprising twenty-three observations.

THE CHAMAELEON COMPLEX IS A rather enigmatic feature of the sky. It's a large star-forming region some 65 light-years wide. Named Chamaeleon I, Chamaeleon II and Chamaeleon III, these clouds – also known as dark nebulae – are molecular clouds whose molecules are packed so closely together that visible light cannot escape. That means that any stars forming within or behind the clouds are invisible. Well, not to Webb's infrared detectors.

In the recesses of the Chamaeleon I dark cloud, ice is forming on the dust particles. In the central part of the image, the stars glow orange against an ethereal backdrop of gas and dust. Using spectroscopy, the light from these background stars can be employed to ascertain the chemical composition of the ice crystals in this cold, dark nebula.

Depending on the chemical makeup of the ice, some wavelengths of infrared light are absorbed as the light passes though the crystals. Using this technique with Webb data, detection of water ice as well as frozen forms of carbonyl sulphide, ammonia, methane and methanol have been found in this cloud. The ice on these dust particles will grow over time and may even seed the cores of planets one day.

This technique is possible thanks to Webb's extreme sensitivity. Data such as this gives us insight into how more complex molecules can be synthesized – molecules that are the first step in the generation of building blocks that could in the future lead to new life.

BELOW: An illustration of how star systems form within nebulae.

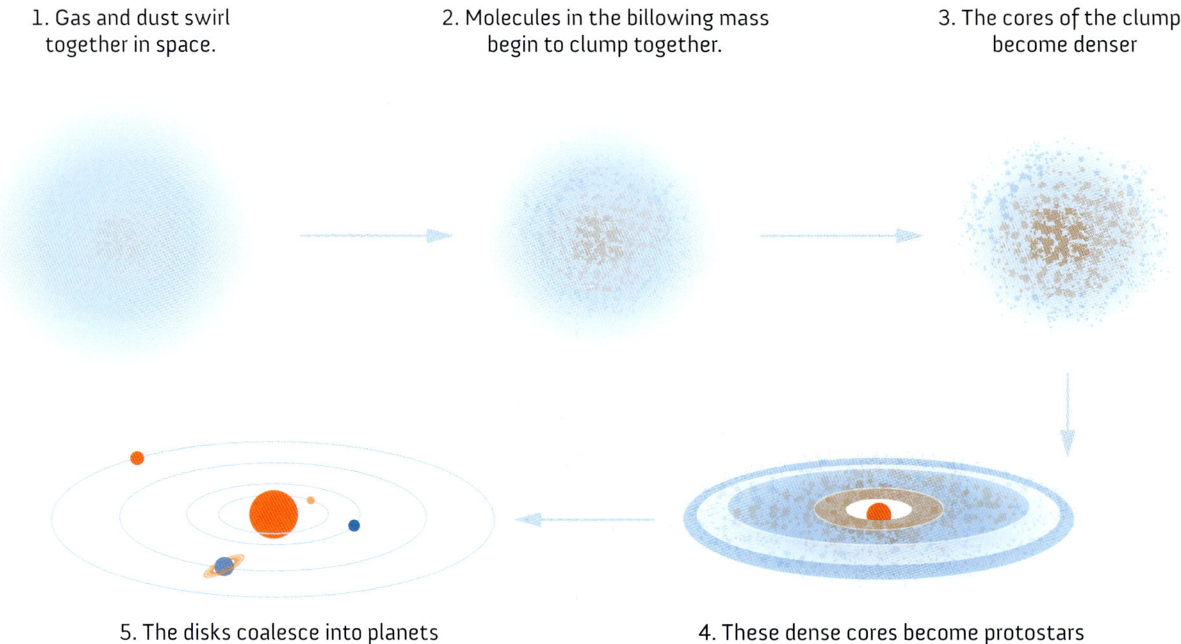

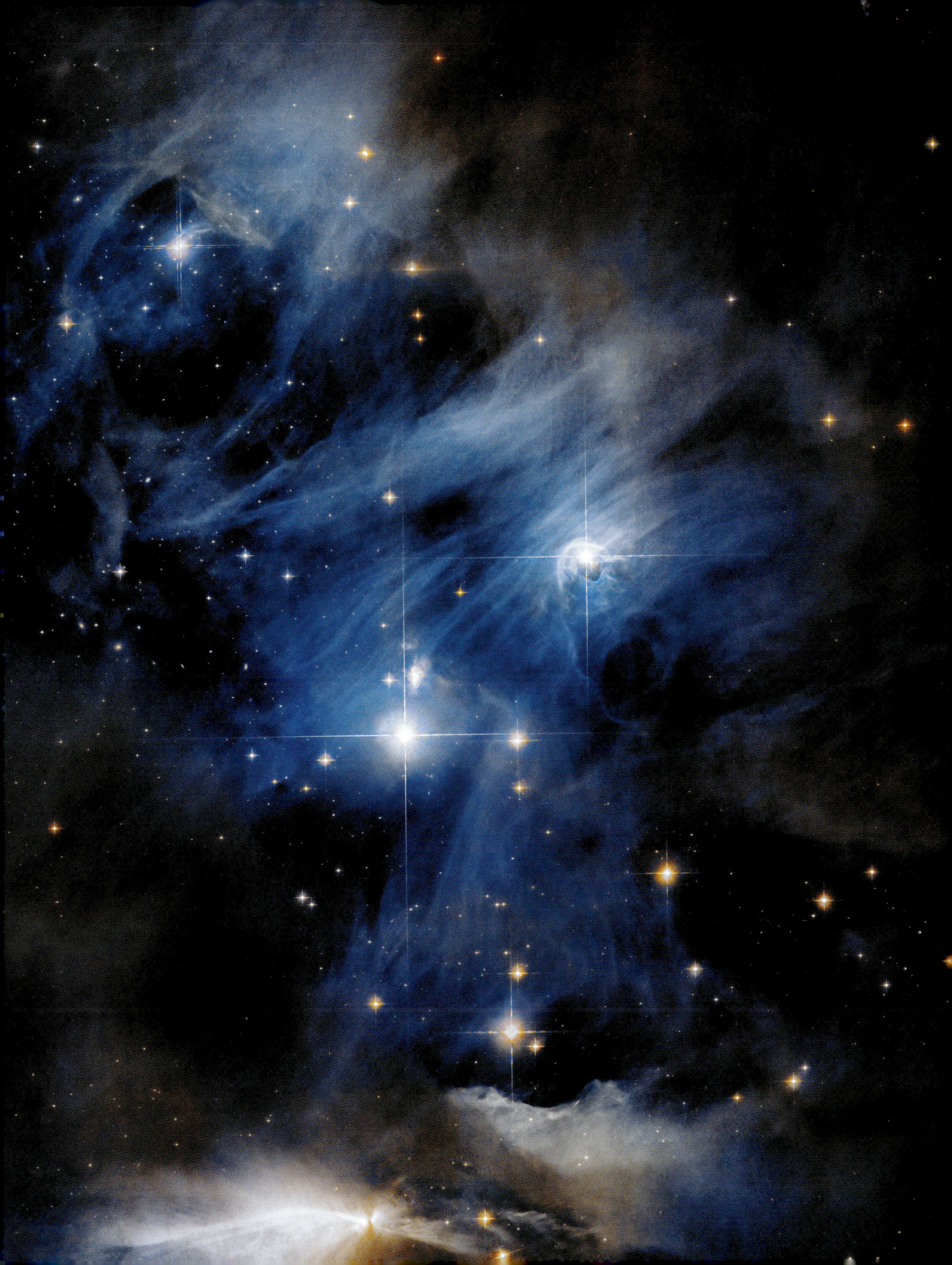

HERBIG-HARO 46/47

Constellation: Vela
Distance from Earth: 1,470 light-years

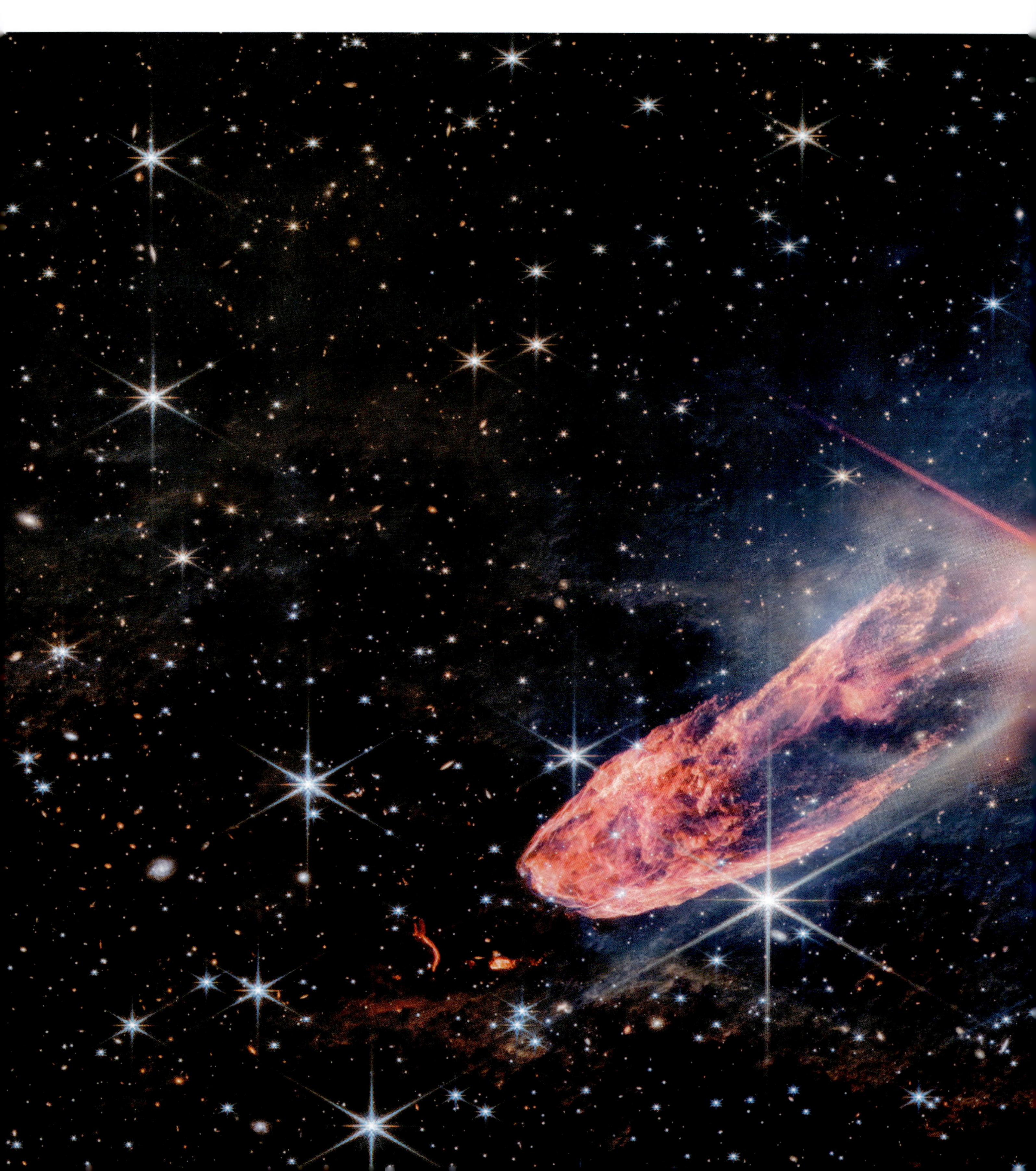

New stars are being born in this Webb image of Herbig–Haro 46/47, which is about 1,470 light-years away in the Vela constellation.

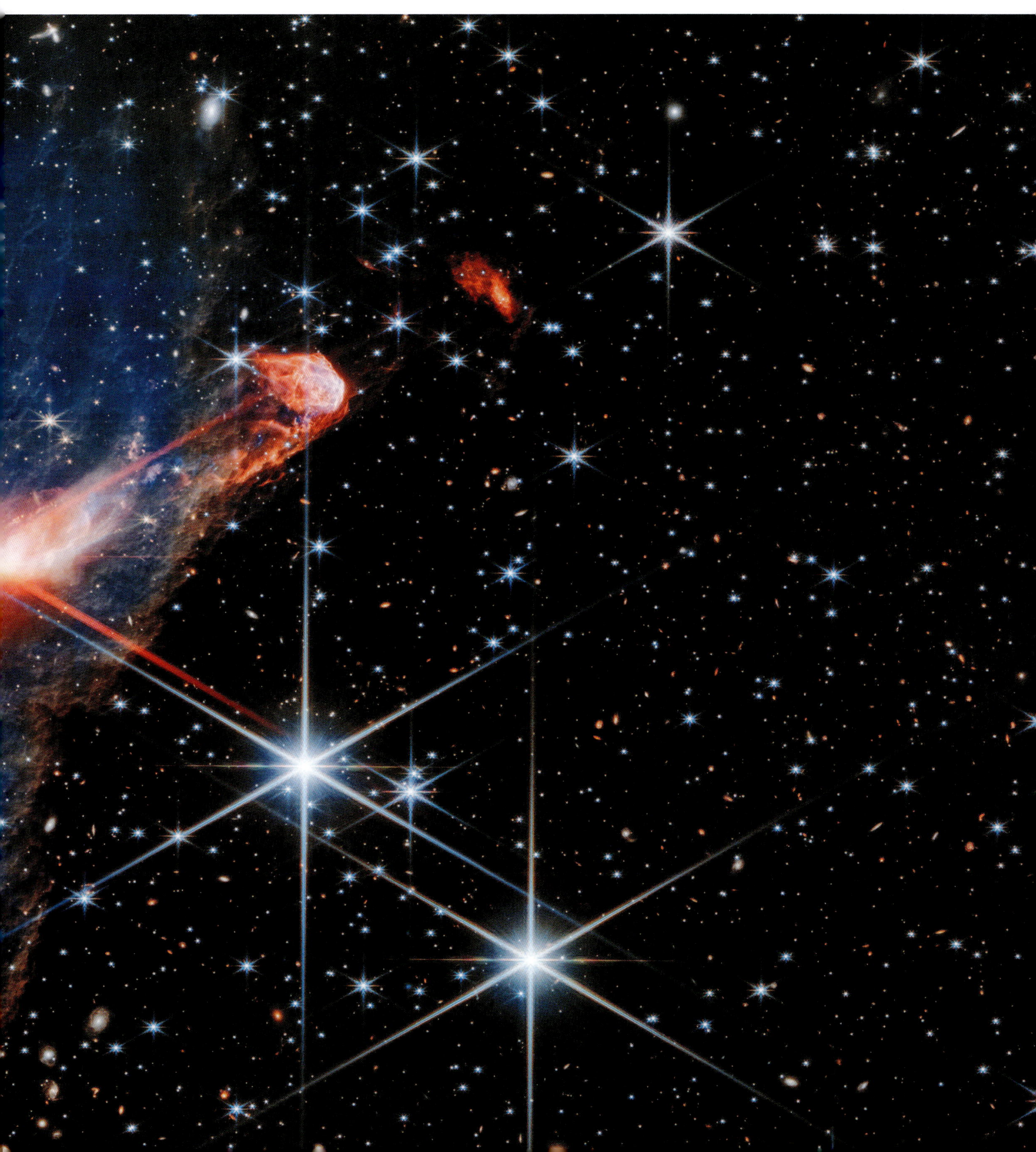

IN THE CENTRE OF THIS burst of colour (on the previous page), two tightly bound stars are forming. To find them, follow the brilliant pink and red diffraction spikes until you reach the centre: the stars are in the orange-white blotch. They are deeply buried within the oval, in a disk of gas and dust that fuels their expansion as they gather mass.

The disk is not visible, but its shadow can be observed in the two dark, conical areas that encircle the core stars. What we are witnessing is the birth of new stars and the cosmic fallout in the region around them. It is not possible to see the shenanigans of these two stars in the visible spectrum, as shown in this image (see opposite) taken by the ground-based La Silla telescope. This is because they are blanketed in dust and gas that blocks the visible components of the electromagnetic spectrum. Webb's infrared capabilities, however, allow us to stare into the middle of this energetic stellar genesis.

THE BIRTH OF NEW STARS

When stars form, they accrete gas and dust from the volumes around them – but just like human babies when they have ingested too much food, they eject the matter. Such phenomena are called Herbig–Haro (HH) objects, and they are the bright glowing patches around newborn stars. Two twentieth-century astronomers – George Herbig and Guillermo Haro – were independently studying star formation in the Orion Nebula when they recognized that these bright areas were linked to new stars being born. The pair met in 1949 and HH objects are named after them.

Because HH objects are associated with the evolving process of star birth, they are transient, lasting for tens of thousands of years, and also prone to rapid (that's 'rapid' in cosmic terms!) change as they move away from their parent stars and into the surrounding gas clouds. The Webb image provides a striking example of Herbig–Haro objects – in this case, Herbig–Haro 46/47. The two orange-sided lobes that flank the stars show the remnants of previous ejections, with the new ejection appearing blue and thread-like. It is through this process, over millennia, that stars manage their masses. It will take millions of years for these two stars to reach maturity.

The technicoloured activity is set against a faint blue backdrop, which is actually a nebula. When viewed in visible light, this whole area looks black – because most of the visible light is scattered, with just a few background stars shining through the gloom. But with Webb's near-infrared detection, we can peer through the cloud to see the celestial objects that lie beyond.

RIGHT: An image of Herbig–Haro object HH 46/47 taken from the ground with the European Space Agency's New Technology Telescope at the La Silla Observatory in Chile.

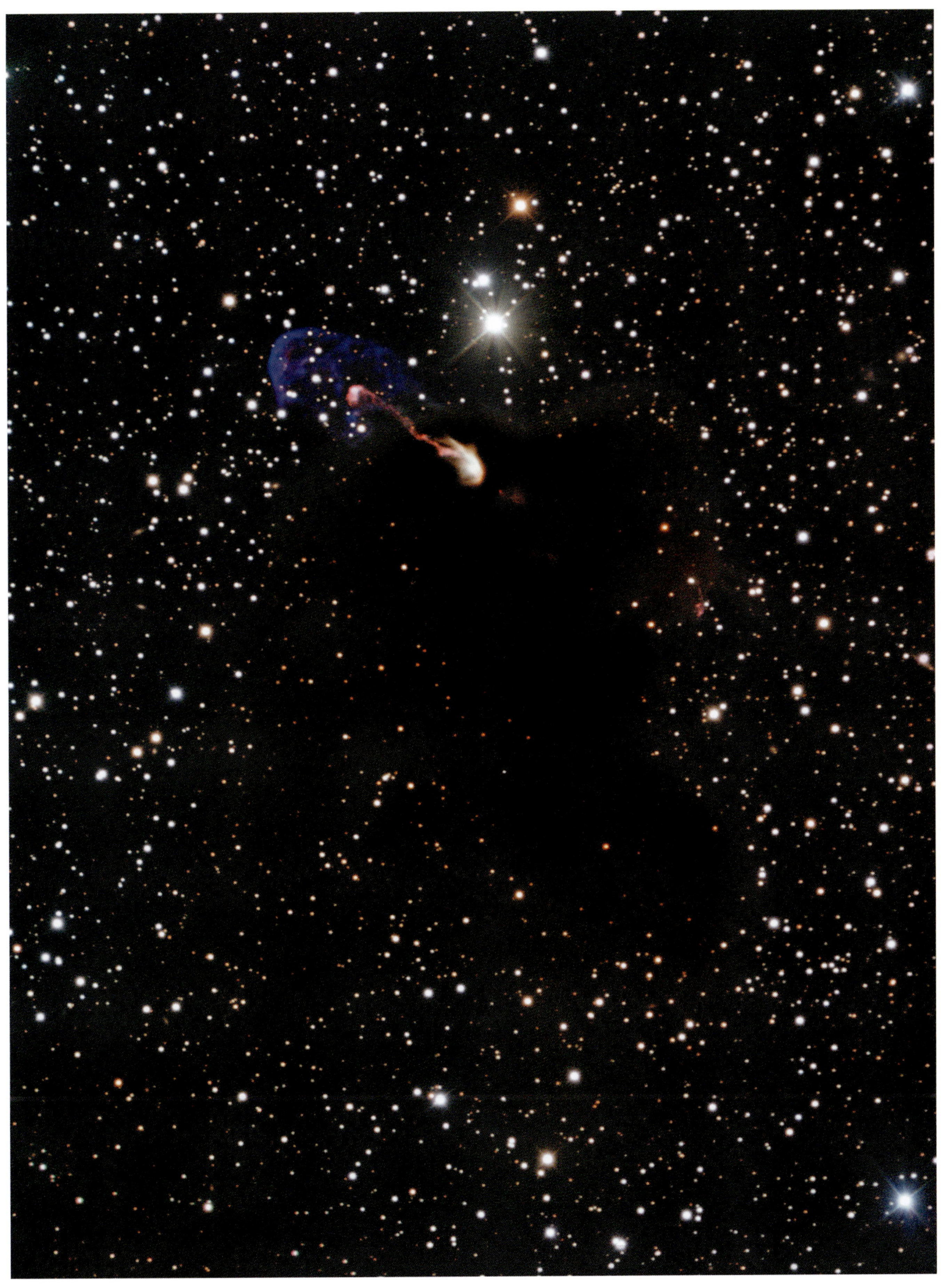

HERBIG–HARO 797

Location: near the open star cluster IC 348
Constellation: Perseus
Distance from Earth: 1,000 light-years

Herbig–Haro 797 streaks across the bottom of this Webb image like a firework. Captured with NIRCam, these types of objects surround newborn stars.

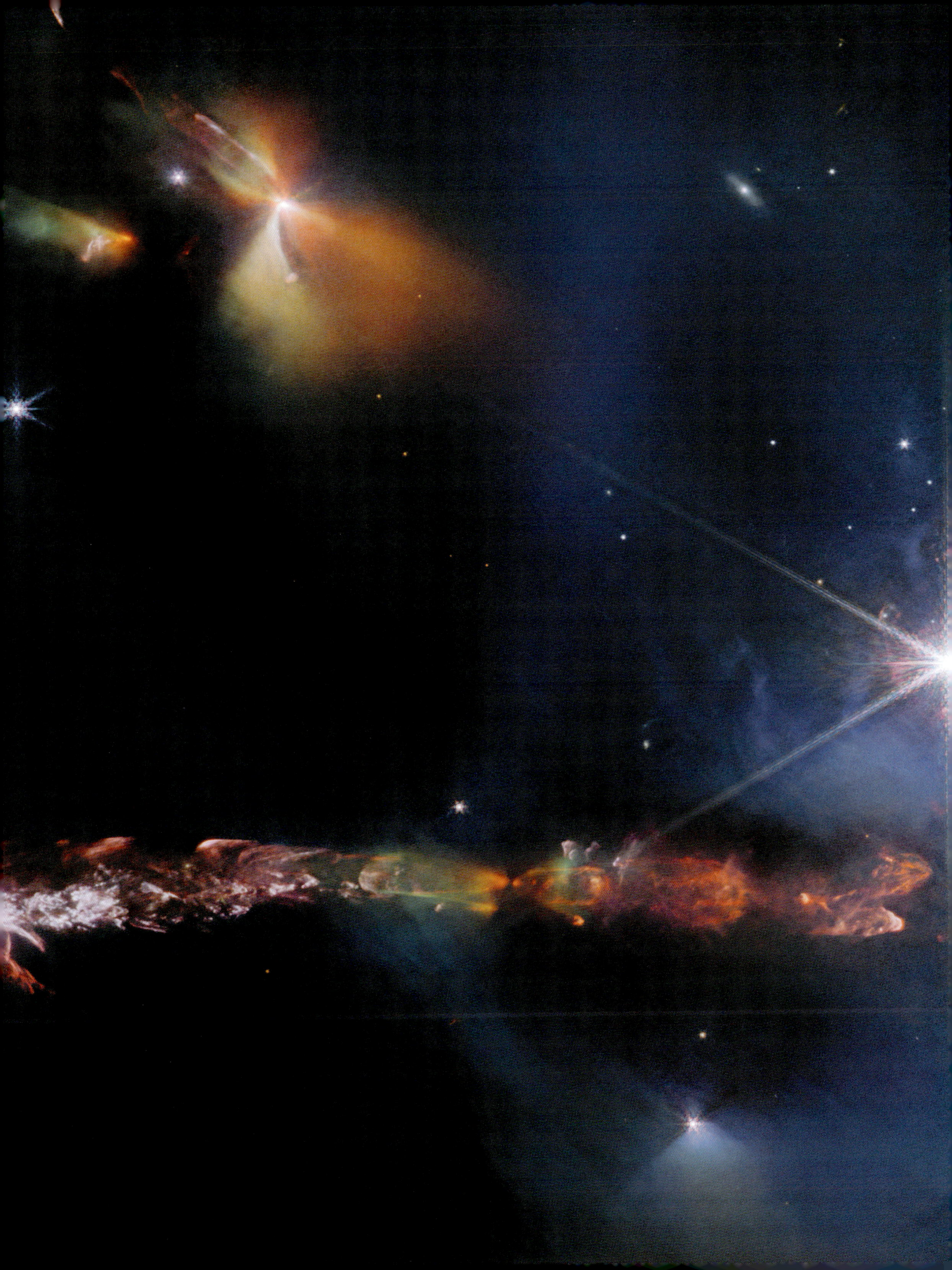

THE DARK-CLOUD COMPLEX IN THE star cluster IC 348 is also a great place to hunt for protostars. These pre-fusion stars are in the energetic and breathtaking stage of their evolution, drawing in material to reach their critical mass but also, surprisingly, expelling it. The expulsion is caused mainly by the protostars' rotational energy, magnetic fields and radiation pressure. These expulsive forces counteract the degree to which gravity draws matter into the stars (their gravitational accretion) and therefore the stars' size.

The layers of dust and gas that envelop the stars make them optically opaque to visible-light telescopes, but Webb's infrared sensitivity – particularly when using the NIRCam instrument – can pierce these clouds and view the stars forming within.

Herbig–Haro 797 dominates the lower half of this image, as though someone loaded a paintbrush with paint and streaked it across the picture. Slightly to the right of the centre, there is a small dark region; in the past, astronomers thought there was just one star inside this area and that its rotation was manipulating its outflow. But there are actually two protostars here, both of them ejecting streams of matter into the surrounding space. This matter crashes with the encircling dust and gas. When Webb's observed wavelengths are coloured in different hues, a kaleidoscopic explosion is produced. This data is not only spectacularly beautiful but also full of scientific revelations.

But Herbig–Haro 797's protostars aren't the only stars in the image. In the top half, there are even more stars being born. Scientists think that these are two more protostars, with energetic jets erupting from them and interacting with particles in the dark cloud. A very bright star, with Webb's distinctive eight diffraction spikes, shines on the far right of the picture.

NEARBY HH 211

The IC 348 cluster and surrounding regions are awash with interesting phenomena. A tiny fraction of a degree north of Herbig–Haro 797, yet another star is being born in a glorious outburst of gas and energy. Herbig–Haro 211 is a protostar and currently a fraction of the size of our sun. However, it gives us an idea of what our sun would have looked like billions of years ago, before it had grown into the luminous behemoth we now recognize it to be.

Astrophysicists are fascinated by these early sun analogues because they allow us to study how our sun formed and evolved over time – as well as the conditions it may have nurtured that allow life to exist on the small blue dot we call home. Observing protostars is quite difficult because they are enveloped in visible opaque clouds of dust and gas; but the violent outflows emit infrared radiation that Webb can detect.

In the image below, you can see energy and matter erupting from the protostar and crashing into the surrounding dust cloud. Within the bow, two red streams explode from the black centre, like arrows through the heart. These are the narrow jets of stellar gas the protostar is expelling. The red streams appear to wobble, which has led to the hypothesis that the protostar is actually two nascent stars that are tightly orbiting each other.

They create bow-like shock waves as they encounter the nearby dust and particles. The innermost outflows are moving at about 90 kilometres (55 miles) a second. However, the researchers looking into this data found that the velocity of the shockwave itself was much slower. This led them to deduce that the jets emerging from the centre of these young stars are actually made up of molecular material rather than simple atoms. Due to the stars' low shockwave velocity, the material is not broken down into atoms. More evolved stars, on the other hand, tend to have speedy outflows of charged atoms.

Haro–Herbig 211, captured here by NIRCam, is near the open star cluster IC 348 in the Perseus constellation and is about 1,000 light-years away.

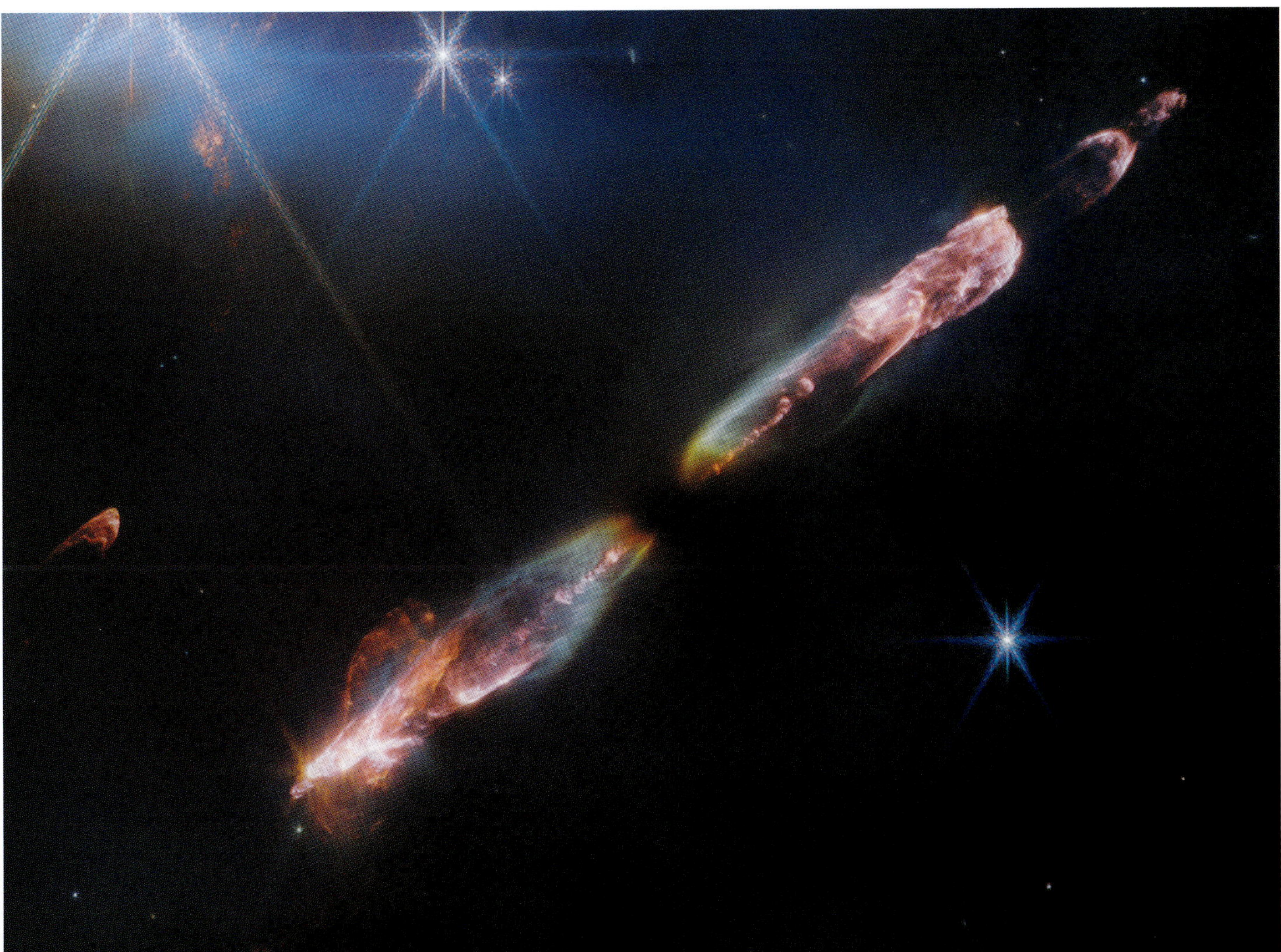

PROTOSTAR L1527

Constellation: Taurus
Distance from Earth: 460 light-years

A new star is forming in the neck of this fiery hourglass, which was imaged using Webb's NIRCam instrument.

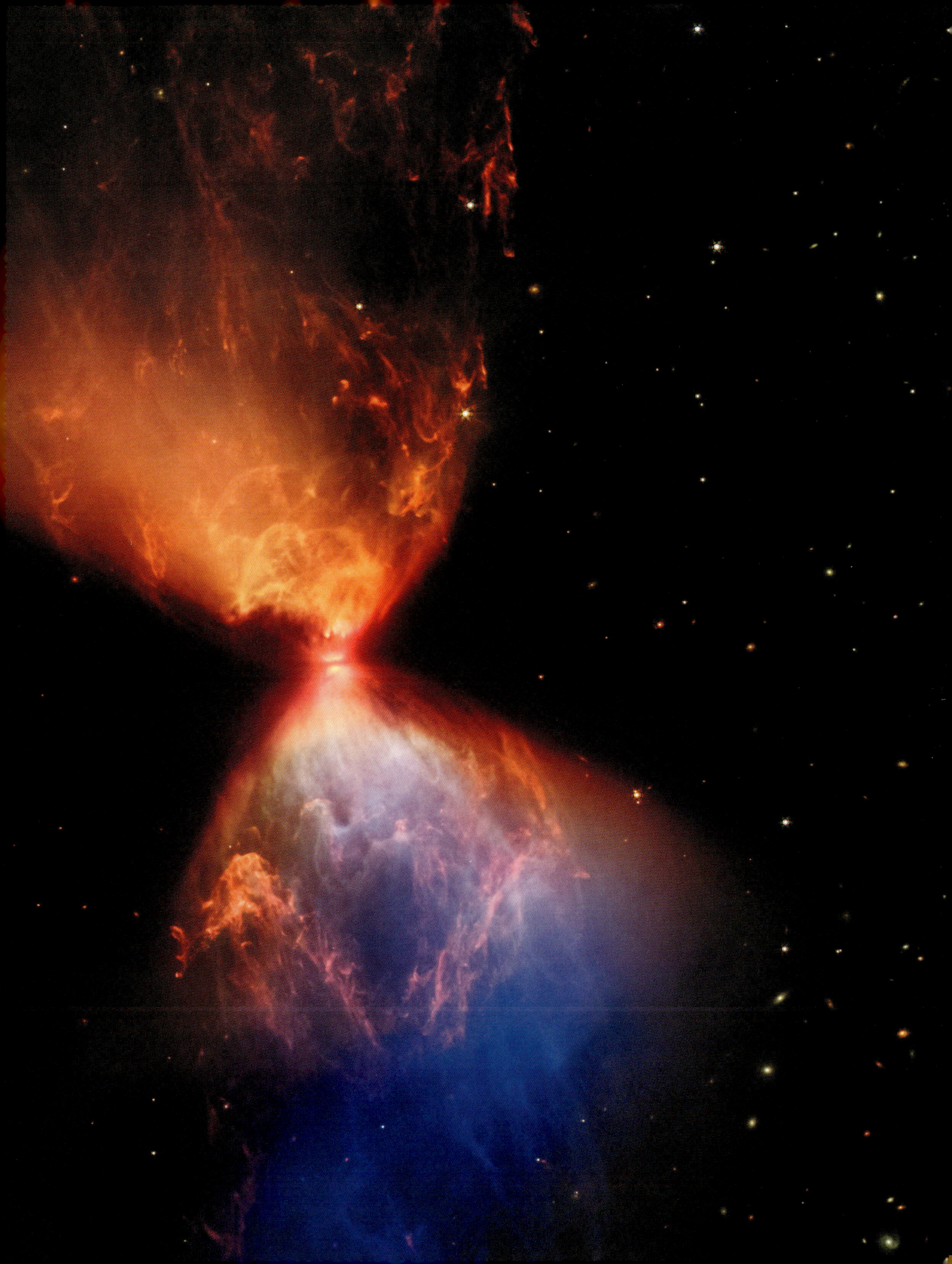

IN THE EYE OF THIS fiery storm, in the neck of the hourglass, a new star is emerging. This protostar is relatively young, about 100,000 years old. The early stages of star formation are normally diffucult to see and thus tricky for astrophysicists to study; Webb's infrared eyes allow us to see the star forming within the dust and gas of its cradle nebula. It is important for scientists to unravel the secrets of infant stars because they can give us an idea of what once happened in our own solar system, when the sun was in its infancy.

Between the two wings in this image there is a faint dark line, like a butterfly's body. The star's gravity is pulling the surrounding dust and gas into a ring around it, which ultimately becomes fuel for the voracious developing star. This is an accretion disk (see page 68) about the size of our solar system.

A new star, L1527, is being born.

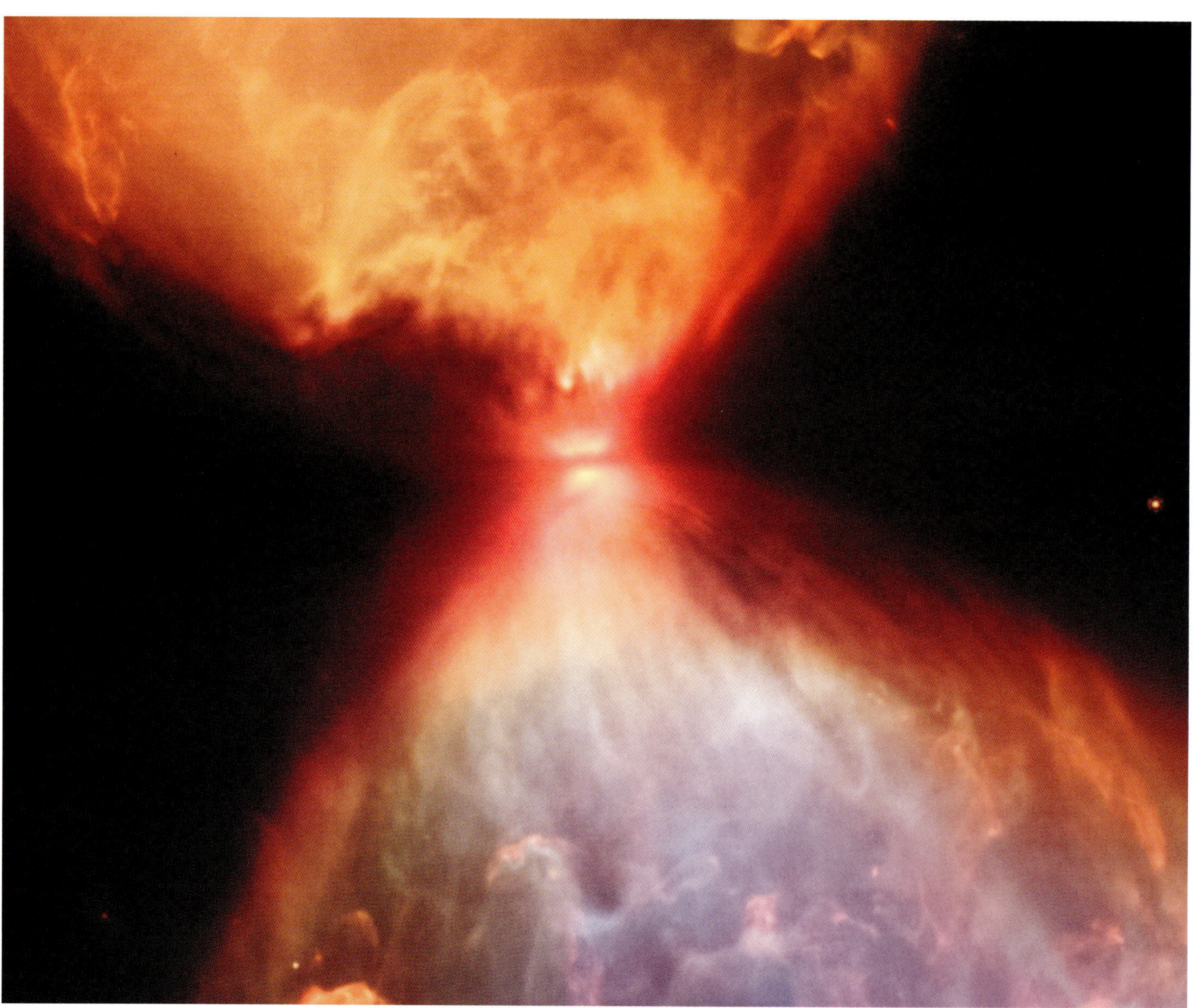

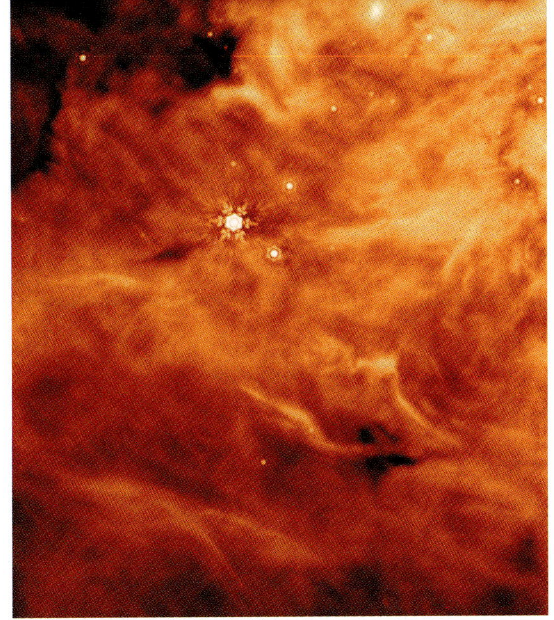

A MIRI image of the region near the protostar IRAS 23385. There is one bright star, and a handful of dimmer ones, with short diffraction spikes.

The environment around an infant star is a turbulent and exciting place. L1527 is drawing material to itself and simultaneously ejecting mass and energy from its poles. These stellar jets fan out into space, interacting with the surrounding dust like giant orange and blue wings. The blue colouring is where the dust is thinnest, spreading out into interstellar space. In the orange plumes, the dust is much thicker. And the young star continues to pump matter and energy into its vicinity, which interacts with the atoms and molecules that it had previously jettisoned into its surroundings. The bubbles within the wings show the sporadic ejections of energy and matter from its core. These shockwaves play a very important role in the star's evolution and disruption of the space around it. For one thing, these energy ejections disrupt other stars from forming the molecular billows. As molecules begin to cluster and clump together, the shockwaves slam into them like waves at the seaside smashing into sandcastles.

A VERY YOUNG PROTOSTAR

L1527 is considered a Class 0 protostar, which is the earliest stage of star formation. For example, they don't generate their own energy just yet. Our sun produces energy through the fusion of hydrogen in its core. Class 0 protostars haven't yet 'switched on' and gone nuclear – literally. Over time, L1527 will gather more material and gravitational forces will compress all that mass into the star. When that happens, its temperature will rise until eventually the elements in its core will begin to fuse and the fusion process will begin. Even so, it will be many years before L1527 begins forming helium in its heart and generating its own energy.

It is also still enveloped within its surrounding dust cloud. If we were looking at it in the visible light range, it would be completely obscured by its dusty cocoon. It's thanks to Webb's infrared eyes that we were able to capture this fiery butterfly.

It will be many millions of years before L1527 matures into a stable star like our sun. And even longer before its planets seed and clump together in orbit around their star. As things stand, it is an unstable ball of plasma, less than half the sun's size. Like Herbig–Haro 211, this protostar gives astronomers a glimpse of what our sun may have looked like when it was young. We can't go back in time to when our sun was a volatile mass, gobbling up everything in its vicinity while throwing matter around like a toddler having a tantrum. But by observing protostars like L1527, we can get an idea of how our life-giving sun was born and evolved as it matured.

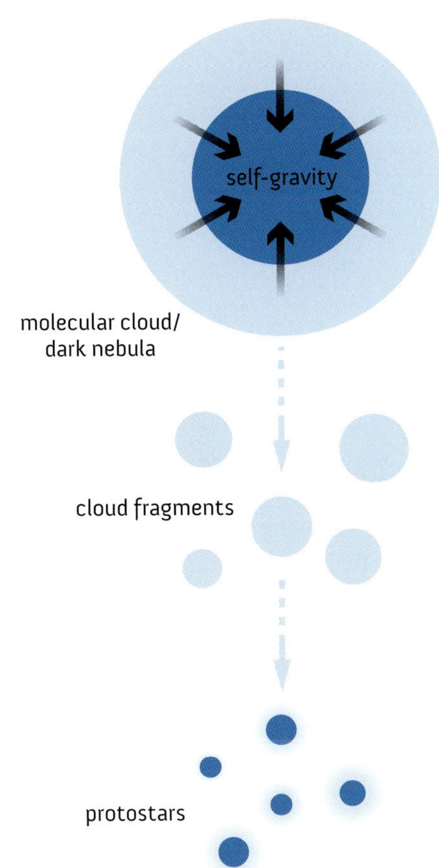

An illustration of how stars form.

WR 124

Constellation: Sagitta
Distance from Earth: 20,000 light-years

A composite image using NIRCam and MIRI data of Wolf–Rayet 124, which is 15,000 light-years away in the Sagitta constellation. This special class of objects are in the brief stage before a star goes supernova.

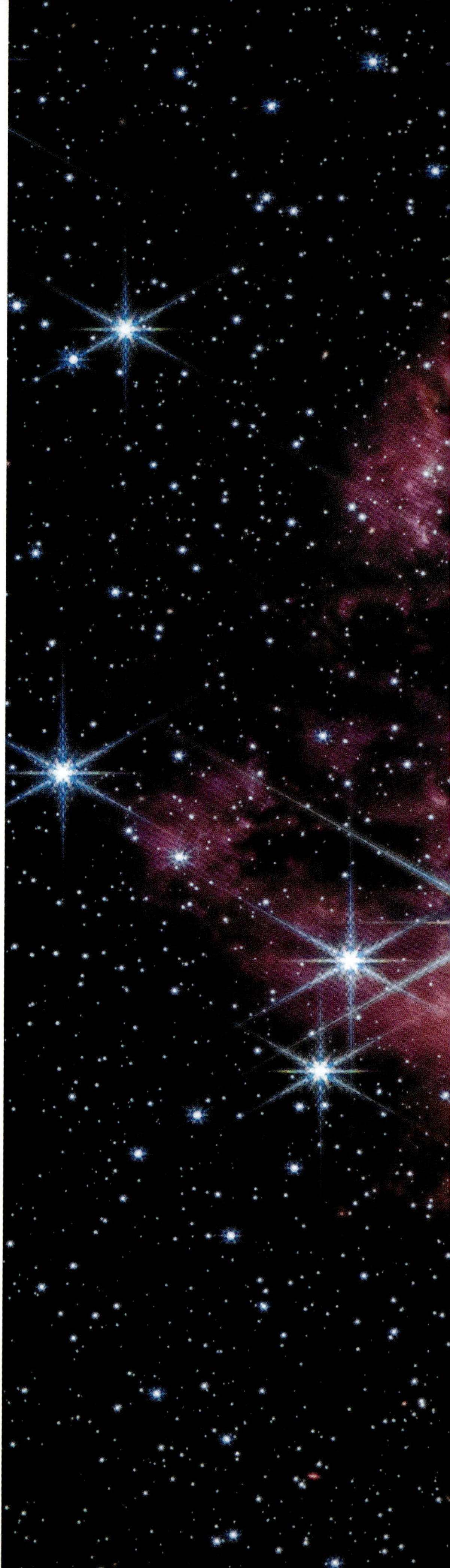

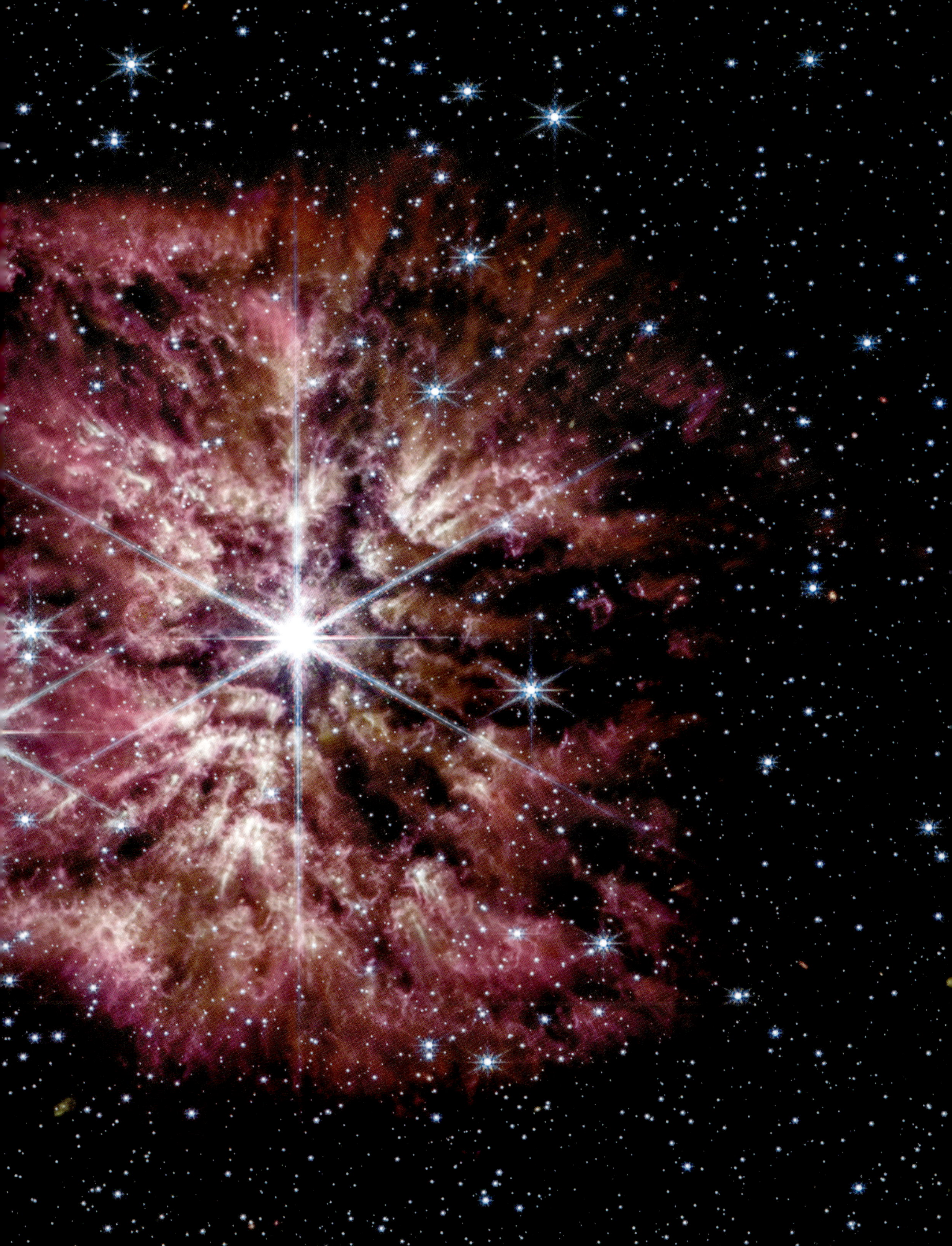

IN THE PREVIOUS IMAGE, WE are seeing a special type of star just before it explodes in a supernova. In the centre of the blossoming M1-67 nebula is a Wolf–Rayet star known as WR 124. These stars are about twenty-five to thirty times the mass of our sun, have an interesting chemical signature and are some of the hottest stars we know of. We can see WR 124 here with Webb's distinctive eight diffraction spikes. Stars become Wolf–Rayet stars towards the end of their stellar evolution: at this stage they have used up most of their simple fuels and start to fuse the heavier elements together.

At the same time, they eject gas from their surface, which cools to form a halo of cosmic dust around the star. These areas of cosmic dust are of great interest because this dust serves as a sort of petri dish for the birth of new stars and planets and is the medium in which molecules cluster and form, possibly giving rise to life.

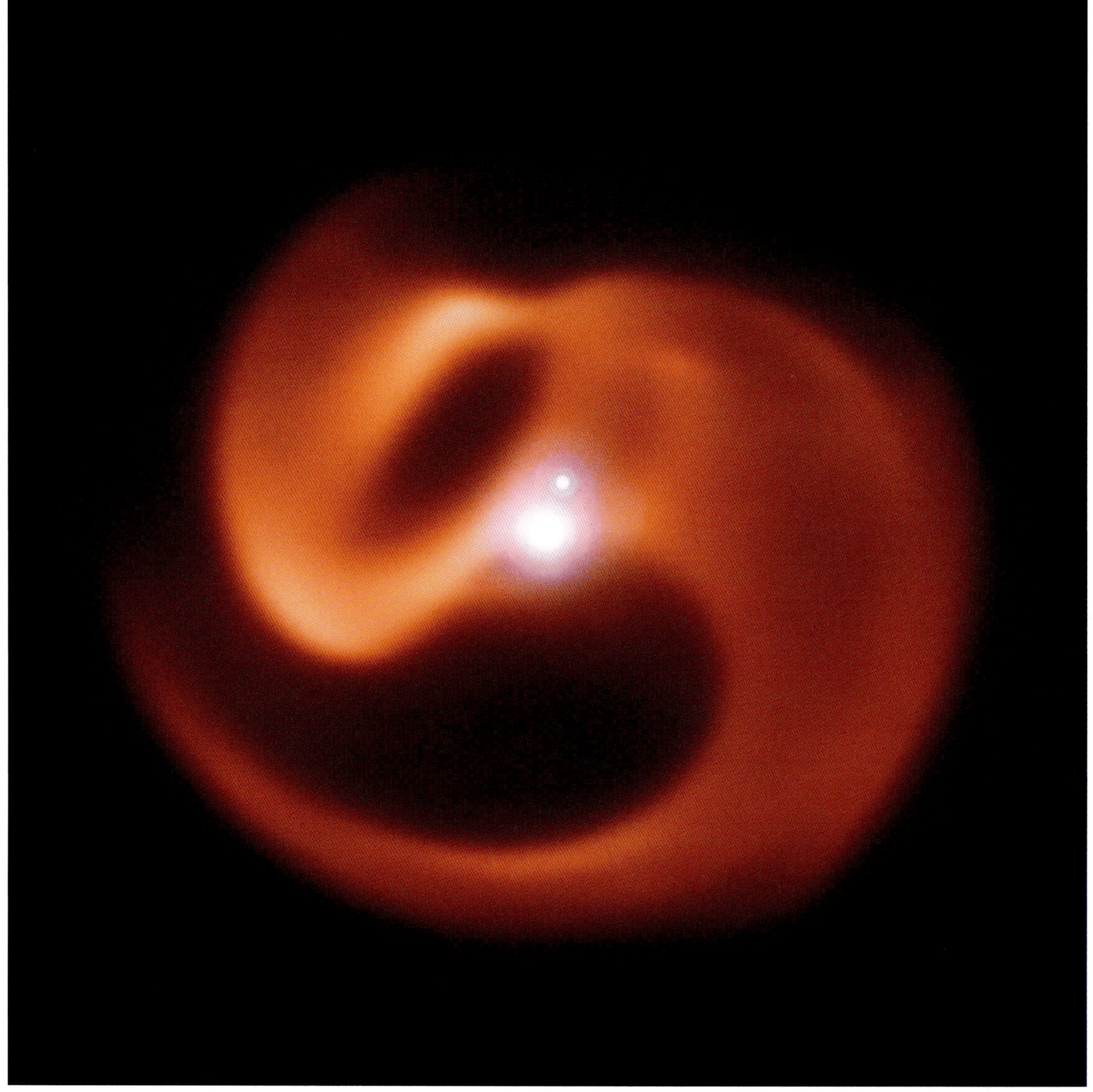

The Apep Star System, captured by the European Southern Observatory, is a triple-star system and showcases the powerful forces that stars can exert on each other and the material around them.

So far, WR 124 has expelled about ten suns' worth of matter. This is the mixture of yellow, pink and reds that resemble a blossoming flower. The resulting dust cloud is a nebula – the imaginatively named M1-67 – which is expanding at more than 150,000 kilometres (93,000 miles) per hour. (For some context, the Earth has a circumference of around 40,000 kilometres/25,000 miles.) At the speed these particles are travelling at, it would be possible to circumnavigate the Earth in about fifteen minutes.

One of the reasons we scientists are fascinated by regions of space like these – aside from the fact that they are awe-inspiring, humbling and filled with spectacular and interesting phenomena – is because in them our laws of physics are pushed to the brink of what we believe to be possible. Wolf–Rayet stars are a case in point. The sheer amount of energy released on such a grand scale is impossible to recreate here on Earth. Theoretical physicists and mathematicians theorize about how matter behaves under such conditions, but we struggle to study it in a laboratory. In these vast energetic cauldrons, though, we are able to get an idea of what is actually possible and understand the universe at its limits.

UNIQUE CHEMICAL SIGNATURE

The name Wolf–Rayet is an amalgamation of the two astronomers, Charles Wolf and Georges Rayet of the Paris Observatory, who first discovered this very hot type of star in 1867. Not all massive stars will become Wolf–Rayet stars. The primary determiner is the chemical makeup of the star. Spectroscopy (see page 44) enables us to identify the chemical composition of stars by analysing their spectral lines, allowing us to find Wolf–Rayet stars and identify stars that are due to go supernova in the near future. To date, only 200 Wolf–Rayet stars have been found in our galaxy, but it is estimated that there may be as many as 2,000, mostly hidden in clouds of dust. With Webb's capabilities, we may find many more in the future.

The Webb image is a composite of MIRI and NIRCam data, combining near-infrared and mid-infrared wavelengths. NIRCam showcases the star's brightness and the dust and gas around it, while MIRI uncovers the petal-like structure of the surrounding nebula. Bright clumps of gas and dust coalesce amid the powerful stellar winds emanating from WR 124.

WR 140

Location: within binary star SBC9 1232
Constellation: Cygnus
Distance from Earth: approximately
5,600 light-years

In this image, MIRI has captured the ripple-like rings of cosmic dust surrounding Wolf–Rayet 140.

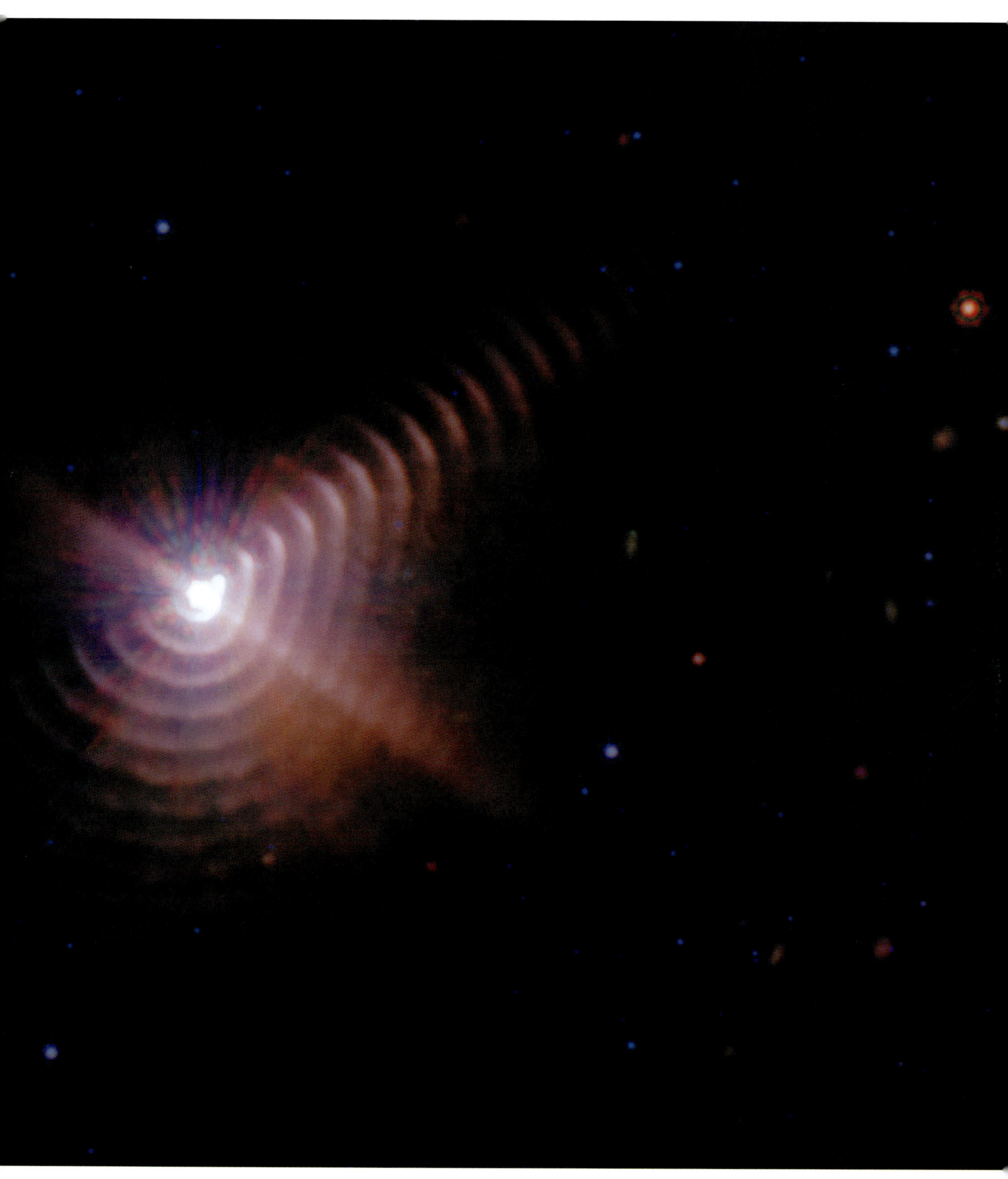

THE HALO EFFECT IN THIS Webb image is not due to a grimy lens or the star's brightness. This Wolf–Rayet star really is encased in gigantic spheres of dust. Previously, we knew only a handful of the star's seventeen shells; but Webb has illuminated them all, giving us a new perspective on this remarkable star system.

As astronomers, we have been studying WR 140 for decades because it's an incredible example of dust production. These types of dying stars experience huge mass loss due to their rapid gravitational collapse. But WR 140 seems to have a particularly high concentration of dust. Scientists are not sure whether this is because of the nuclear processes within the star or because it is not alone.

WR 140 is paired with a massive star, and the two are locked in a cosmic dance around each other. The other star (which is not on the brink of death) is bigger and brighter than WR 140, but the pre-supernova star's broad range of emissions make it very interesting to the likes of Webb. Every eight years, these two stars are brought together by their orbits. When they are furthest apart, they are

As Wolf–Rayet 140 and its companion star near each other, their stellar winds collide and produce large quantities of dust.

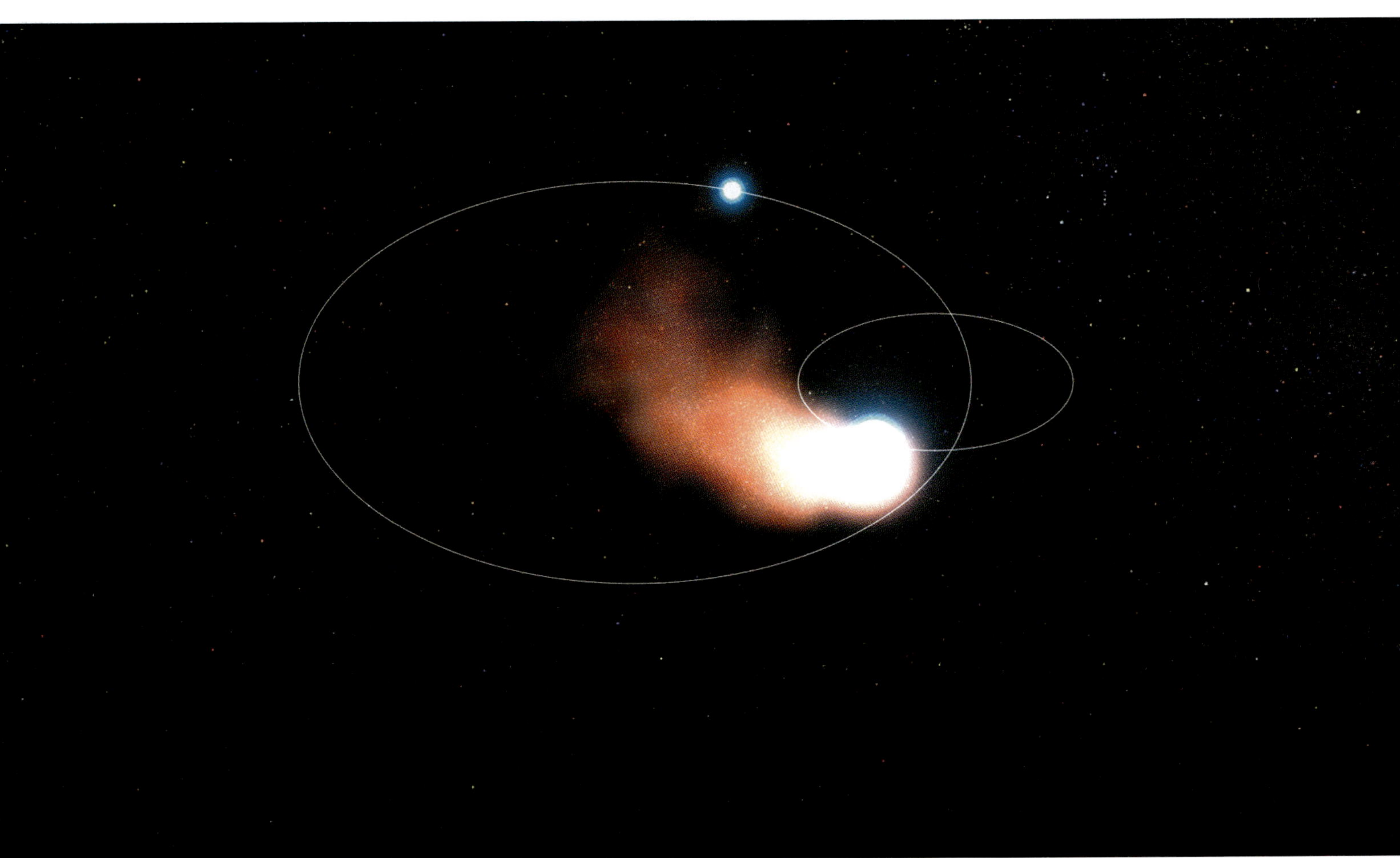

slightly more distant from each other than the sun and Uranus – just under 3 billion kilometres (1.9 billion miles). At their closest, which happens every eight years, they are as near to each other as the sun and Earth. And when they are close together, exciting things happen.

DUST SHELLS

The stellar wind from the larger star collides with the Wolf–Rayet dust emissions and creates the rhythmic pulse that generates the concentric dust shells around the pair. They are like three-dimensional tree rings, with the seventeen shells chronicling more than 136 years of dust production.

During those years, the outer ring has traversed millions of kilometres (about 70,000 times the distance from the Earth to the sun). Using the MIRI instrument, Webb was able not only to see these rings but also to investigate their chemical composition. Scientists suspect that the rings contain PAHs, the carbon-rich molecules that are thought to be vital for planet and star formation – and life.

The lifecycle of stars.

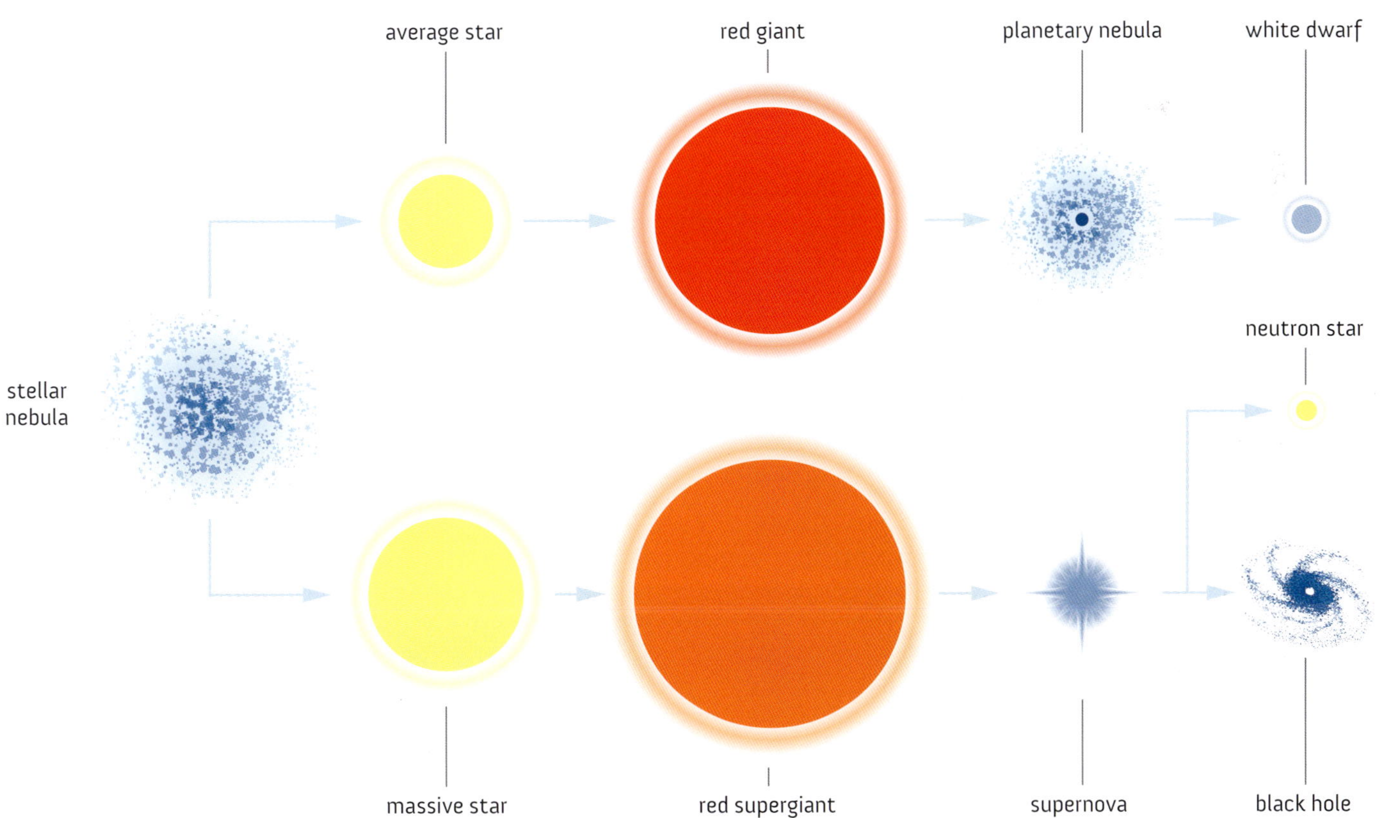

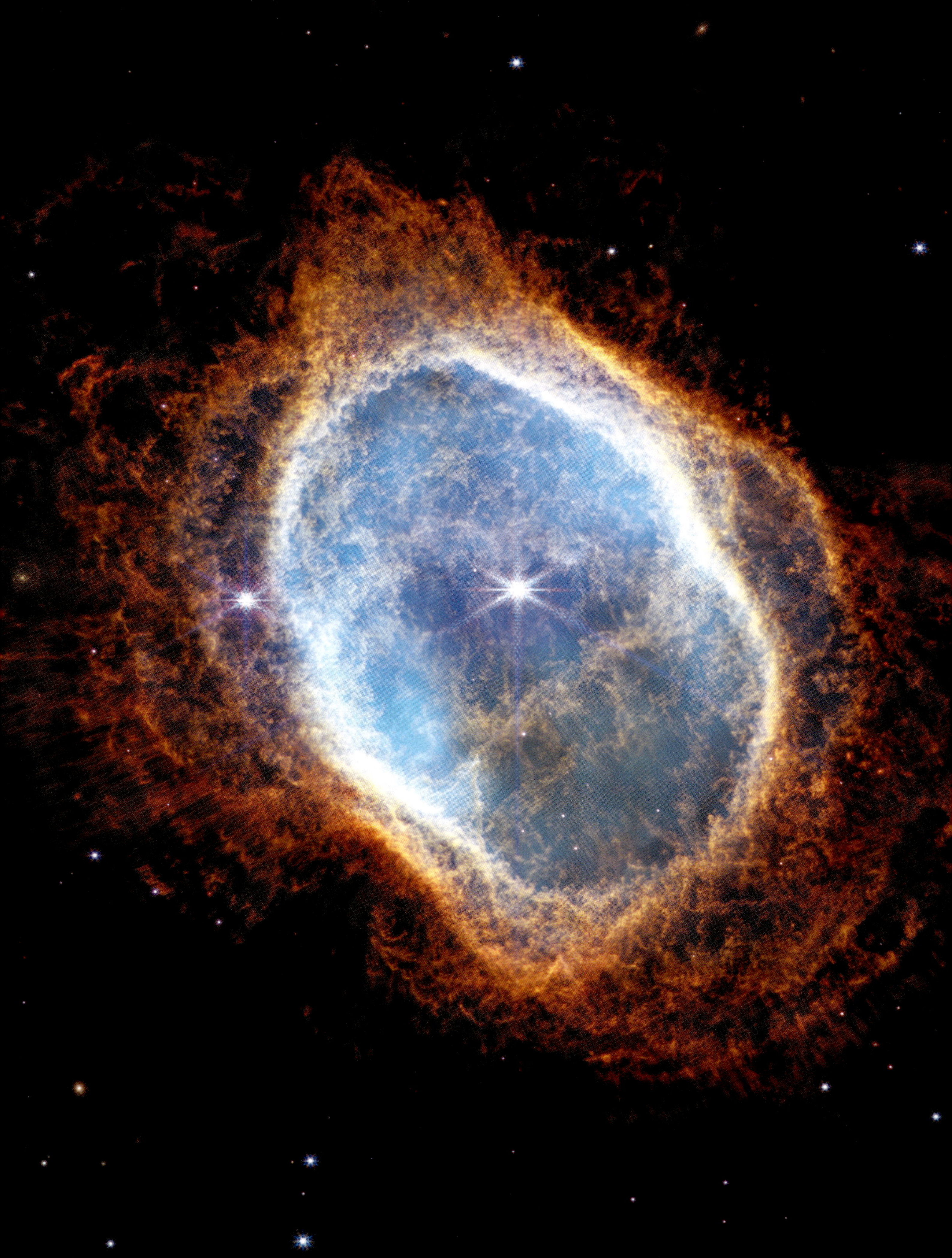

SOUTHERN RING PLANETARY NEBULA

ALSO KNOWN AS NGC 3132

Constellation: Vela
Distance from Earth: about 2,000 light-years

The Southern Ring Planetary Nebula, also known as NGC 3132, is about 2,000 light-years away. In this MIRI image, it is possible to see one of the massive stars in its centre.

TWO STARS SIT IN THE centre of this pond of rippling radiation, and one of them is dying. It has been emitting its last breaths of dust and gas for thousands of years. NGC 3132 is what is known as a planetary nebula – a phenomena that in fact has little to do with planets. Planetary nebula is the name given to nebulae with this distinctive ring shape, which is actually huge shells of gas that have been ejected by expiring stars. Such celestial phenomena give scientists important insights into what happens when stars die, and how the material they eject transforms the environment around them. Cataloguing the various molecules present also feeds into our understanding of how new stars form.

DANCING STARS

In the centre of the gas rings, the two stars are in different stages of their stellar evolution. As they continue to orbit each other, they shape the cosmic landscape around them. The Webb image on the bottom right shows, for the first time, that the second star is enveloped in dust. This star, coloured in red in the MIRI image, is a white dwarf. During its transformation from active star to white dwarf, it ejected pulses of matter. This matter moved out into the surrounding space like ripples in a pond.

Scientists believe that the star should have shed its last layers by now, and so are curious as to why it is still shrouded in dust. The other star, in blue, continues to interact with the white dwarf's old emissions. Their dance has warped the emissions ripples and forced the white dwarf to spray ejected material in a haphazard way. And this startling cosmic tug of war takes place against a backdrop of multicoloured stars and galaxies. In this image, stars look like tiny triangles while galaxies resemble squashed circles, ellipses and jagged scrawls.

ESCAPING LIGHT

When viewed through NIRCam (previous page) the white dwarf is basically invisible, but the other active star shines out like a pearl in an oyster. It has Webb's signature eight-point diffraction spikes. This star's light streams through holes in the nebula. However, parts of the dust cloud are too dense for light to escape, and in this image those regions are teal. Meanwhile, the transparent red parts of the rippling nebula contain distant galaxies.

TOP and BOTTOM LEFT: The Southern Ring Nebula (NIRCam and MIRI composite images).

BOTTOM RIGHT: A view with MIRI.

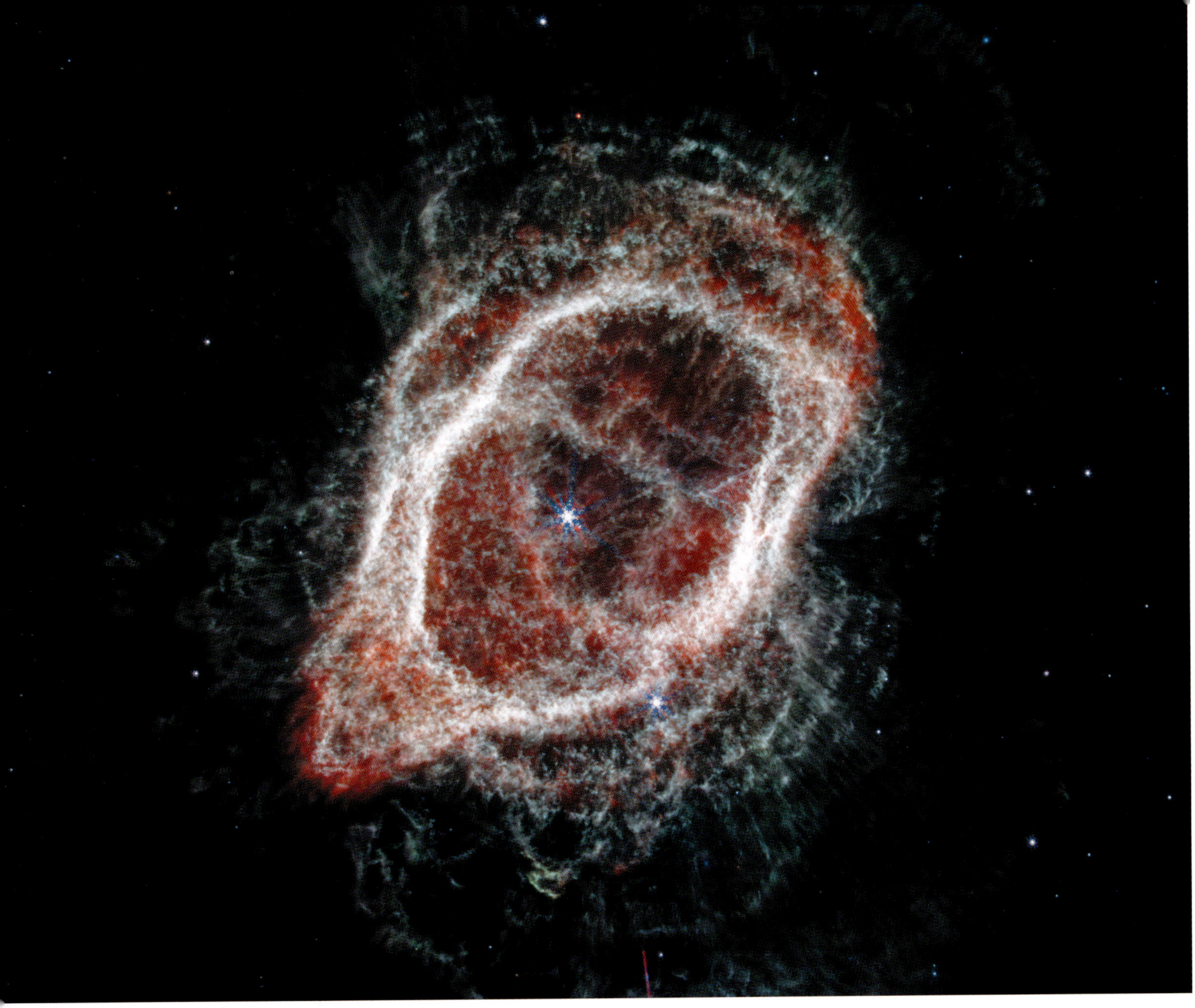

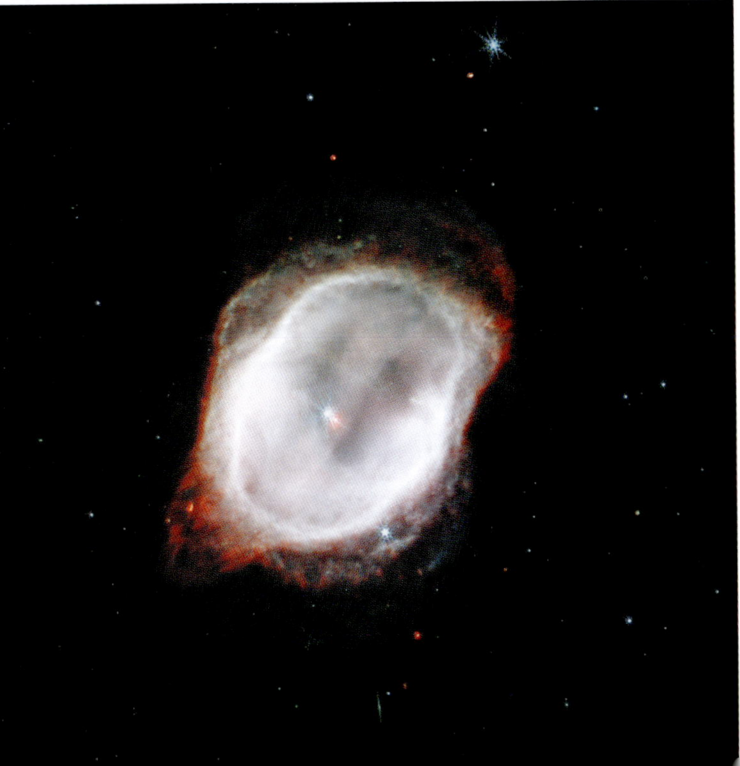

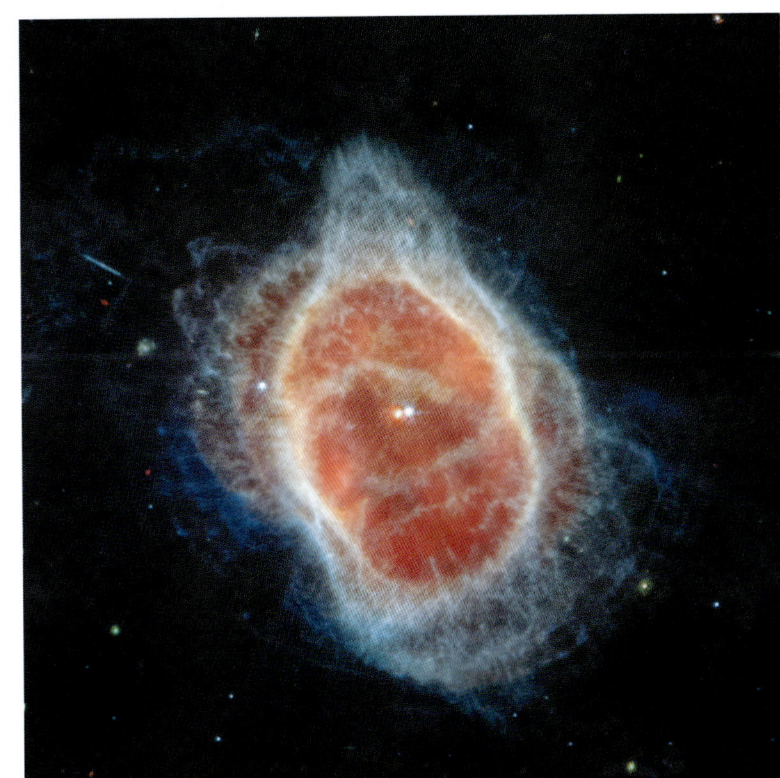

CASSIOPEIA A

Constellation: Cassiopeia
Distance from Earth: about 11,000 light-years

This NIRCam image of Cassiopeia A highlights new features in the well-known supernova remnant.

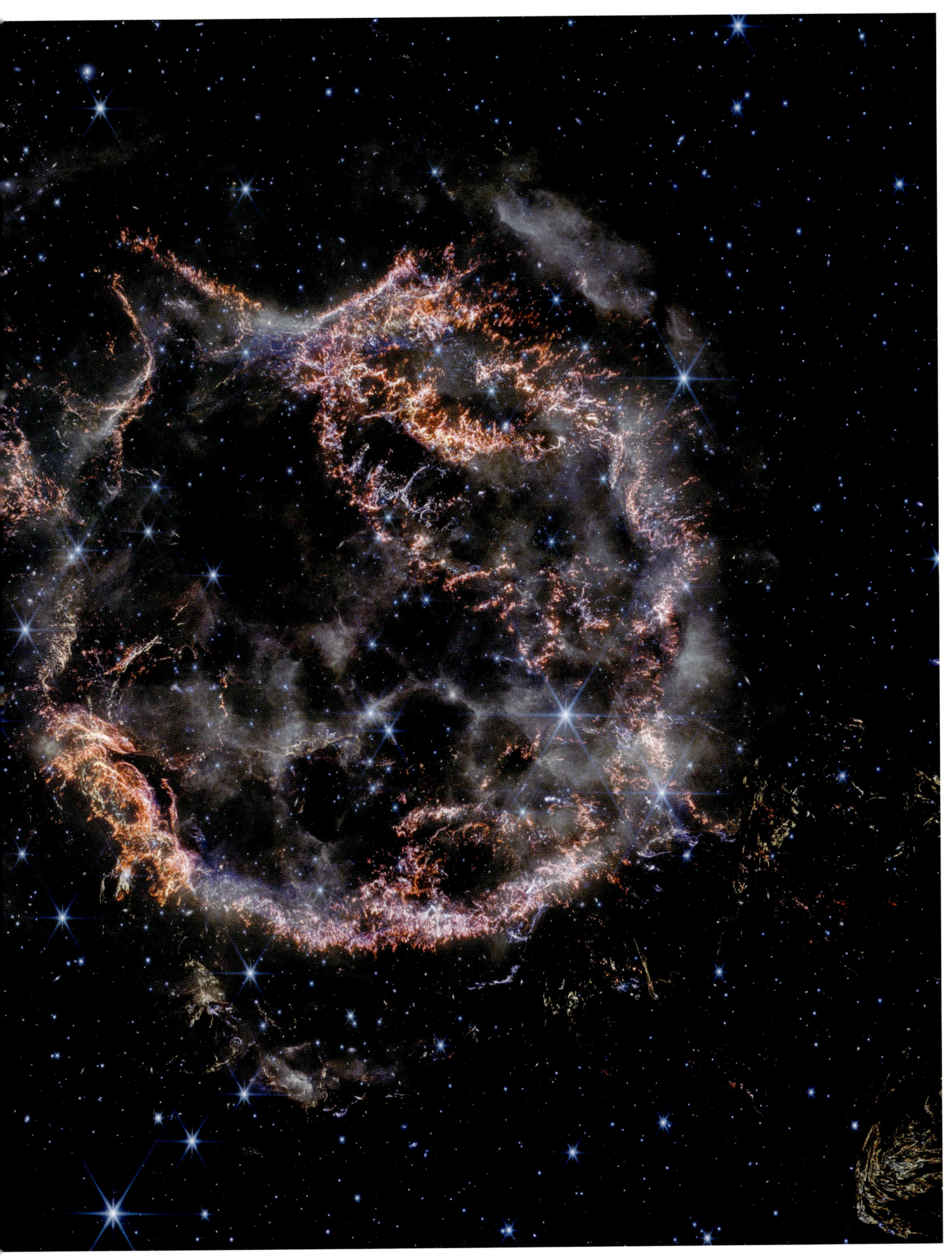

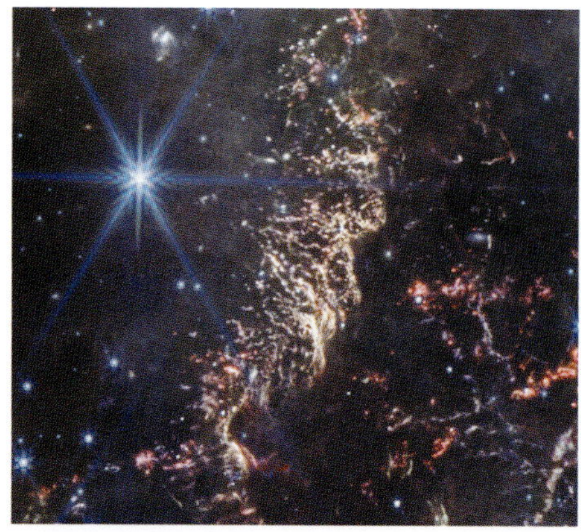
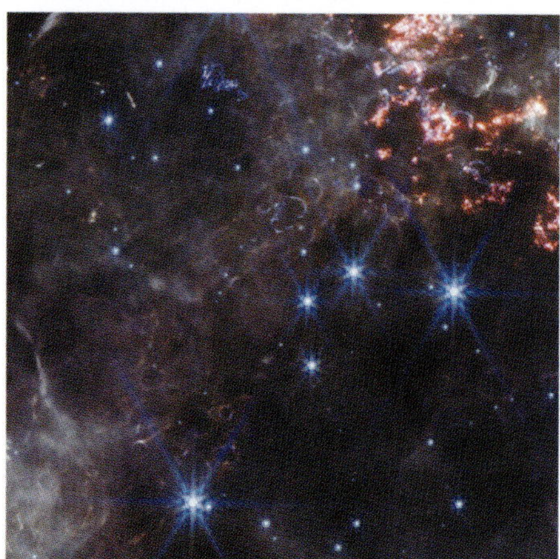
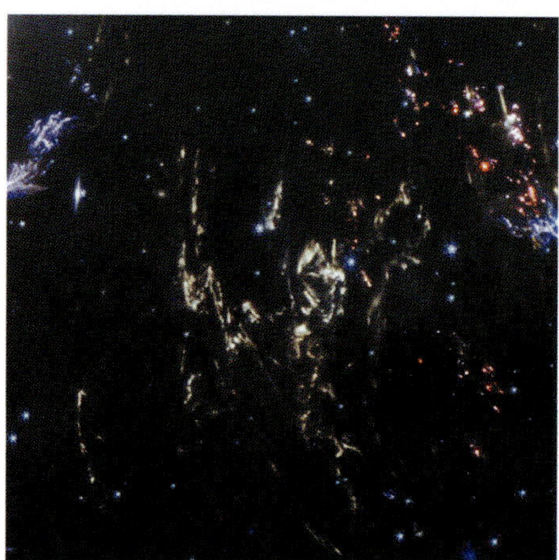
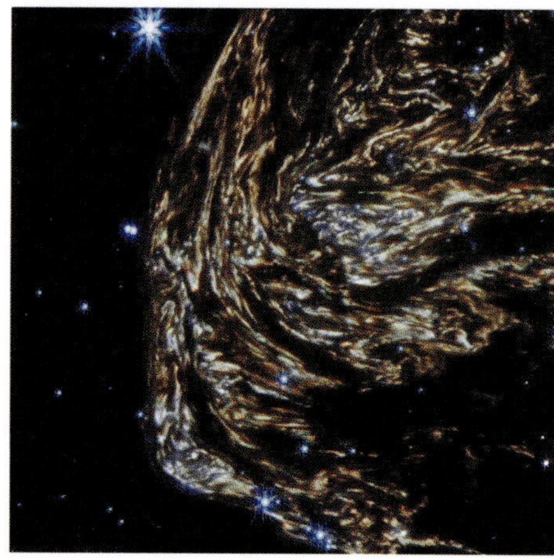

ABOUT 340 YEARS AGO, THE light from the supernova that was the star Cassiopeia A reached Earth for the first time. Supernovae can throw off several solar masses of material in a very short time and this material can be ejected at velocities of around 5 to 10 per cent of the speed of light. This material leaving the star in all directions so quickly creates an expanding shockwave into the surrounding interstellar medium.

Yet through the death of one star, others may be born. The remains of a star are rich in higher atomic number elements and the expanding shell can cause clumping in the surrounding medium, triggering star formation. Scientists estimate that Cassiopeia A's expansion shell is about 30 million°C (over 50 million°F) and that it is expanding at between 4,000 and 6,000 kilometres (2,500 and 4,000 miles) per second. That's roughly the distance from London to New York – *in one second*. However, in this case the remnants are not expanding uniformly – there are two nearly opposing effects on the energy and matter that are exploding into space at speeds of up to 14,500 kilometres (9,000 miles) per second, which is absurdly fast.

In this image, the supernova's expanding shell is smashing into previous gas eruptions. The supernova's inner shell is made up of heavier elements erupting from the star itself, namely sulphur, oxygen, argon and neon. These clusters of elements are coloured bright pink and orange. They exploded into the surrounding space when their star went supernova.

The white wispy clouds within the Cassiopeia A's shell are actually light from synchrotron radiation, which is emitted by charged particles as they rocket into magnetic fields and dance around them at high speeds. Webb can see this synchrotron radiation with its infrared detectors.

A LIGHT ECHO

In the far bottom corner of the NIRCam image, scientists stumbled upon a surprising find: a light echo. These echoes occur when light reflects off a distant source; in this case, light from the ancient explosion has reached a clump of gas, heating it up. Researchers have dubbed the large, stripey clump Baby Cas A – even though it is actually quite far away (about 170 light-years) from the supernova remnant. There are other light echoes in the image, swimming through the supernova remnants like striped fish.

LEFT: A pullout of interesting features from the NIRCam image on the previous page. These include the light echoes and Baby Cas A (bottom left).

RIGHT: A MIRI view of Cassiopeia A.

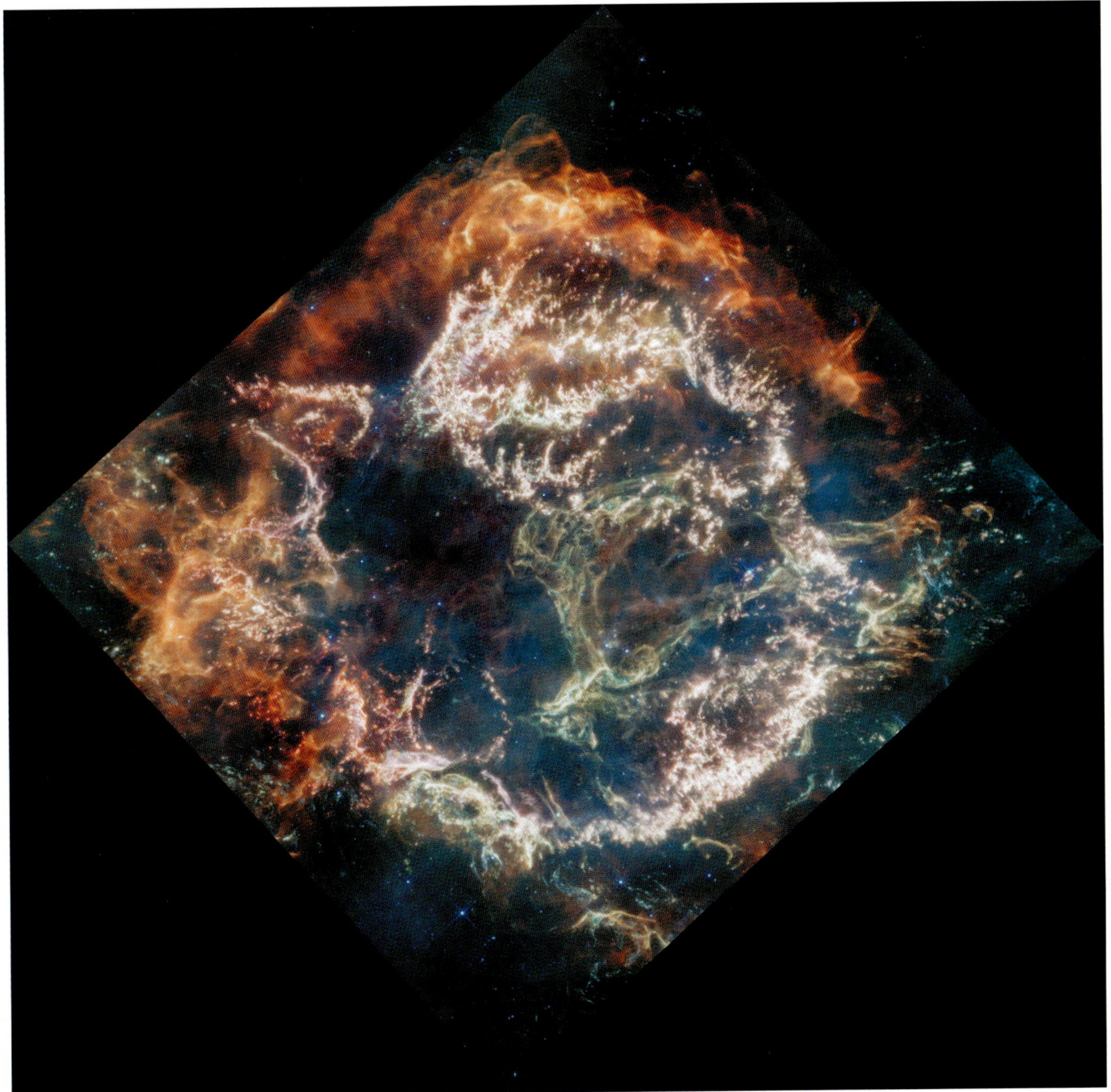

SUPERNOVA REMNANTS

In the above image, Webb is yielding new details about the remnants of this exploded star. The orange marks that resemble flames show the ejected material encountering the surrounding gas and dust. The bright-pink scrawls and dots flag matter from the star itself – a mixture of heavy elements such as oxygen, argon and neon.

Scientists nicknamed the green loop the Green Monster in honour of the left-field wall at Fenway Park in Boston, home to the Boston Red Sox baseball team, and are still uncertain about what it represents and why it looks like that. For example, the green loop is pockmarked with mini-bubbles, but they are not yet sure what these bubbles are.

RING NEBULA

ALSO KNOWN AS M57 OR NGC 6720

Constellation: Lyra
Distance from Earth: about 2,500 light-years

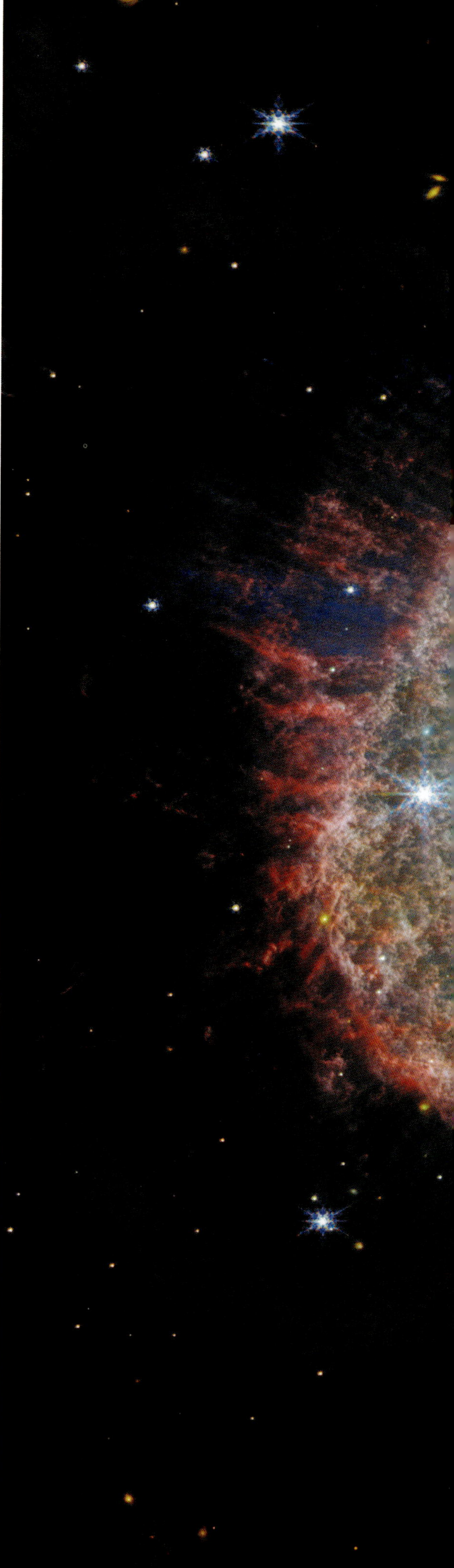

The Ring Nebula, as seen with NIRCam, is a planetary nebula. Astronomers are sifting through the data for clues about the complex process that led to its creation.

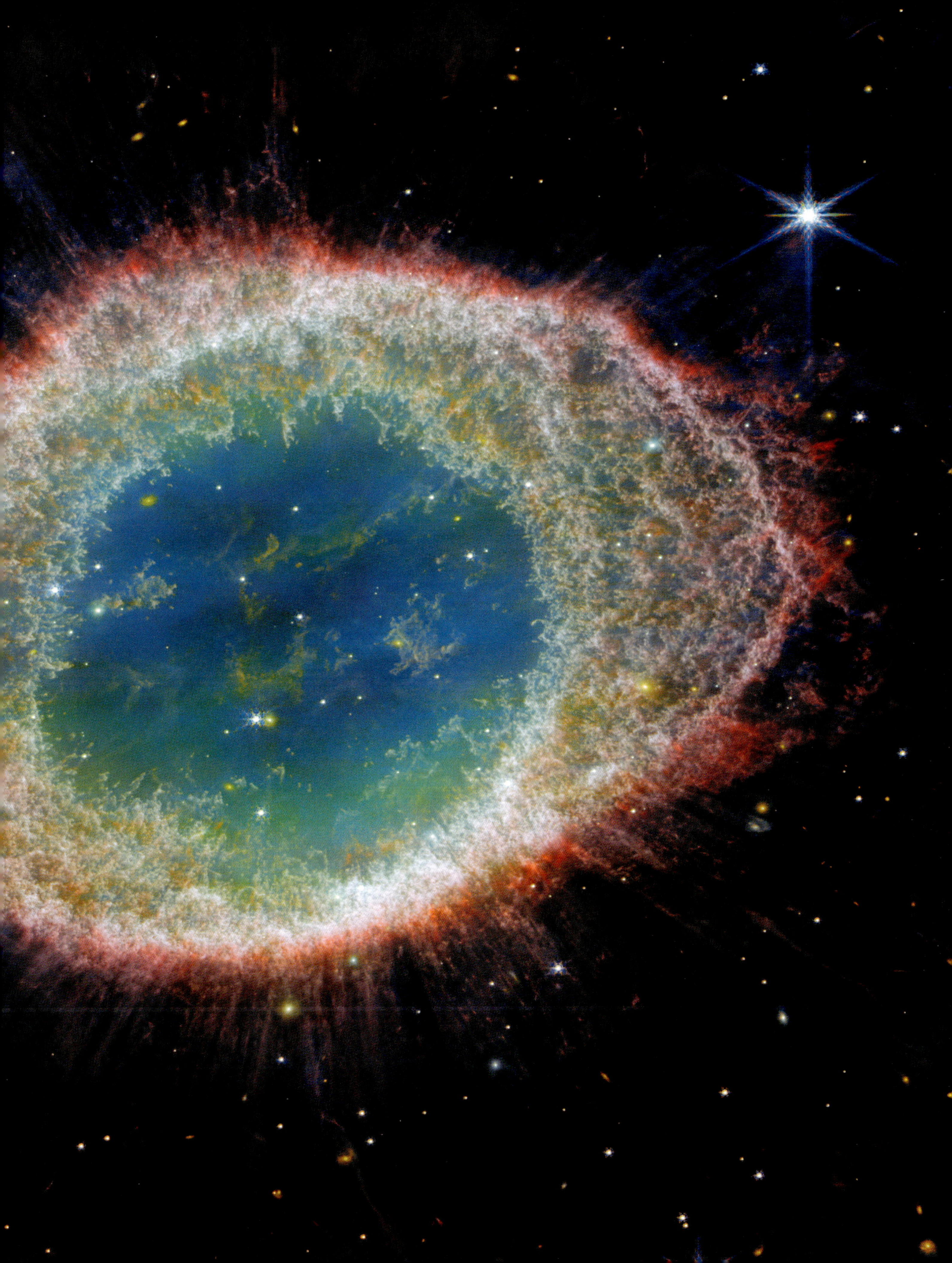

THE RING NEBULA IS ANOTHER example of a planetary nebula. A few thousand years ago, a star was dying and turning into a red giant. In around 5 billion years' time our own star, the sun, will turn into a red giant, with its matter engulfing Mercury, Venus and even Earth. As these stars expand, their surface temperature drops from around 5,600°C (10,000°F) to between 2,000 and 3,000°C (3,600 and 5,500°F). These cooler temperatures cause stars to produce light in the redder part of the spectrum, and these are known as red giants.

Stars shine for most of their lives, until they eventually start to run out of fuel. When this happens, the outward pressure caused by the fusion process reduces and the force of gravity dominates, causing the core to start collapsing in on itself. Average stars, like our sun, tend to become white dwarfs. They are roughly the size of Earth, but substantially denser, and don't give out much radiation. Scientists hypothesize that over time white dwarfs will crystallize, ultimately becoming black dwarfs. But the amount of time this would take is longer than the maximum possible age of the universe, so they're not sure what eventually happens to white dwarfs.

A different view of the Ring Nebula with MIRI.

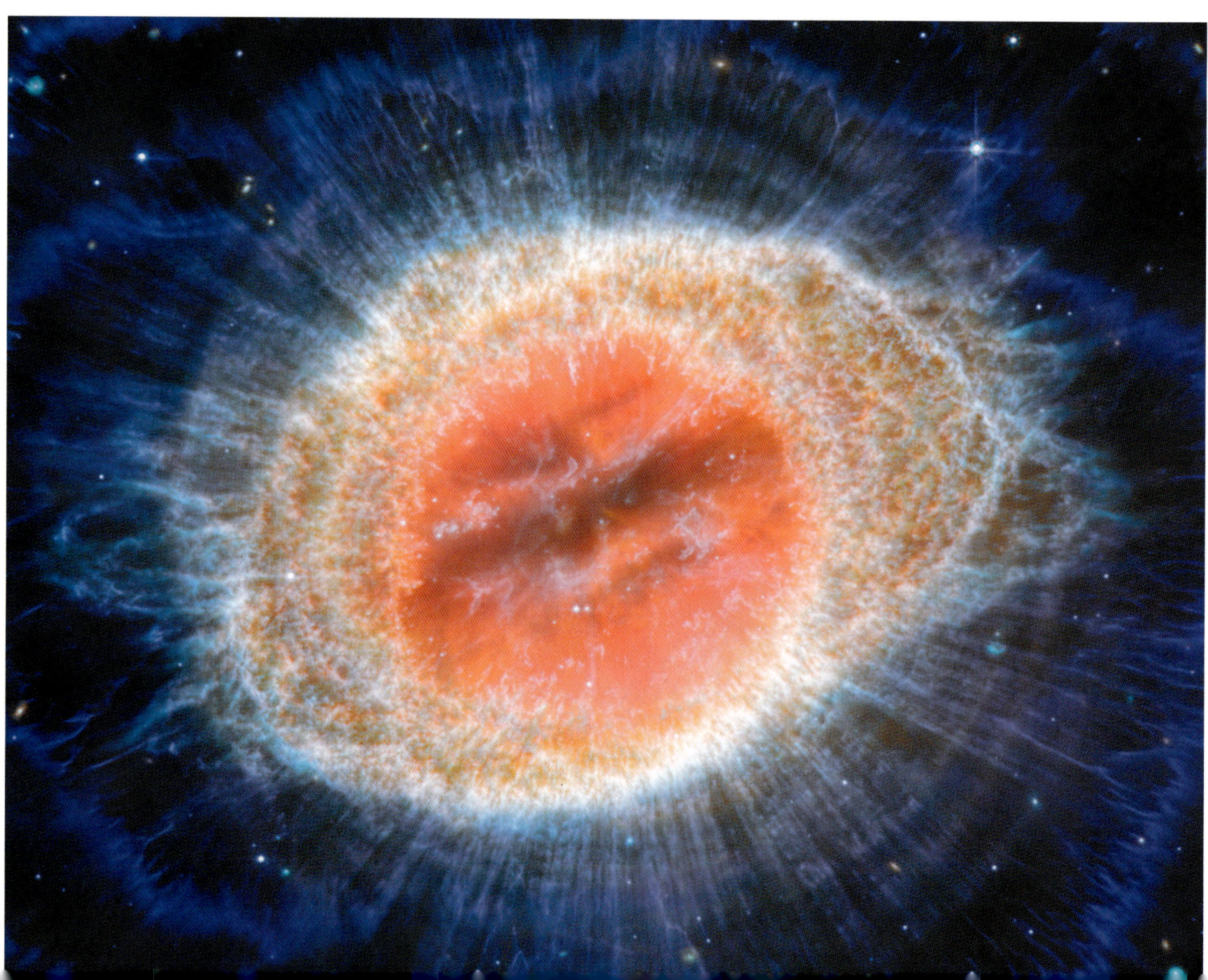

Much larger stars can end up going supernova, in which their cores collapse and then explode. After they have consumed all their internal fuel, these stars can no longer support the gravitational pull towards the centre; the result is so spectacular that their demise can be seen across the universe. These stars' giant stellar volume shrinks to a fraction of its former girth and their temperature skyrockets. The exploding star can form either a neutron star (one of the smallest and densest objects in the universe) or a black hole (a region of space whose gravity is so strong that not even light can escape), or can detonate into a diffuse nebula (and create a breeding ground for new stars).

So, as this star died, it shed most of its mass. With each pulsing ejection, it sent matter and energy into the space around it, forming these distinctive watermarks of emissions. Using MIRI and NIRCam, Webb has given scientists an unprecedented view of this well-studied planetary nebula.

CLUMPS OF GAS

In the NIRCam image, we can see the delicate structure of the nebula's rings, particularly the hot gas in its heart. It gets its name from the bright ring that circumscribes it, and Webb has shown us that the bright ring is made up of about 20,000 clumps of dense molecular hydrogen gas. Each of those clumps is about as big as Earth. Just within that ring, there are emissions from PAHs. Scientists were surprised to find these biologically important molecules in the Ring Nebula. The NIRCam image also shows 'spikes' pointing away from its centre – a bit like fur on an angry cat – a prominent feature in the infrared. These spikes are thought by some to be molecules that can form in parts of the nebula that are shielded from the central star's radiation, but more evidence is needed to verify this theory.

THE HEART OF THE RING NEBULA

While NIRCam gave us a glimpse into the complex heart of the Ring Nebula, MIRI shows us how far its ripples extend. In the image on the left, there are about ten circular rings beyond the main halo. Calculations indicate that these circles formed every 280 years as the dying star sloughed off material. But that's not usually what happens when a single star implodes and forms a planetary nebula. The glorious rainbow rings surrounding the dead star suggest the likely presence of another star orbiting somewhere nearby.

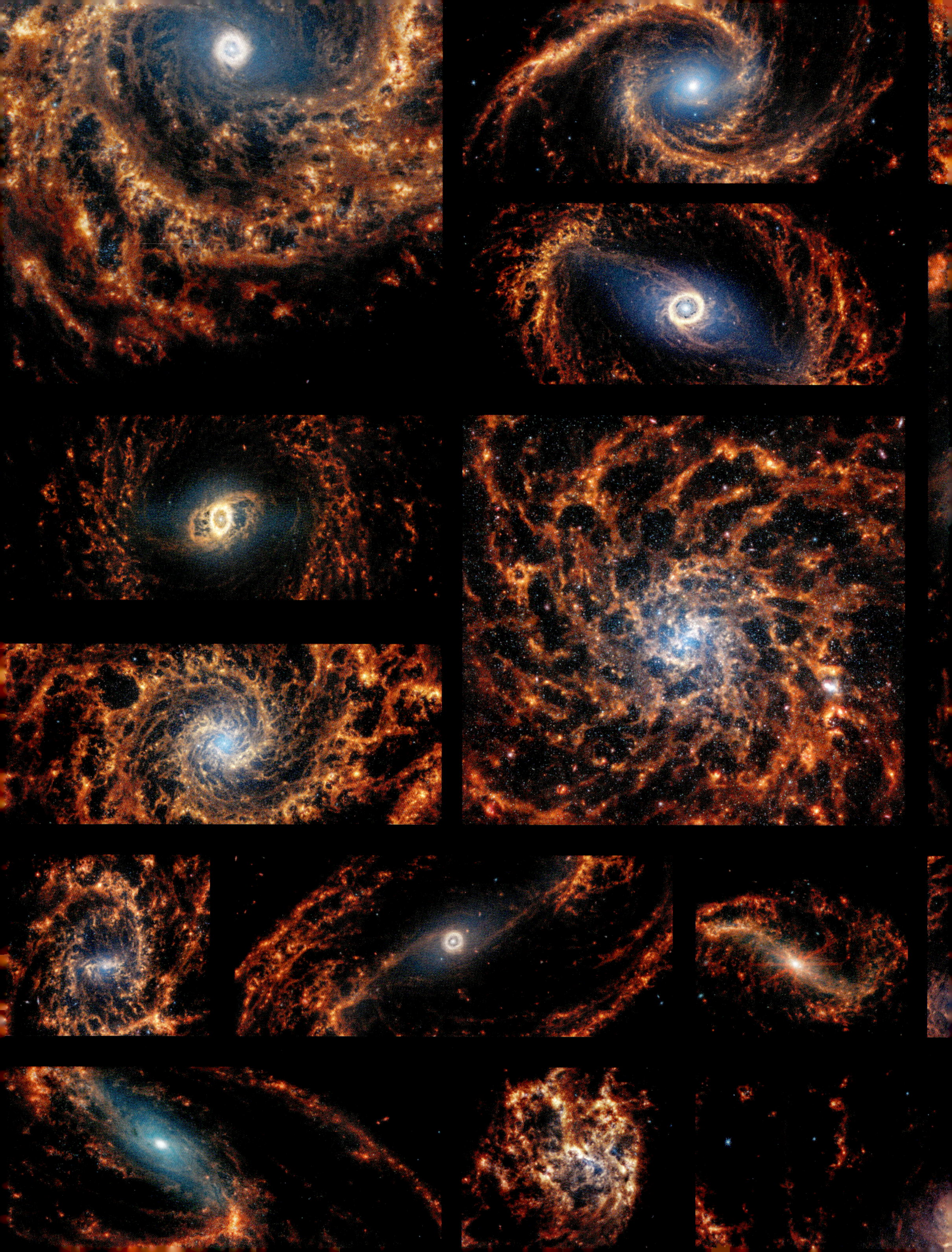

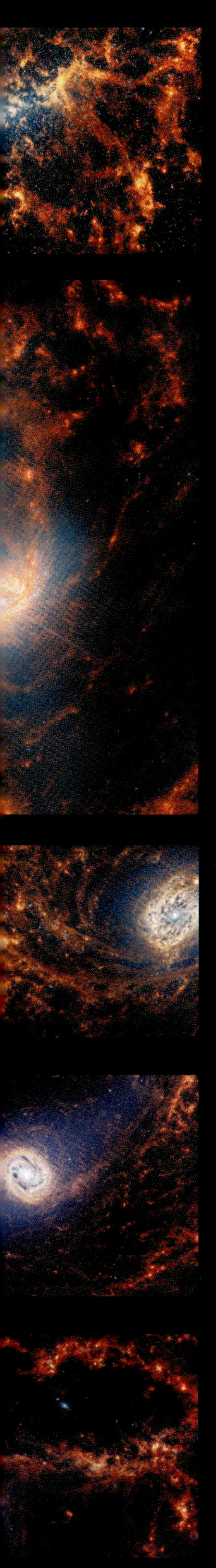

GALAXIES

Galaxies are giant cosmic systems made up of stars, interstellar gas, dust and dark matter, all held together by gravity. They come in a variety of shapes and sizes, and there is still a lot we do not know about them. We don't actually know exactly how many galaxies are out there. Some of the first indications were gained using Hubble's Deep Field imaging, on the basis of which it is estimated that there are between 100 billion and 200 billion galaxies in the observable universe – with each of those galaxies potentially hosting as many as a trillion stars.

Galaxies are divided into three main categories: elliptical, spiral and irregular. Elliptical galaxies account for around a third of the known galaxies. They contain little gas and dust, and tend to no longer have active star-forming regions. They mainly look circular or a little squashed. Spiral galaxies, such as our own Milky Way galaxy, look like the classic flat disks of stars, dust and gas, with a bulge in their centres. These galaxies contain actively forming stars and are very common. Irregular galaxies fall into neither of these categories, and are more common when observing the early universe before spiral and elliptical galaxies formed.

There is another galaxy type that deserves a mention: dwarf galaxies are relatively small compared to the others, with only a few billion stars. They often orbit larger galaxies – our Milky Way, for example, has at least a dozen. These include the Small and Large Magellanic Clouds, which are among my favourite things to spot whenever I am in a clear-sky area in the southern hemisphere.

Webb showcases the marvellous diversity of galaxies in the cosmos. As part of the Physics at High Angular resolution in Nearby GalaxieS (PHANGS) programme, Webb observed nineteen nearby face-on spiral galaxies.

SAGITTARIUS C

Galaxy: The Milky Way
Distance from Earth: 26,000 light-years

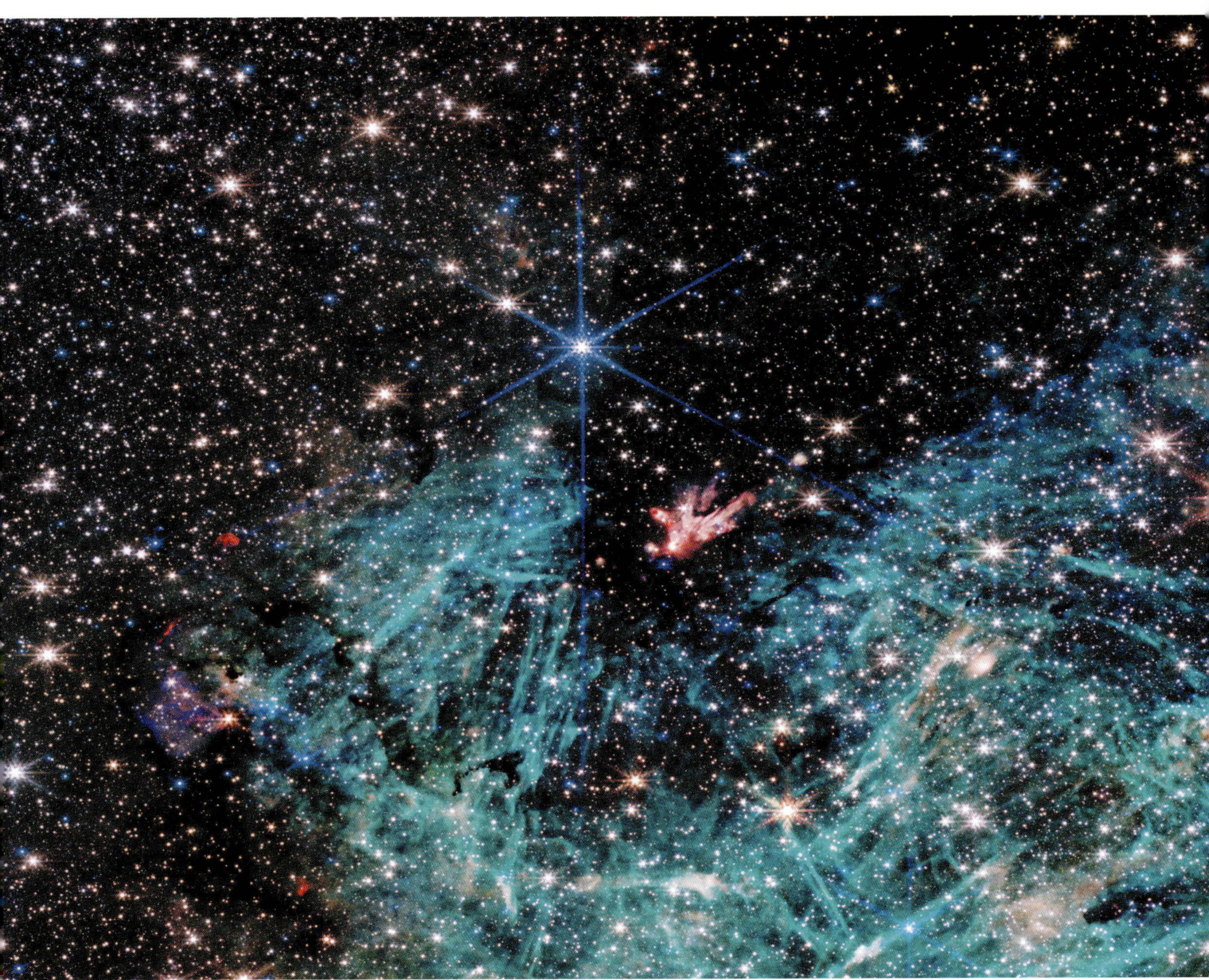

Sagittarius C, captured here with NIRCam, is part of our own Milky Way galaxy. About half a million stars shine in this image.

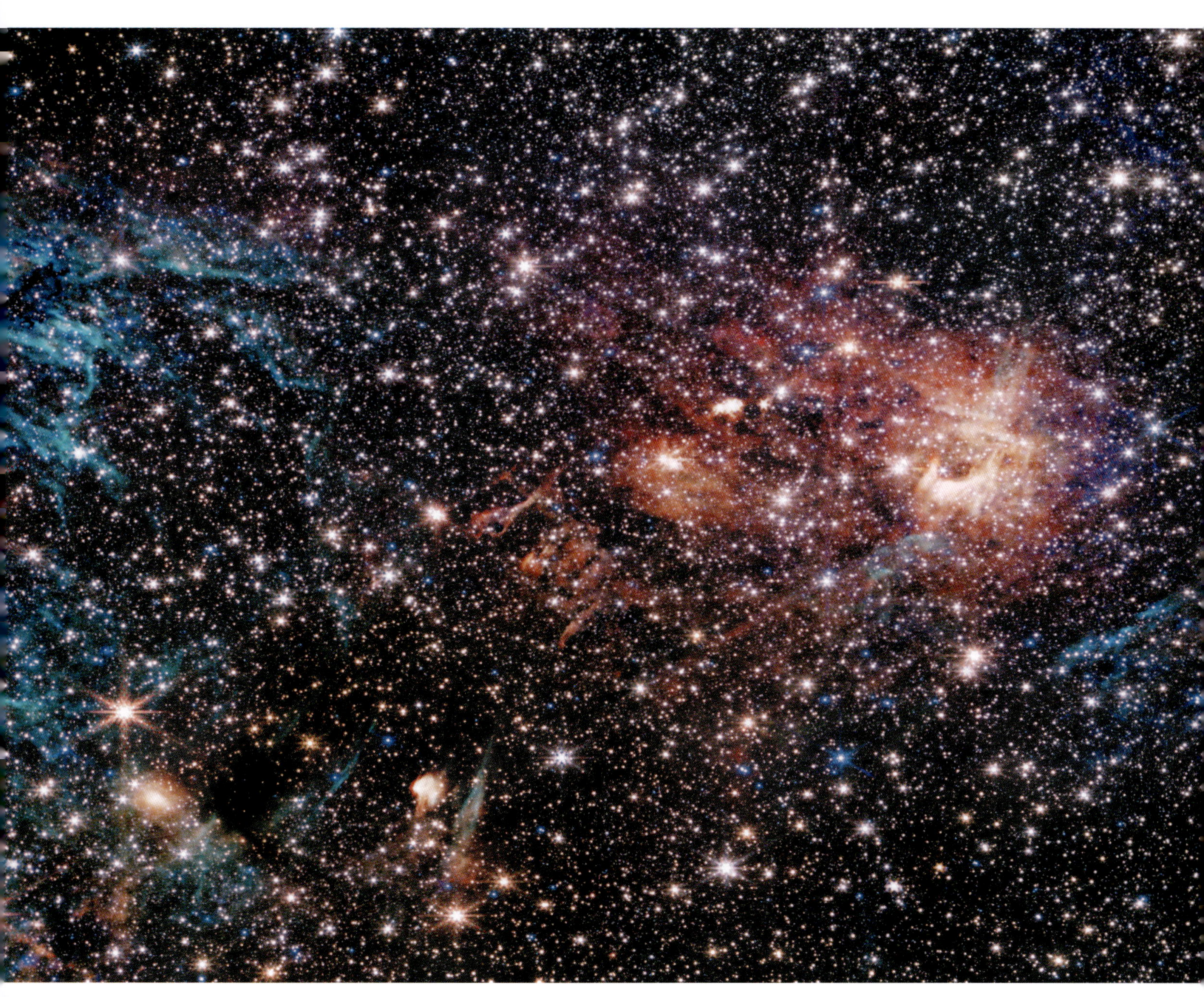

THIS SUMPTUOUS KALEIDOSCOPE IS OUR cosmic backyard. Sagittarius C is right here in the heart of the galaxy we call home, our Milky Way. It is a very active star-forming region close to our supermassive black hole, Sagittarius A*.

A large cyan cloud dominates the Webb picture. This is a giant cloud of ionized hydrogen, which is typically a site where stars have recently formed. The shape and topography of such areas are often determined by the star formation that is happening within them, but this image of Sagittarius C has some puzzling features that have generated more questions than answers. Needle-like structures of ionized hydrogen have formed within the cloud, looking like ancient human scratchings on a cave wall. But they aren't uniform and don't seem to follow any particular pattern.

BLACK HOLES

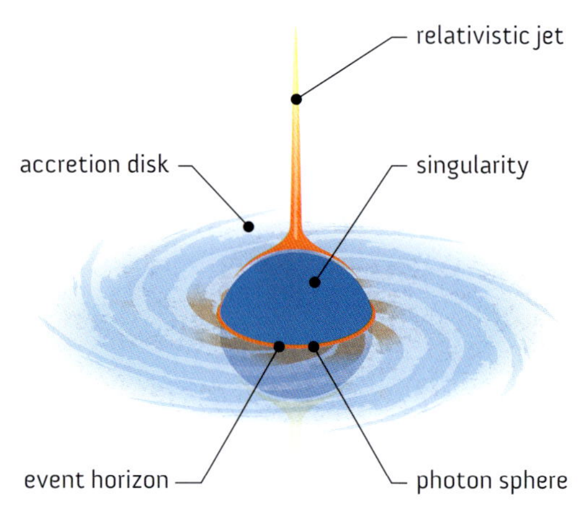

An illustration of a black hole.

At the centre of many galaxies, including our own, are supermassive black holes – celestial objects with such strong gravitational fields that not even light can escape. Everything that happens inside black holes is hidden from us. The sorts of forces they exert are unfathomable and, as such, matter behaves very strangely around them. Black holes push the boundaries of our laws of physics and are useful to test our theories of the universe.

Your average black hole is about ten to twenty-four times more massive than the sun. Supermassive black holes are a different category altogether, with a mass up to a billion times that of the sun. Their gravity is so strong that it enables them to act as a galactic pivot around which the millions or even billions of stars of a galaxy orbit. We are still unsure how the goliath black holes form. They may coalesce out of many stars at the heart of a galaxy, or they could have been there from the beginning.

A COSMIC CAULDRON

It's worth saying that this region of interest is a mere 300 light-years away from the Milky Way's voracious active galactic nucleus – and interesting things tend to happen in the vicinity of black holes, especially supermassive ones. That's mainly because this is such an extreme environment, where immense pressures, sizzling energies and strong magnetic fields push matter to the boundaries of what is possible. In other words, a good place to produce new stars. The mesmerizing Webb image contains about half a million stars.

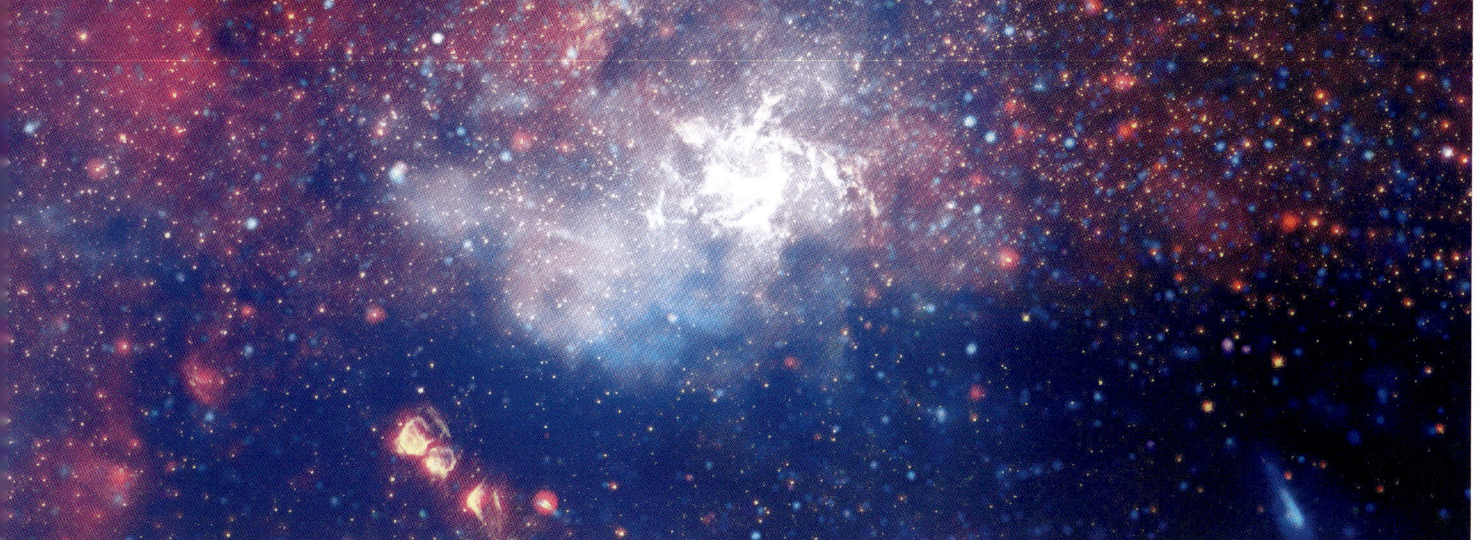

A composite image of our galactic centre, which combines Hubble, Chandra and Spitzer data.

As Sagittarius C sits in our galaxy, it is close enough that we can observe individual stars, allowing us to understand how they form, live and ultimately die. In the image on the previous page, a protostar cluster flares red to the right of centre, just above the cloud of ionized hydrogen. The cluster of protostars, some of which you can see in bulbs of yellow, looks like a bushel of fireworks that have just been ignited.

INFRARED-DARK CLOUDS (IRDCS)

The image may also hold even younger stars. The protostar cluster is an island in a sea of relative darkness. Compared to the rest of the image, that section looks much darker. A smaller, similar example sits to the right of the cyan cloud like an inky thumbprint in an otherwise busy sky. These are infrared-dark clouds. We've only known about IRDCs since 1996, and we are still trying to figure out exactly what they are. We do know that they are cold, dense regions within molecular clouds. Astronomers suspect that in their dark hearts new stars at the earliest stage of star formation are coalescing. This naturally makes IRDCs even more intriguing, because there are still many questions about how stars actually form, where they do it and why.

STELLAR EVOLUTION

The Webb image shows various stages of star evolution, from their very beginnings right through to bright, mature stars. Some of the brightest objects look like cartoon stars, with Webb's characteristic diffraction spikes. Having such a clear image of Sagittarius C could give astrophysicists some of the information they need to better understand these luminous objects and how their environment influences their genesis and evolution.

NGC 346

Constellation: Tucana and part of Hydrus
Distance from Earth: 210,000 light-years

NGC 346, imaged here with NIRCam data, is in the Small Magellanic Cloud. It's a star cluster about 200,000 light-years away.

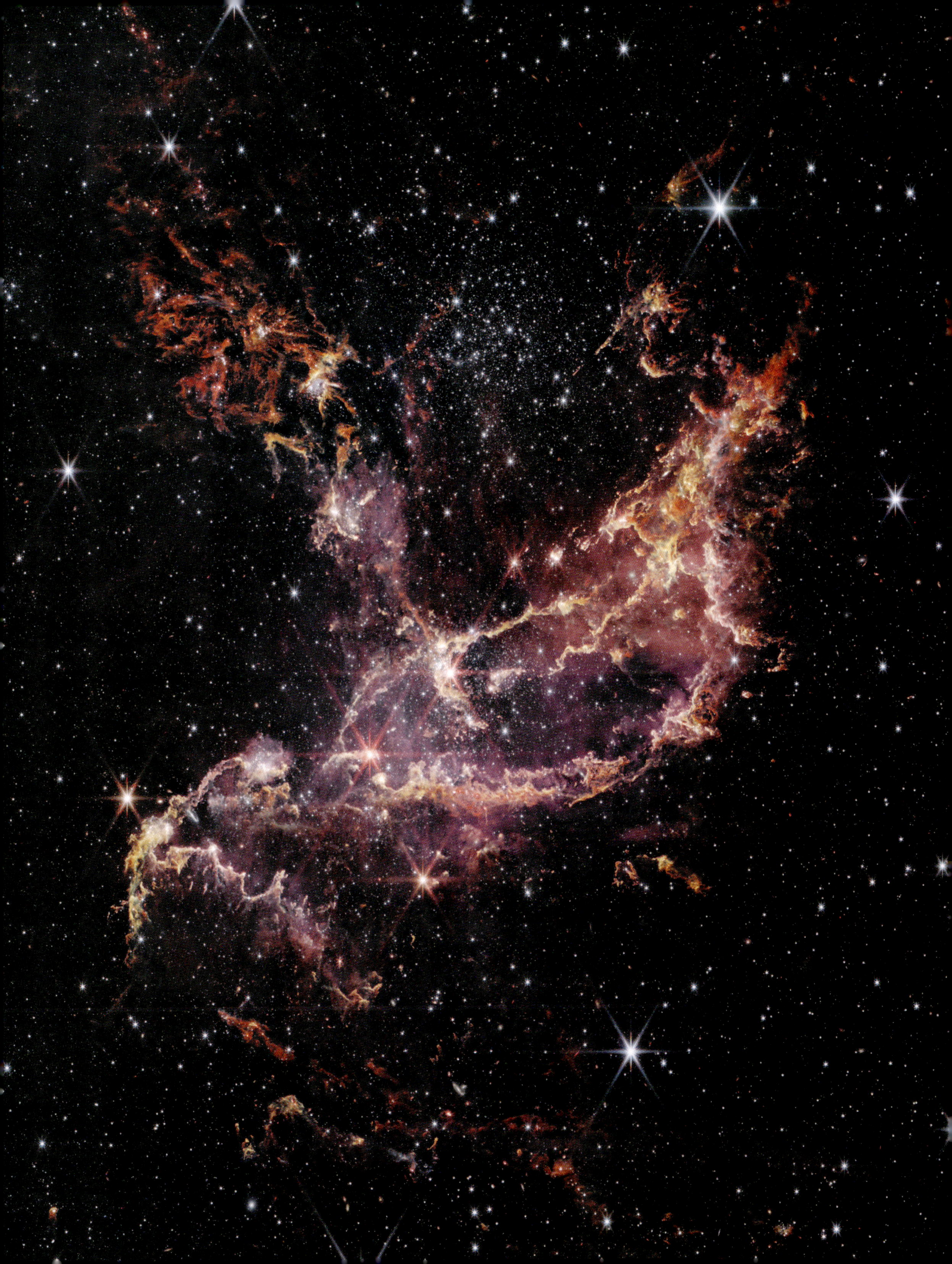

THE MAGELLANIC CLOUDS – BOTH Small and Large – can be viewed in close proximity to each other. As with many faint astronomical objects that can be viewed with the naked eye, to see them it is often best to look at a spot close to them rather than directly at them. This allows their image to fall on the part of the retina at the back of the eye with the most rods – the detectors that pick up low light levels. To the uninitiated observer, the Magellanic Clouds just look like small wisps in the sky, hence their name.

The Small Magellanic Cloud (SMC) NGC 346 is a mesmerizing galaxy for us to study. It's technically a dwarf irregular galaxy, which means that it is relatively small and doesn't have a typical shape like a spiral or ellipse. It is one of the dwarf galaxies that is orbiting our own Milky Way. But the reason that this galaxy, with its several hundred million stars, is so interesting is that it mimics the types of galaxies that existed in the early universe.

The Large Magellanic Cloud, observed by Hubble, is a satellite galaxy of the Milky Way.

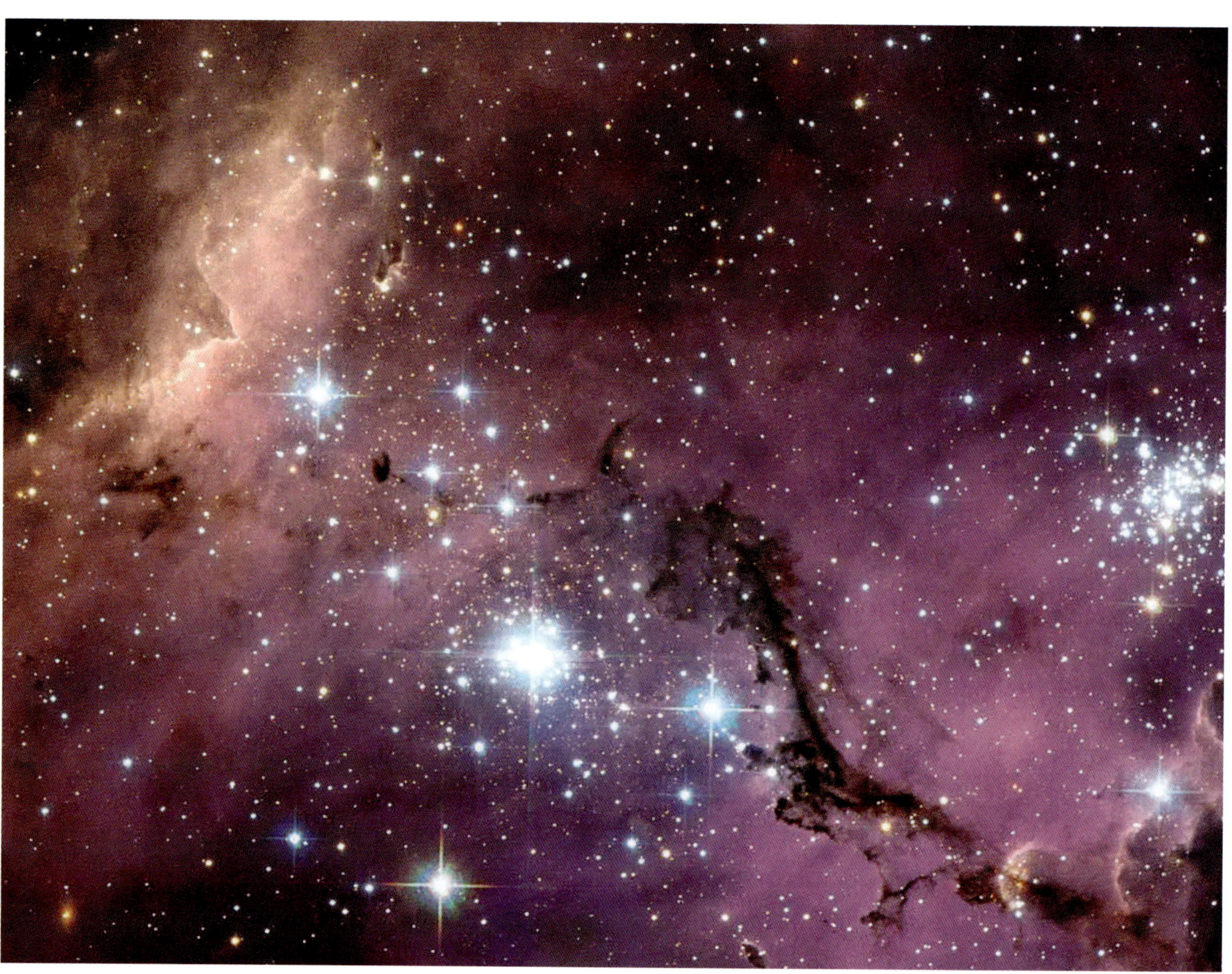

This captivating corner of the Small Magellanic Cloud was imaged using data from Hubble, Chandra and Spitzer.

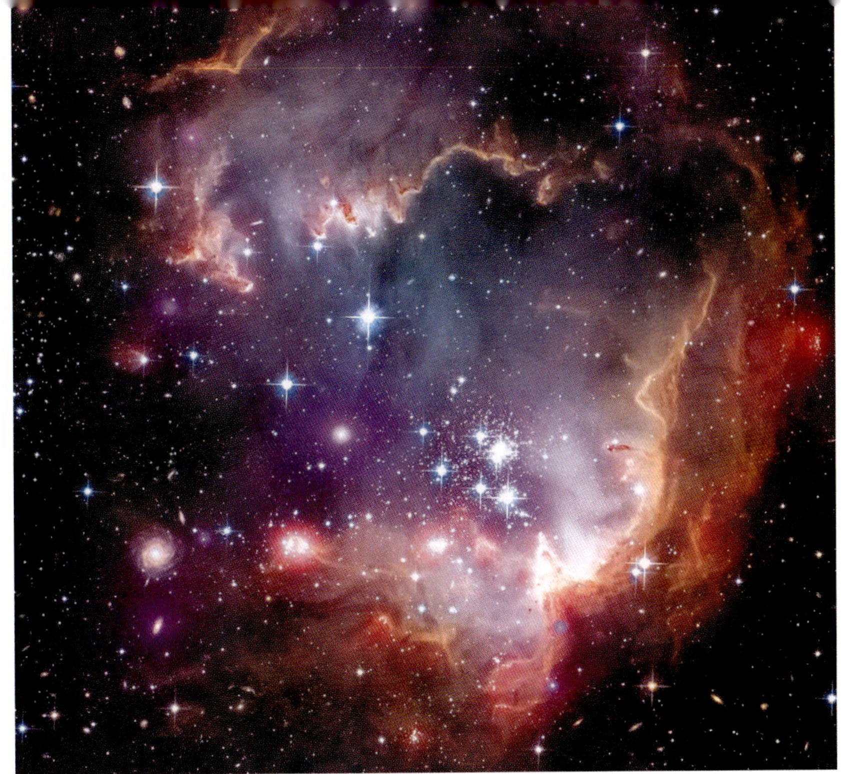

THE COSMIC NOON

Back in the early years of the universe, galaxies were mainly made up of simple atoms like hydrogen and helium. This era, between 2 and 3 billion years after the Big Bang, was known as the Cosmic Noon (coming after the Cosmic Dawn) and saw an intense period of star formation. Over time, stars died and threw their metal-rich hearts out into the vast reaches of space. These heavier atoms became more stars and these grew into the common galaxies we see today, such as our own Milky Way.

The SMC is a bit of a mystery because its structure and chemical composition mimic that of those galaxies and star clusters found during the Cosmic Noon, with low concentration of metals.

The main image shows NGC 346, a young, open cluster of stars in a nebula. Canopies of dust and hydrogen billow through the image, with newborn stars and planets in their recesses. There are actually two different types of hydrogen in the image: the orange plumes are molecular hydrogen (hydrogen in molecule form, with two hydrogen atoms bonded together), while the pink clouds are positively charged, unpaired hydrogen atoms. The orange sections are usually incredibly cold, which makes them a good place for stars to form. The illuminated pillars show places where young stars are eroding the dense clouds surrounding them. By studying new stars forming, astrophysicists can compare them with stars forming in older galaxies, such as our Milky Way, to see how they differ and how star evolution changes in different generations of galaxy.

SMACS 0723

Constellation: Volans
Distance from Earth: 4 billion light-years

When Webb's First Deep Field was released on 12 July 2022, it was the deepest and sharpest infrared image of the distant universe to date. In the foreground, it shows galaxy cluster SMACS 0723, with its thousands of galaxies. But the true magic is in the background, where there are ancient galaxies, the light from which has taken billions of years to reach us.

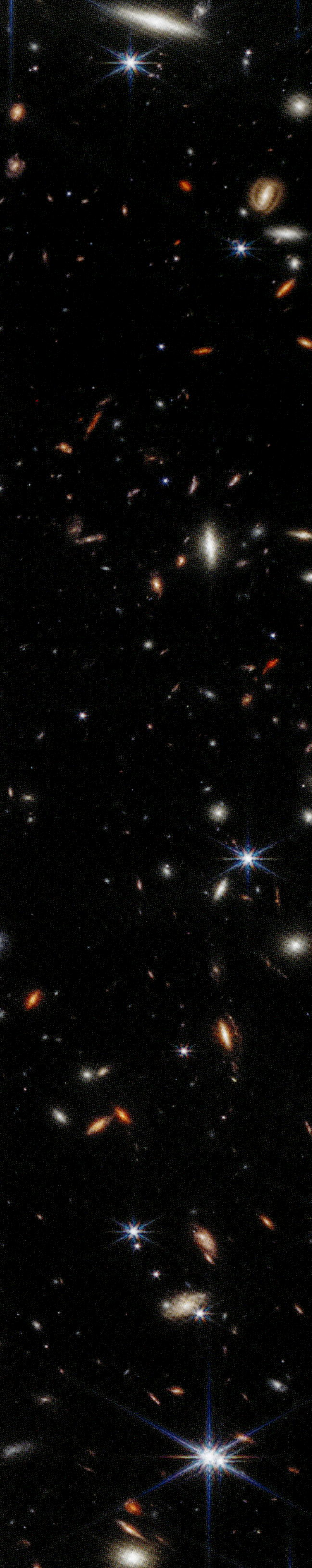

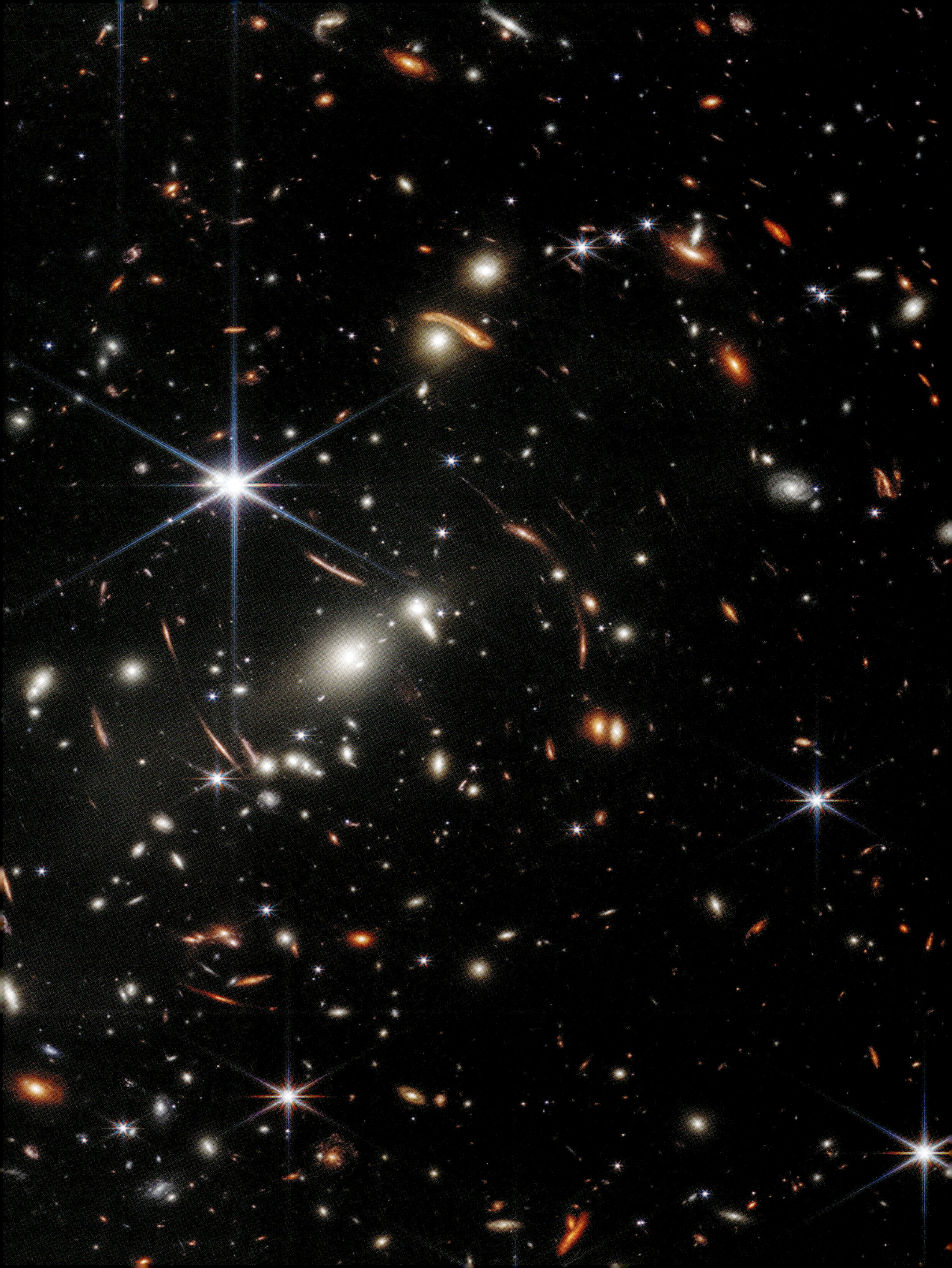

THE LIGHT FROM THE EARLIEST galaxies is incredibly faint, separated from us through both time and space by billions of years. Until recently, it was difficult to image these early galaxies because our technology was just not up to the task. But the latest generation of telescopes – and most recently Webb – have the ability to see deeper into the universe than ever before.

To pick up on these ancient signals and to gather enough radiation to draw meaningful conclusions, deep-field observations are undertaken. A number of years ago, I met Robert Williams, who was the director of the Space Telescope Science Institute in Baltimore. He commissioned the first of these deep-field observations using Hubble – the HDF. In these surveys, the telescope is directed to look at a particular part of the sky for a very long period of time – days rather than the usual minutes – so that a prolonged exposure can be achieved with the onboard detectors. Just as with a standard camera, the longer the exposure, the more photons can be gathered.

Robert told me that many people were shocked when he first suggested this idea. To use a world-class telescope in this way seemed madness. Telescopes like Webb and Hubble are in high demand, with many applications for use being rejected, and yet here was Robert proposing to do a ten-day exposure. What was worse was that he and his team were choosing an area of the sky with no known bright objects – a field of view that seemed to be empty space. But any bright stars in the field would overwhelm the exposure, so empty space was needed.

The observation went ahead during December 1995 and 342 separate exposures were taken. When amalgamated and analysed, the HDF exposures revealed the distance, age and composition of the galaxies imaged. Bluer objects, for example, may include young stars or be quite close together. Redder objects may include older stars or be further away. From this observation an estimation of the number of galaxies in the universe was also inferred.

A RICHNESS OF GALAXIES

The Hubble Deep Field was a game changer in our understanding of the universe and several similar exposures were later taken with Hubble.

With such a legacy, people were expecting a lot from Webb's First Deep Field image. This was also its first operational image to showcase its capabilities. It was taken with NIRCam, and is

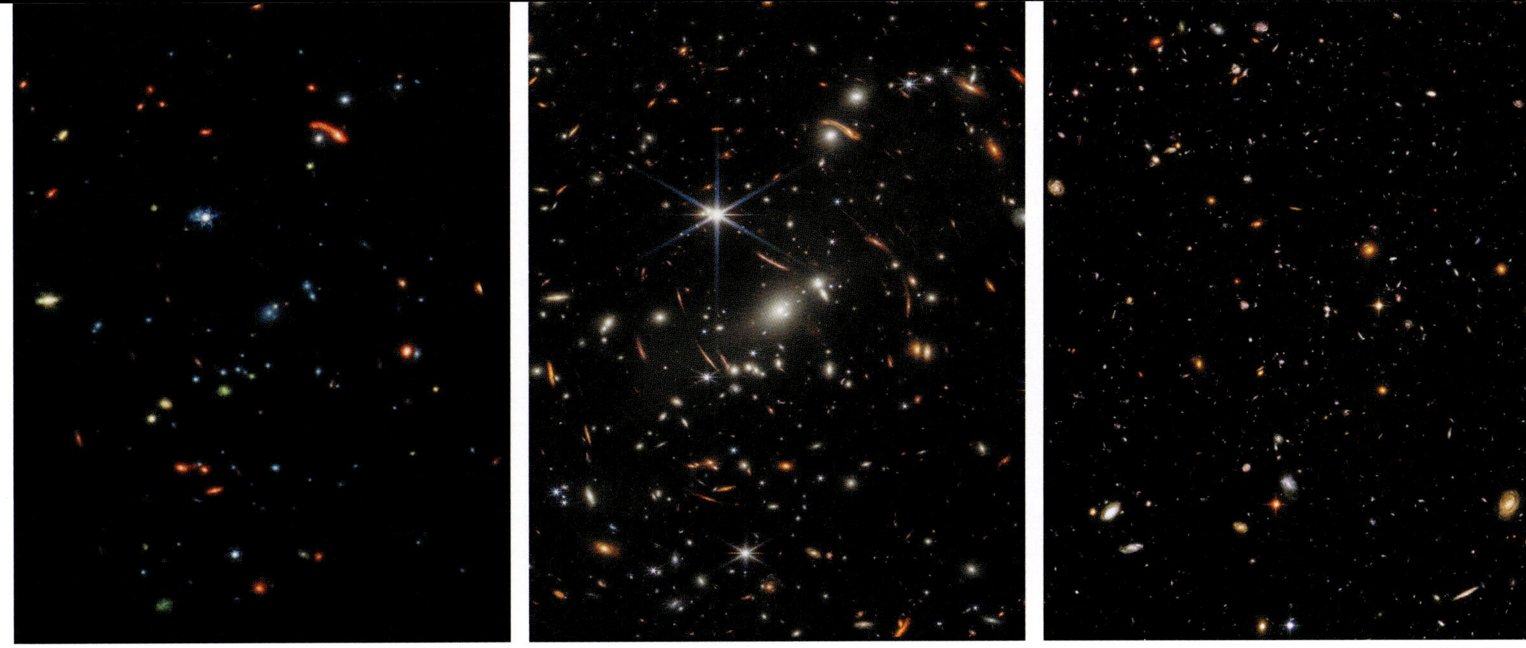

The first and second images show SMACS 0723 from the different perspectives of MIRI (left) and NIRCam (right). The third image is also of the galaxy cluster, but taken with Hubble.

made up of several images snapped at different wavelengths. In total, it's the result of 12.5 hours of exposure time. At the time of writing, Webb's First Deep Field is also the highest-resolution image of the early universe. It contains thousands of galaxies, displaying the sheer diversity of galactic shapes, sizes and ages. And to think – this is only a tiny portion of the sky around us.

In the foreground is the galaxy cluster SMACS 0723 as it appeared more than 4 billion years ago. It is visible only from Earth's southern hemisphere, but Webb's mobility allows it to capture these images from the freedom of space. SMACS 0723's constituent galaxies include bright elliptical bursts of white, which stand out in the image. While this is a gloriously crisp image of SMACS 0723, it is overshadowed by the excitement in the background.

GRAVITATIONAL LENSING

SMACS 0723's galaxies are bound together by gravity, concentrating them in a cluster. Their combined mass and the invisible dark matter contained within them manipulate the light from the galaxies behind them. This natural phenomenon is called gravitational lensing. The giant object – in this case SMACS 0723 – magnifies, distorts or sometimes even mirrors the light of objects behind it.

There are several mirrored galaxies in this deep-field image – you can see them in the orange stripes that flank the brightest cluster galaxy. Each galaxy, with its bright, star-filled centre, appears twice in each stripe. You can see their edges sparking with stars, like sunlight catching the rim of a wave. Not all of the orange galaxies are mirrored, though. They are simply stretched.

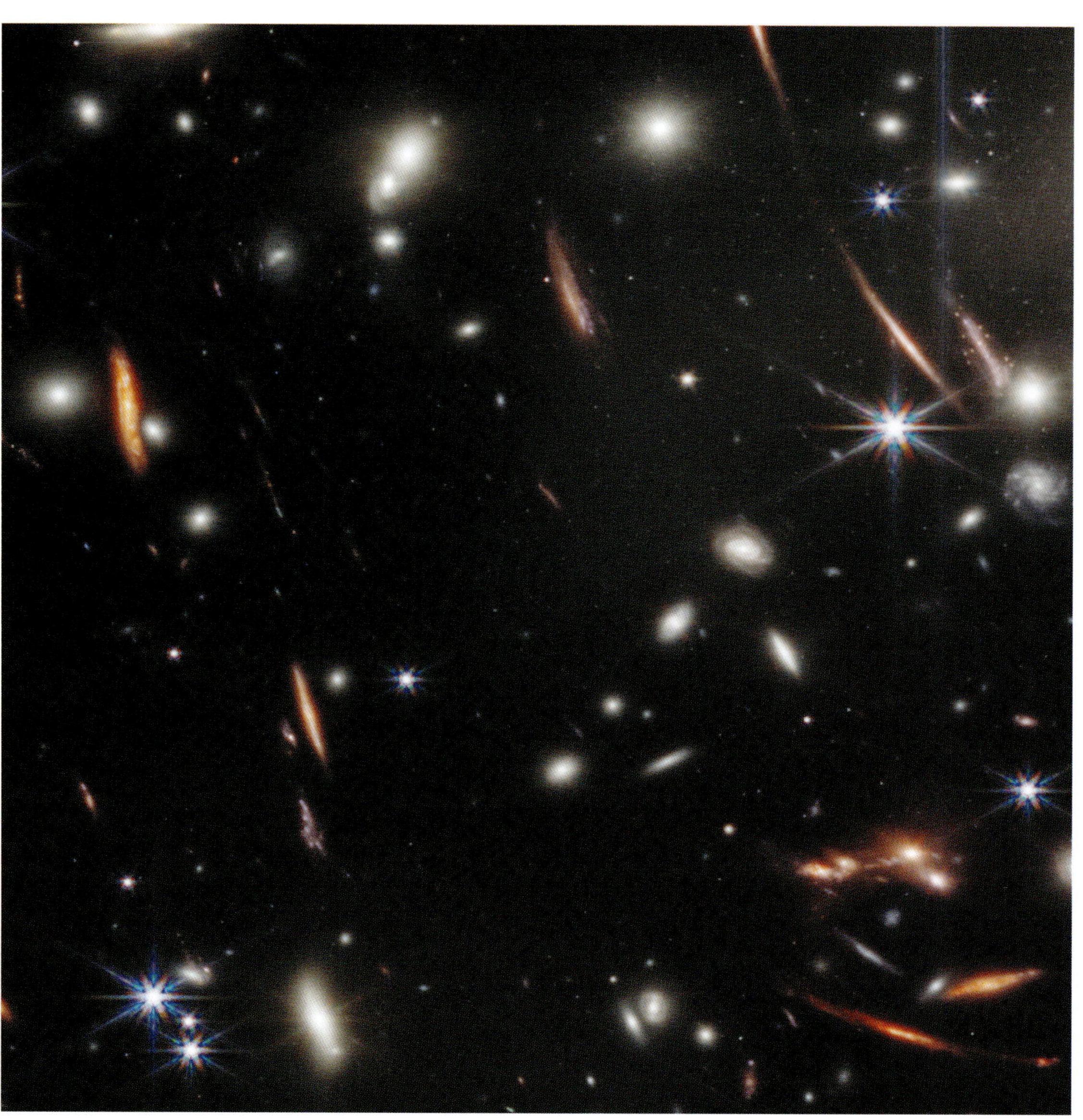

A magnified image of SMACS 0723, with fiery smudges of galaxies.

What this Webb image shows is the sheer diverse richness of galaxies in the universe. You can see orange elliptical galaxies and shining silver swirls. Diffuse galaxies look like piles of glowing jewels. Many of the galaxies – such as the long, orange-red stripes flanking the bottom of the biggest diffraction spikes – are so clear and crisp that you can see incandescent pockets of star formation. And there are other galaxies so bright that they themselves appear as luminous eight-spoked stars.

ANCIENT GALAXIES

The real wonder, however, is tiny red blobs in the background. They look like nothing in the eye-grabbing circus playing out in this portion of the sky. A minute red-yellow splodge to the bottom left of the biggest diffraction spike is actually one of the oldest galaxies that we have imaged to date. Light from that galaxy travelled 13.1 billion years before it bounced off Webb's mirrors and was sent to the telescope's instruments.

Webb's NIRSpec instrument, with its microshutter array (see page 58), can shut out the light from distractingly bright closer objects and home in on specific objects of interest. It's also able to observe up to 150 objects simultaneously, while ignoring the dramatic galactic spectacles playing out in the rest of its field of view. When Webb was observing SMACS 0723, the NIRSpec microshutter array was simultaneously viewing forty-eight individual background galaxies and observing their spectra.

Spectral analysis is an incredibly powerful tool that can give a detailed breakdown of the different wavelengths received from an object. This data contains information about the radiation that might have been generated by the object as well as about chemical reactions that may be taking place. These come out as dark bands where specific wavelengths are absorbed by the reaction. If there are any substances between the observer and the object – for example, clouds of dust or gases – these too will show up in the spectrum because these bodies can absorb some of the radiation emitted by the object. Analysis of known absorption bands can also indicate whether objects are moving away from the observer (redshift) or towards them (blueshift) (see page 46). Using this technique with Webb data allows us to determine how old these objects are.

In the graph below, you can see the spectral lines follow the same pattern: a hydrogen spike followed by two oxygen spikes. Depending on where these spikes are located (i.e. to what extent red or blue shifted) indicates how far away the galaxies are. If those emission lines are found in signals with longer wavelengths, the signal is older and the galaxy is further away.

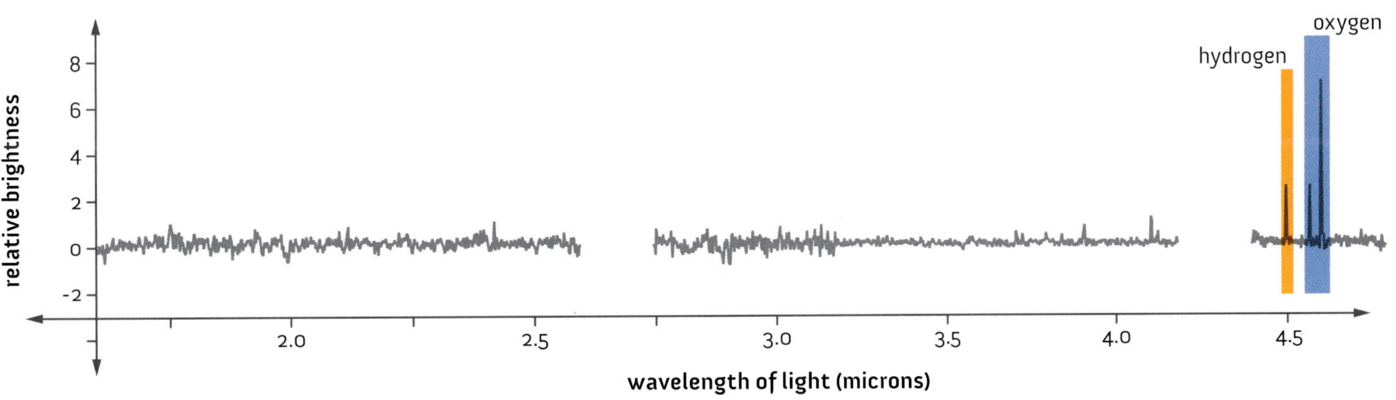

CARTWHEEL GALAXY

ALSO KNOWN AS ESO 350-40 OR PGC 2248

Constellation: Sculptor
Distance from Earth: about 500 million light-years

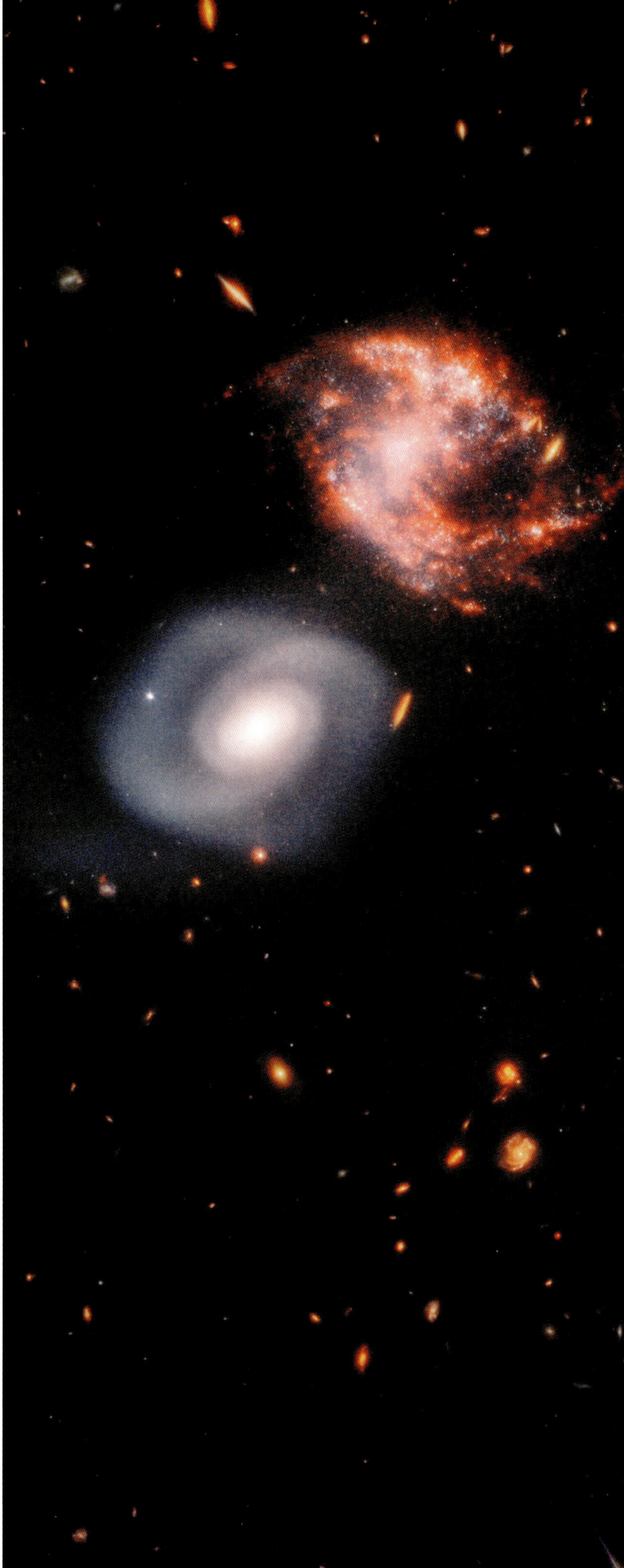

This image of the Cartwheel Galaxy is a composite of MIRI and NIRCam data. Located about 500 million light-years away, the Cartwheel Galaxy is the result of a cataclysmic collision between two galaxies.

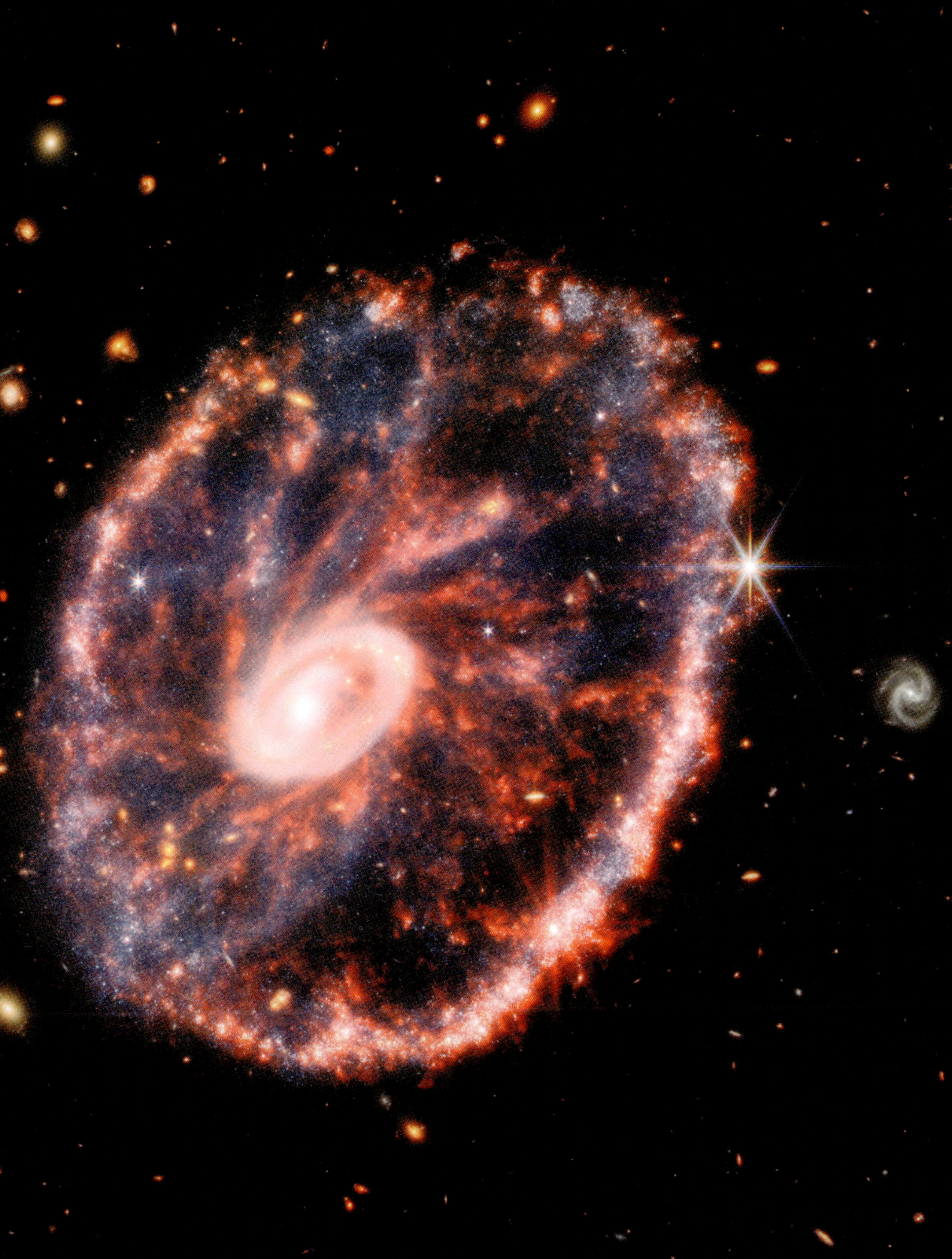

THE CARTWHEEL GALAXY HAS HAD a tumultuous history, and it shows. It's known as a lenticular ring galaxy, which is rather rare. Once upon a time, it was a rather pedestrian spiral galaxy, which are common in the universe. But about 400 million years ago it had a head-on collision with another, smaller galaxy. As this smaller galaxy travelled through the spiral galaxy, it sent shockwaves reverberating through the Cartwheel Galaxy, creating its unique structure. While it's classified as a 'small' ring galaxy, the Cartwheel Galaxy is not really small at all – it's about 1.5 times the size of our Milky Way.

Webb's infrared detectors have been able to peer into the dusty chaos and give us new clues as to how it responded to this cataclysmic event and perhaps shed some light on what will happen in the future. The main image is a composite of NIRCam and MIRI data. NIRCam's wavelengths have been coloured blue, orange and yellow, while MIRI's are red. Many telescopes, including Hubble, have imaged the Cartwheel Galaxy over the years, but by looking at it in the infrared Webb is able to uncover many details previously shrouded in dust.

TWO RINGS

Using MIRI, it is possible to see the dusty regions within the Cartwheel Nebula and the young stars within them.

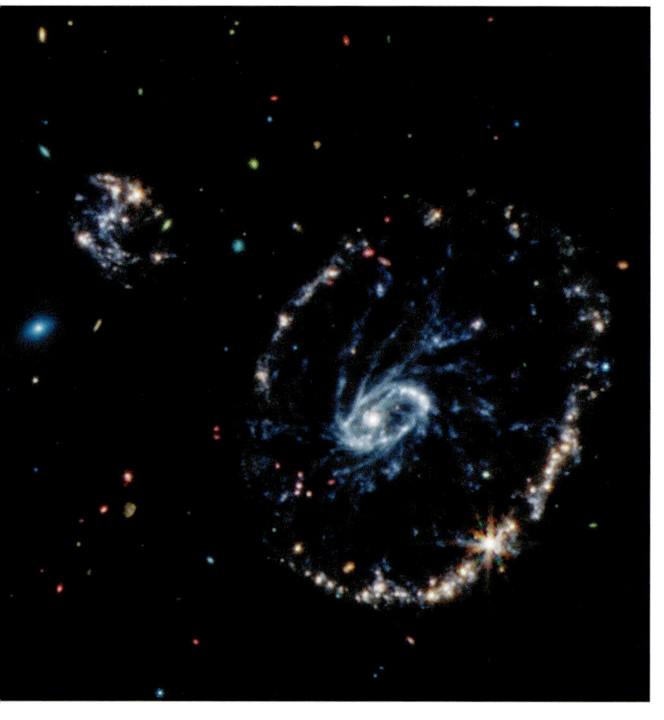

The Cartwheel Galaxy has two rings – the glowing white inner ring and the star-speckled red-pink one that circles it. The outer ring is the leading edge of the shockwave from the ancient collision and has been expanding for millions of years. At its frontier, dust and gas smash into the surrounding interstellar gas, creating extremely high pressures and becoming a rich crucible of star formation and supernovas.

In the image on the previous page, it looks as though the rim of the Cartwheel Galaxy has been liberally salted with stars. We can see so many stars thanks to the precision of NIRCam. They appear as blue and white dots. Webb's sensitivity also means we can tell the difference between older star regions, in which the stars are more spread out and defined, and new ones, where the stars clump together in hot clusters.

MIRI, with its ability to look at mid-infrared wavelengths, is revealing the secrets of the dust that characterizes this galaxy. The red dust that forms the spokes of the 'cartwheel' comprises complex molecules, such as hydrocarbons, as well as silicate dust, which is found all over Earth. These spokes form the backbone of the galaxy as seen in streaks of red connecting the Cartwheel's glowing centre to its star-filled rim.

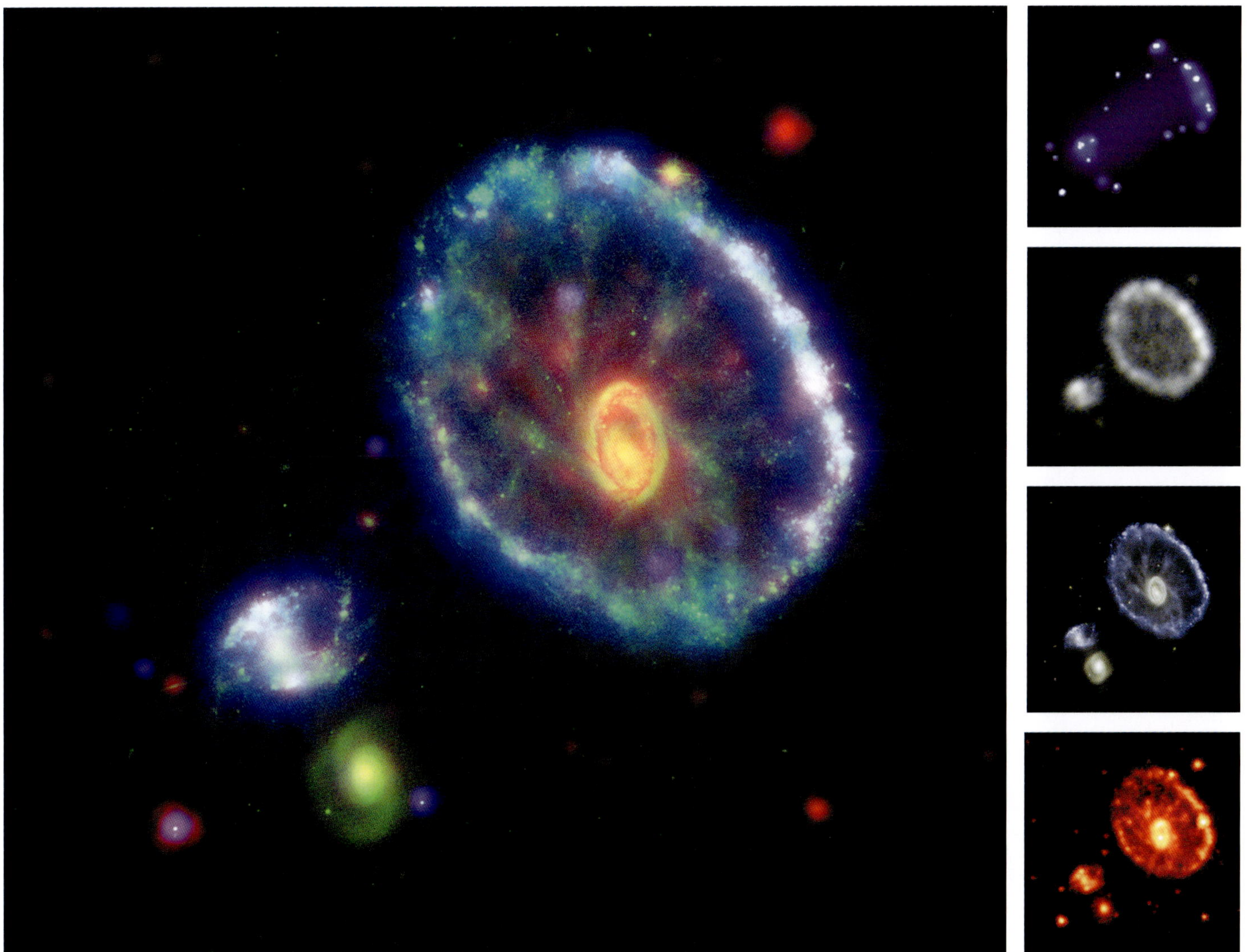

The main image is a composite of Chandra, GALEX, Hubble and Spitzer data. Right, in descending order: the Cartwheel Galaxy imaged by Chandra (X-ray), GALEX (ultraviolet), Hubble (visible) and Spitzer (infrared).

ANOTHER COLLISION

Over time, scientists suspect that the Cartwheel will revert to its original spiral shape and that the dust and stars riding the shockwave frontier will eventually start to fall back towards the galaxy's centre. Already, we can see that its spokes have the tell-tale curve of spiral galaxies' arms.

However, it's possible that the Cartwheel Galaxy could once again collide with other nearby galaxies. The Cartwheel is actually the dominant member of a group of galaxies, known as the Cartwheel Galaxy Group. The main player has three companions, and we can see them in the main Webb image. On the left, two spiral galaxies rotate – one in a skirt of white and pale blue and another in a fiery star-filled red swirl. On the right, you can see the distinct white arms of a tiny turning galaxy.

IC 1623

ALSO KNOWN AS INTERACTING GALAXIES VV 114

Constellation: Cetus
Distance from Earth: about 275 million light-years

Two galaxies are colliding in the centre of this Webb image. Known as IC 1623, the merging galaxies are sparking a frenzy of star formation.

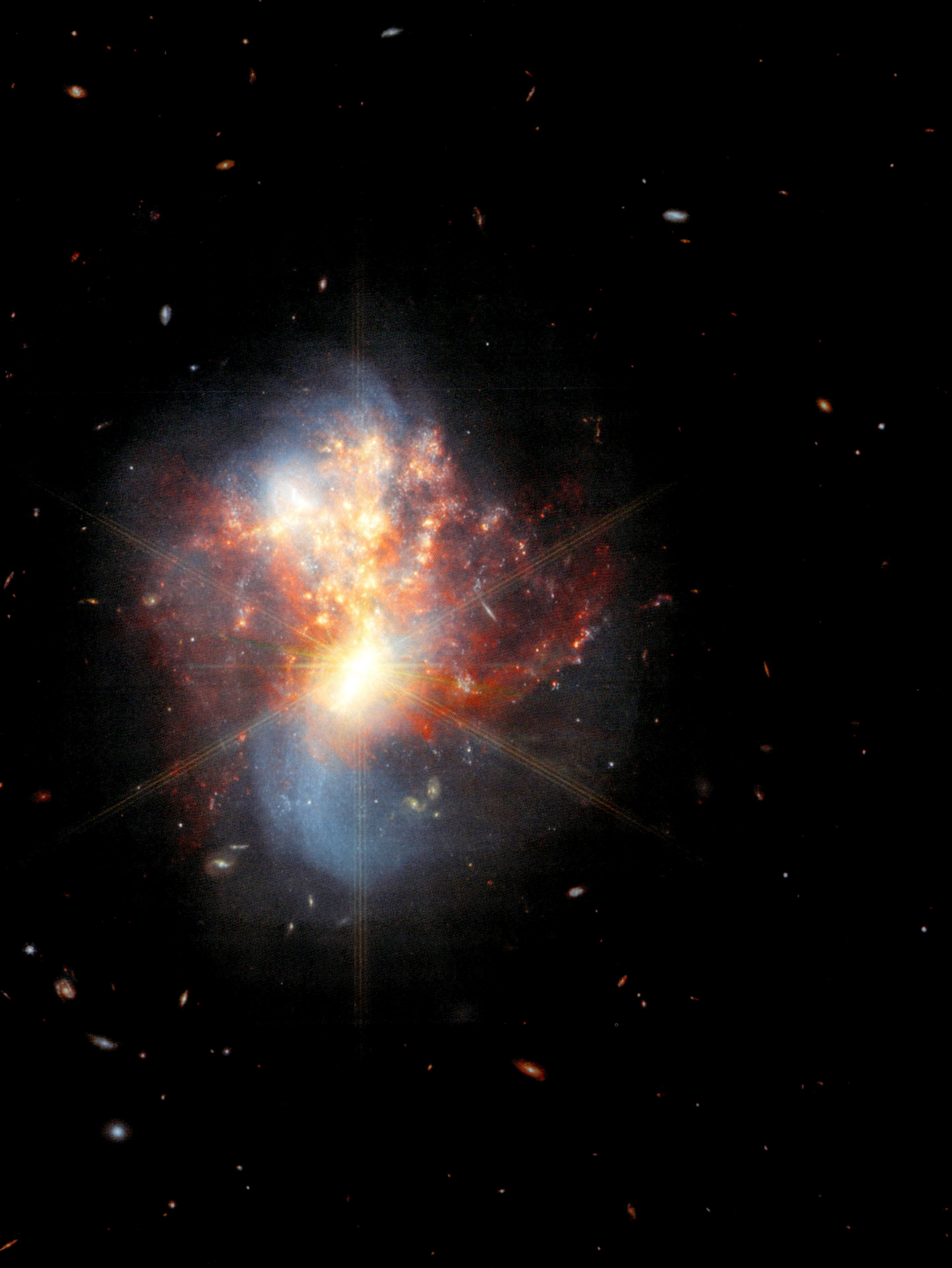

IN THIS VIBRANT IMAGE, TWO titans are colliding. IC 1623 is a pair of interacting galaxies, known as VV 114E and VV 114W, which are in the final stage of a process called a galaxy merger. Their nuclei are separated by 26,100 light-years.

A galaxy merger may sound like a violent event because the distance between the stars is so great, but very few stellar collisions happen. Where the two galaxies are coalescing, there is a flurry of star formation. Scientists call this wild rush of star development a starburst. The collision, about 275 million light-years away, is spawning stars at more than twenty times the rate in our Milky Way.

In this Webb image, you can see the ongoing stellar genesis creating such strong emissions that it forms Webb's distinctive eight-spike diffraction pattern. Scientists suspect that these merging galaxies may be in the process of generating a black hole.

The blue galaxy (VV 114W) appears to swirl in the background like an 'S', with its long arms freckled with the light speckles of stars. In the foreground, the dust centre of the other galaxy (VV 114E) glows like an inferno, with golden embers of new stars. Using the Webb data, scientists have identified forty star-forming knots in that galaxy, more than a quarter of which have not been viewed in visible wavelengths. Complex carbon molecules – PAHs – infuse the whole system.

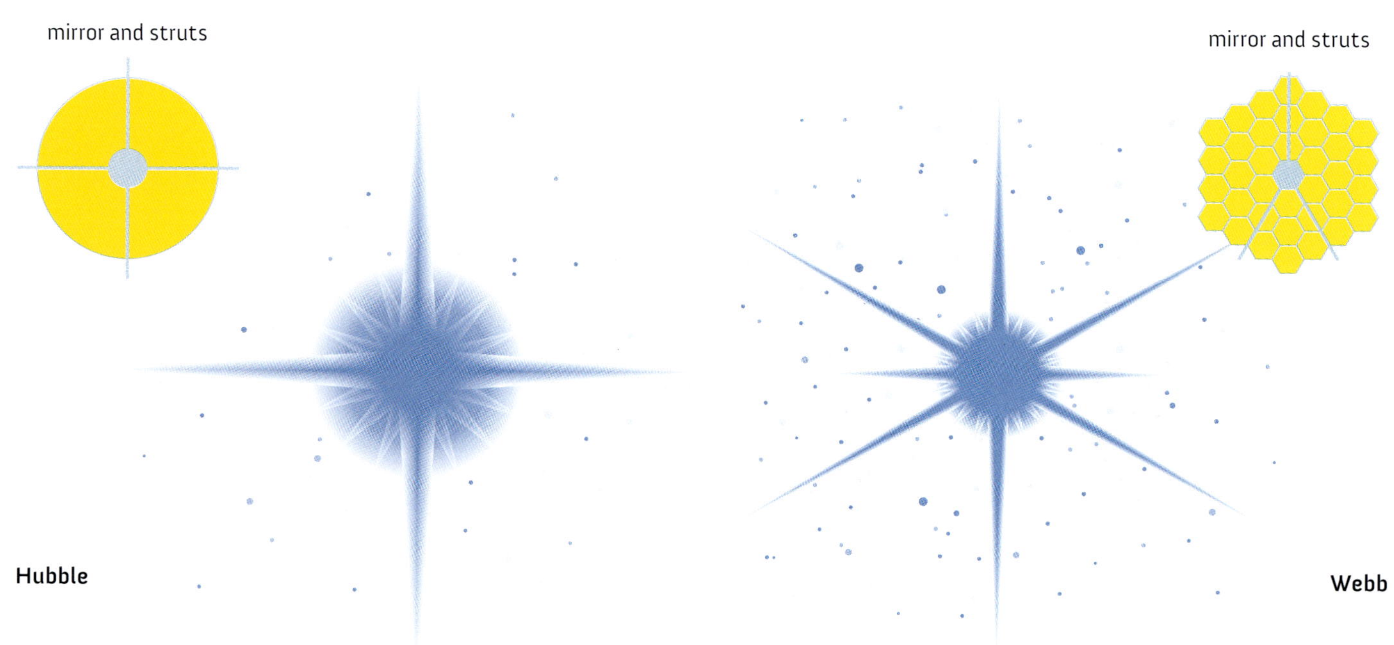

Telescopes have characteristic diffraction patterns. This is why very bright objects appear to have eight diffraction spikes when viewed by Webb.

An image of IC 1623 using Hubble.

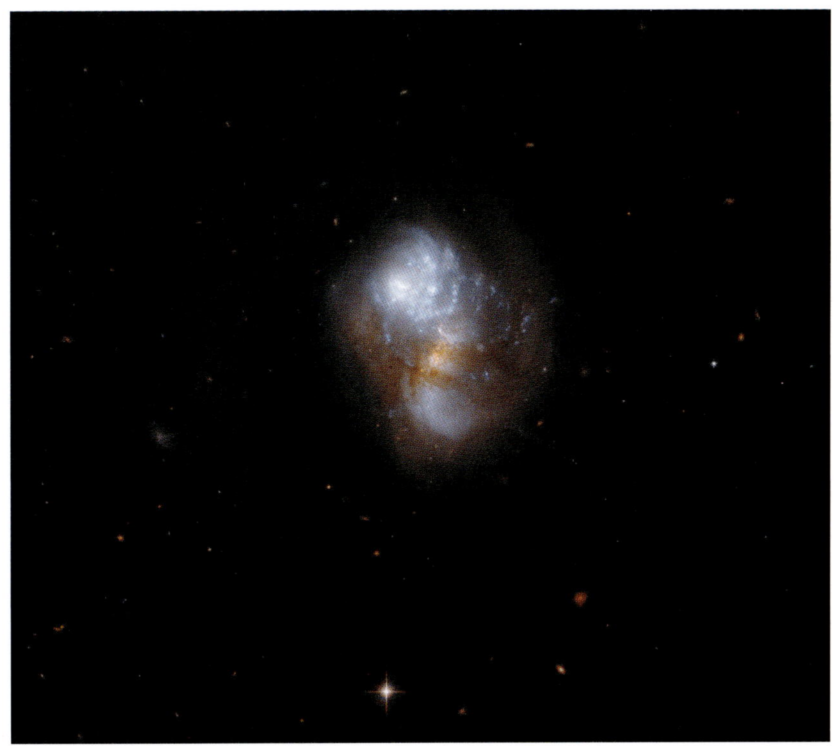

DIFFERENT EYES

This is not the first image of IC 1623, but it is the most sensitive in the infrared. Titanic collisions such as that of this galaxy merger generate a great deal of dust, making it difficult to see what is actually happening when you're looking with optical instruments – or when you're looking at it from Earth, for that matter.

This merger highlights how altered objects appear in different wavelengths. In the Hubble image above, VV 114W takes centre stage and dominates the picture in whimsical wisps of white and pale blue. Meanwhile, VV 114E is a smudge-like afterthought. The Webb image on the main page, however, tells a different story. VV 114E glows like a wildfire across the collision, its hot gas lighting up under Webb's gaze. To generate this image, a hat-trick of Webb instruments was used – MIRI, NIRSpec and NIRCam – and the resulting compound image has created a wealth of data that scientists will be poring over for years to come.

NGC 5584

Constellation: Virgo
Distance from Earth: about 75 million light-years

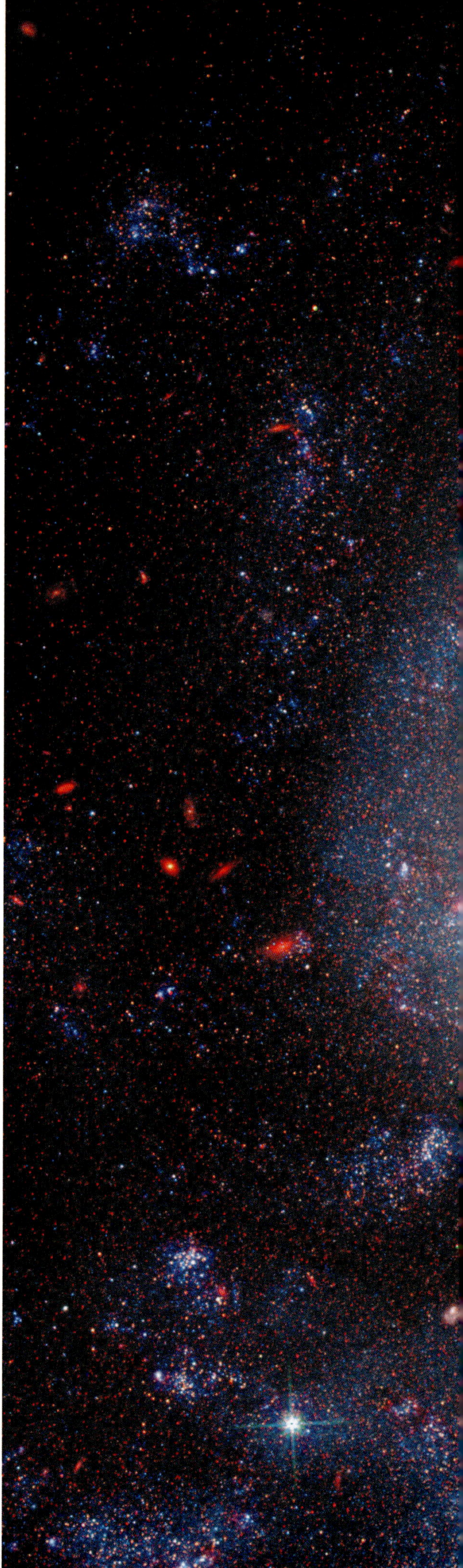

Scientists combined data from Hubble and Webb's NIRCam to create this image of spiral galaxy NGC 5584. In the liberal salting of stars, there are a special type of pulsing stars, known as Cepheids, which can be used to measure distance in space.

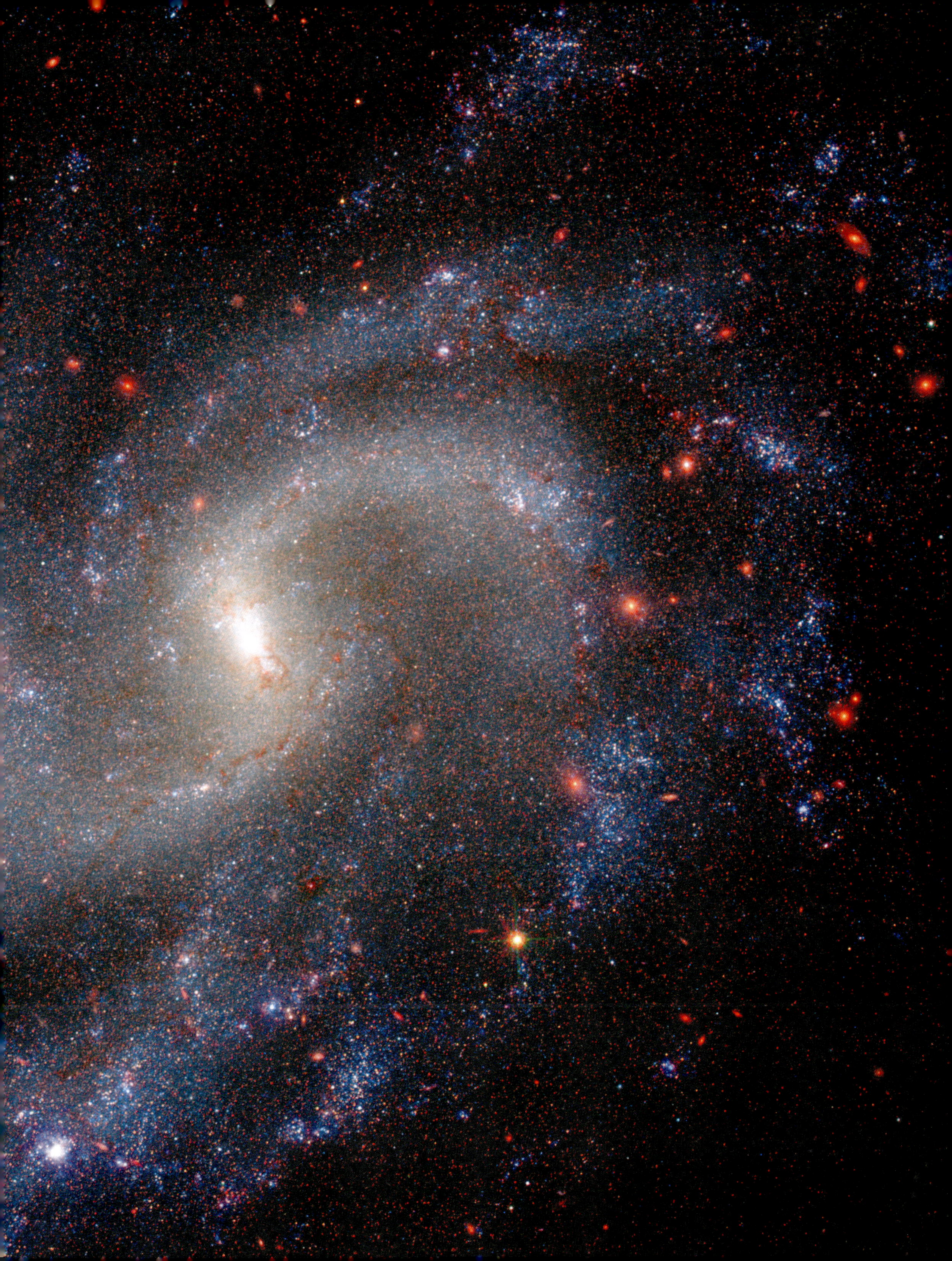

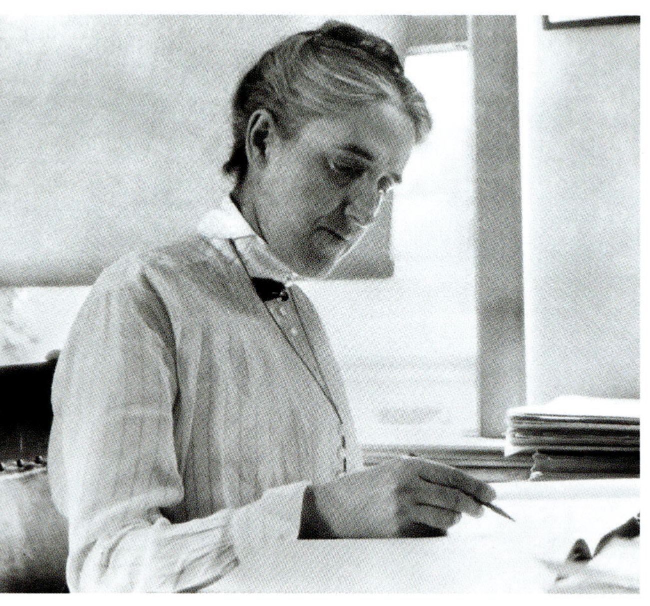

American scientist Henrietta Swan Leavitt.

HERE WE SEE NGC 5584, a galaxy that resembles haphazardly pulled pieces of cotton wool. It's neither a strict spiral nor completely diffuse. This type of galaxy is sometimes called a flocculent spiral galaxy because its arms are discontinuous, flailing out from a compact bright core. Uncountable stars sprinkle the galaxy's arms. This image is a composite of Webb's NIRCam and Hubble's Wide Field Camera 3.

CEPHEID VARIABLES

Some of these stars seen in the image on the previous page are particularly special. They are known as Cepheid Variables. The term 'Cepheid' comes from the name of the constellation Cepheus, where this type of star was first identified by John Goodricke in 1784. The word 'variable' refers to the fact that the brightness of these stars changes periodically as they pulse with a regular rhythm. Although they had been known for centuries, it took the genius of American scientist Henrietta Swan Leavitt in 1908 to discover a relationship between the timing of their pulses and their intrinsic brightness. It turns out that the more luminous the variable stars are, the longer the frequency of their pulses. This period-luminosity relationship is now known as Leavitt's Law.

This discovery marked a major breakthrough in our understanding of the size of the universe. Leavitt's Law gave us the tools to measure distances between our location and galaxies that are millions of light-years away. Because we know that light gets dimmer with distance (something called the inverse-square law), we can work out how far away the Cepheid Variable is by measuring the brightness we observe here on Earth and comparing this to the known actual brightness of the star. As we can spot Cepheids in galaxies far, far away, we can measure the distance between our location and that of the galaxy.

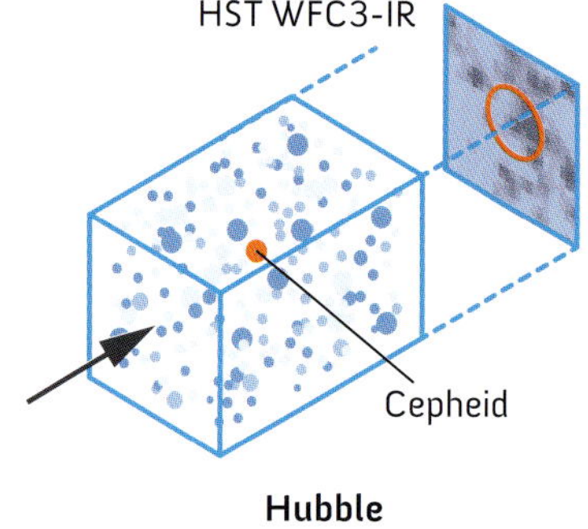

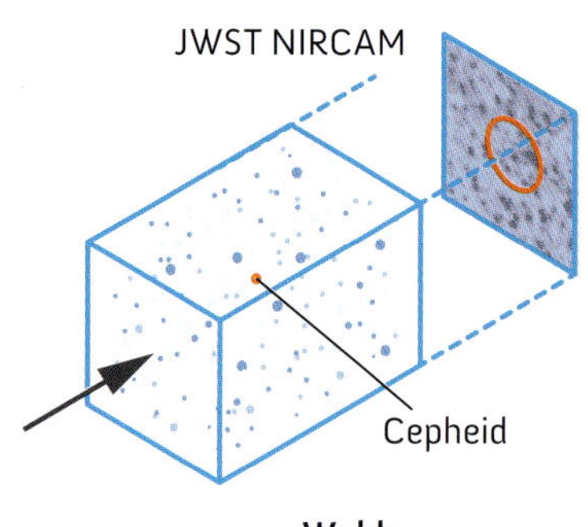

Webb's precision allows it to better distinguish Cepheids from surrounding stars.

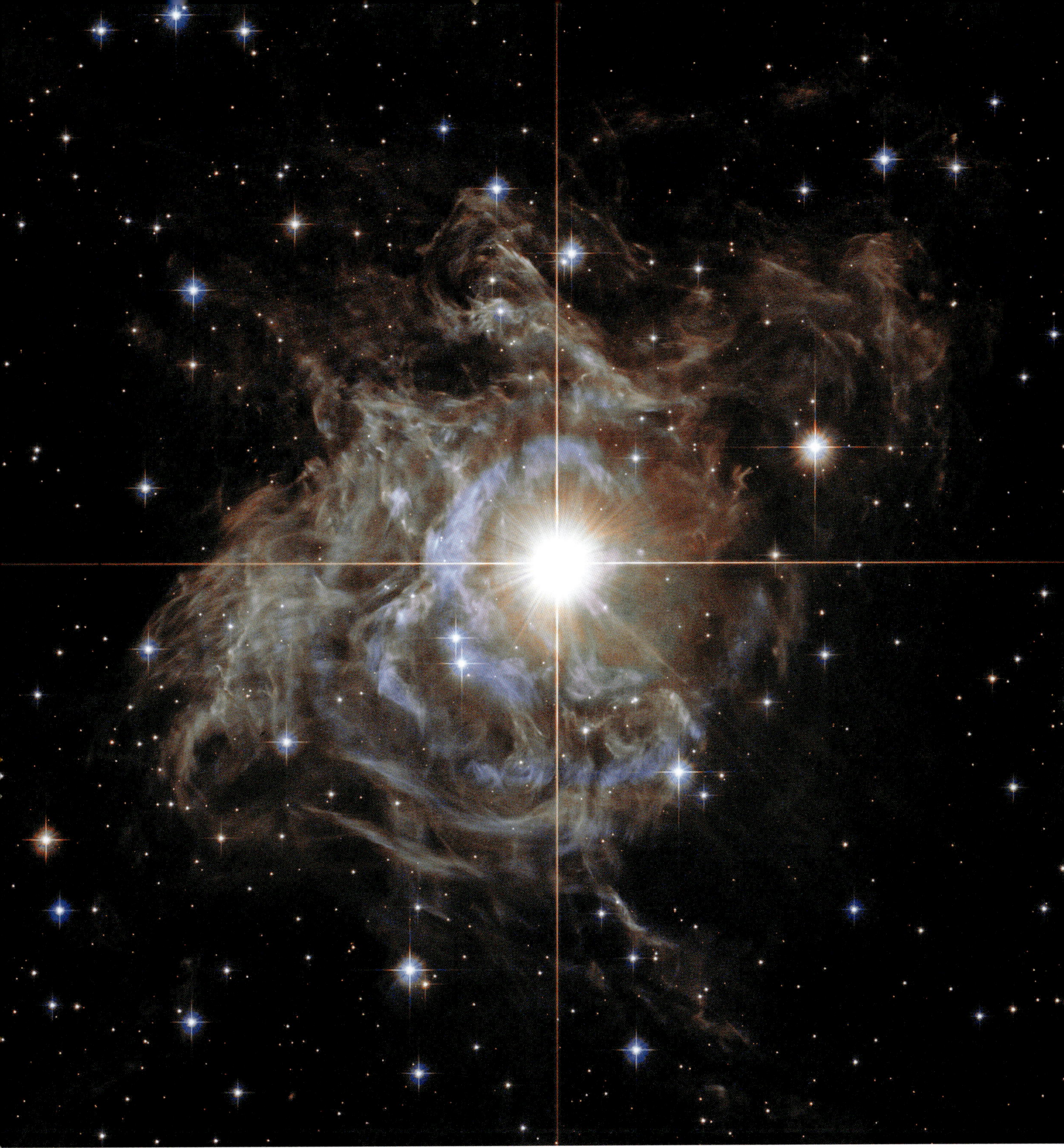

RS Puppis, captured by Hubble, is a Cepheid variable star. It is also a superstar that is 200 times larger than our sun.

As a result, Cepheid Variables are sometimes referred to as a 'standard candle' for cosmological measurement. This is particularly important when you're trying to tease apart what happened in the early universe. But first you need to be able to accurately identify and isolate the Cepheids – which can be tricky to identify and characterize because they are often nestled within fields of stars, overwhelmed by their light. That is where Webb's capabilities come to the fore. In the infrared, it is possible to get images that are clearer in terms of their brightness and locations. Used in conjunction with a special type of supernova called Type 1a, Cepheid Variables can be used as beacons for measuring distances.

THE HUBBLE CONSTANT

In the early twentieth century, Edwin Hubble used spectroscopic techniques to look at the spectra of a range of stars and objects and published evidence that would change the way we viewed the universe. He had found (though others had proposed the idea before him) that the further away a celestial object such as a galaxy was, the faster it was moving away from us. This observation is now called the Hubble–Lemaître Law, after Hubble and French scientist Georges Lemaître. The Hubble constant is the rate at which the universe is expanding. It tells us the speed of any object at any distance, as it accelerates away from the site of the Big Bang. But there's a problem: the different ways of measuring the Hubble constant give differing results.

COSMIC MICROWAVE BACKGROUND (CMB)

The remnants of the Big Bang's first light continue to pervade the universe. This 'fossil' light is like an echo of the birth of the universe. In the 1960s, two American radio astronomers, Robert Wilson and Arno Penzias, accidentally stumbled upon the fact that a constant hum of radio emissions is present everywhere all the time. This ancient signal – which we now call the cosmic microwave background – fills all space in the observable universe and is considered a shockwave of the Big Bang.

Using this Big Bang echo, mathematicians can predict how fast the universe should be expanding. However, there is a big discrepancy between the predicted velocities of this calculation and the ones measured using spectral analysis of the movement of objects. The problem is that no one has been able to work out why the two

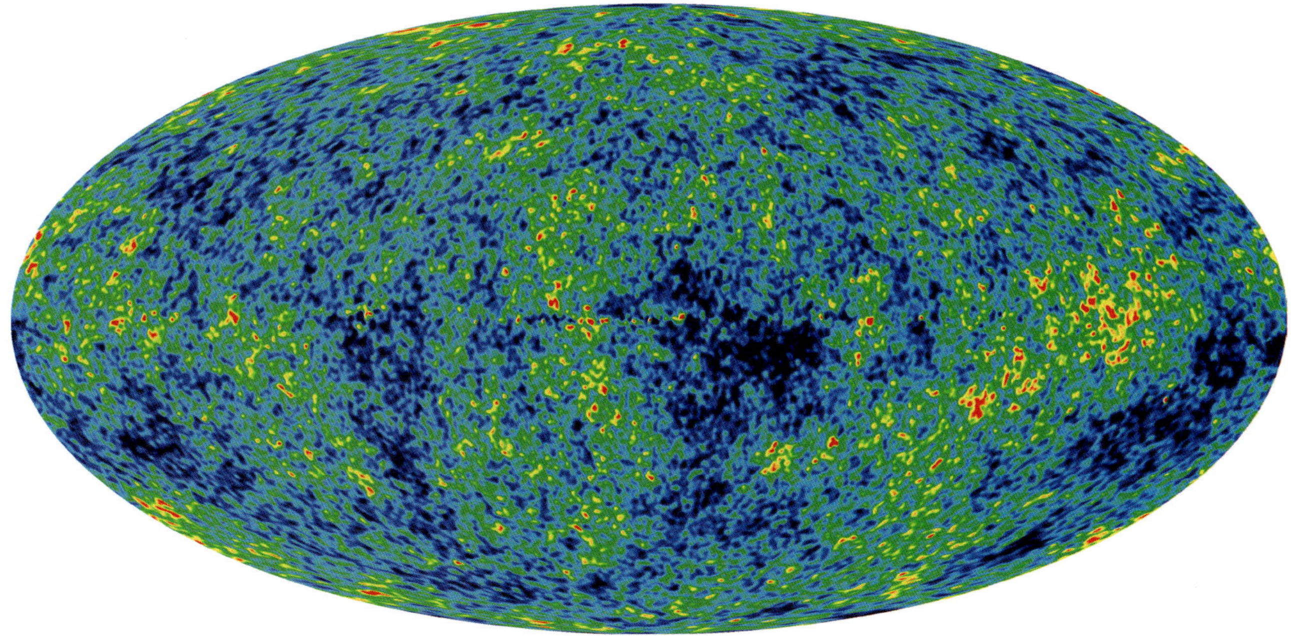

The Cosmic Microwave Background is fossil radiation that permeates the universe.

results don't add up. The universe is expanding at a much faster rate than our models suggest it should. It is clear that we're missing something fundamental about our understanding of the cosmos.

MOST ACCURATE CONSTANT YET

As instruments and techniques have become more sophisticated, we have managed to close the gap between the measurements and calculations. Webb will play an important role in that with even more accurate data. Hubble identified many of the Cepheid Variables in NGC 5584, and scientists have followed up with Webb to make their observations even more precise. However, the significant discrepancy that remains between the model and real measurements demonstrates that there is a lot we have yet to discover about the universe.

PHANTOM GALAXY

ALSO KNOWN AS M74 OR NGC 628

Constellation: Pisces
Distance from Earth: about 32 million light-years

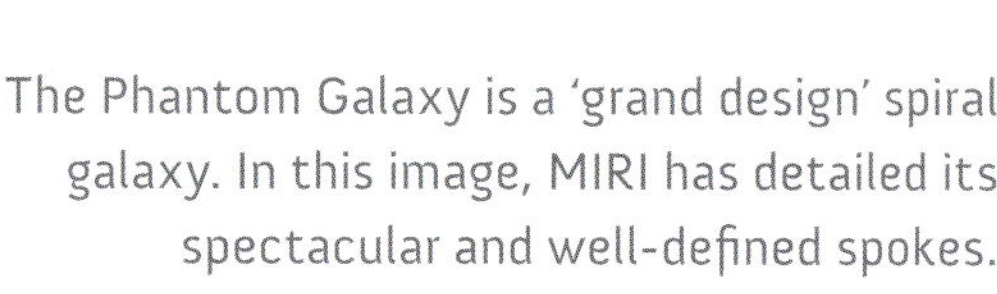

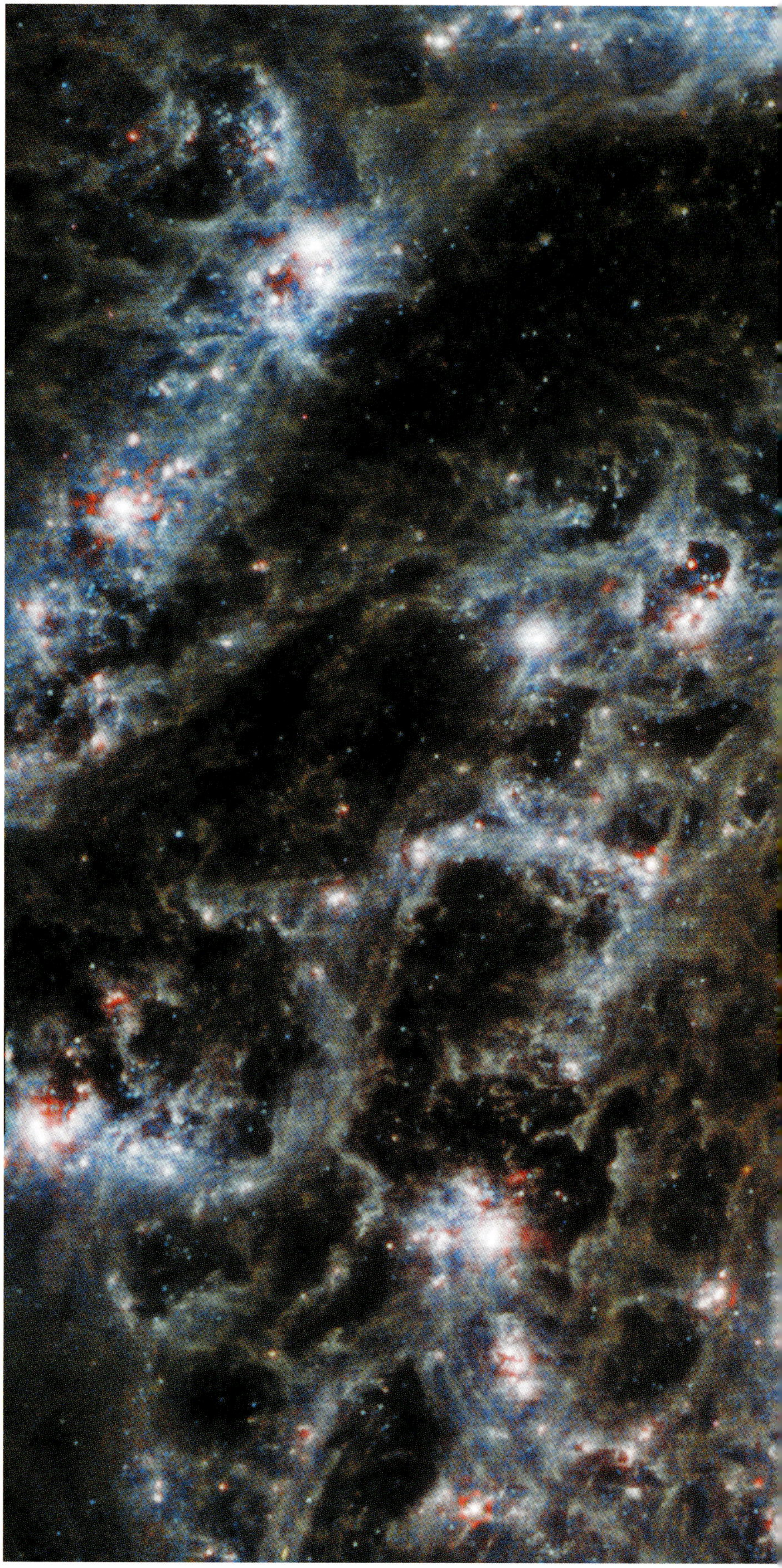

The Phantom Galaxy is a 'grand design' spiral galaxy. In this image, MIRI has detailed its spectacular and well-defined spokes.

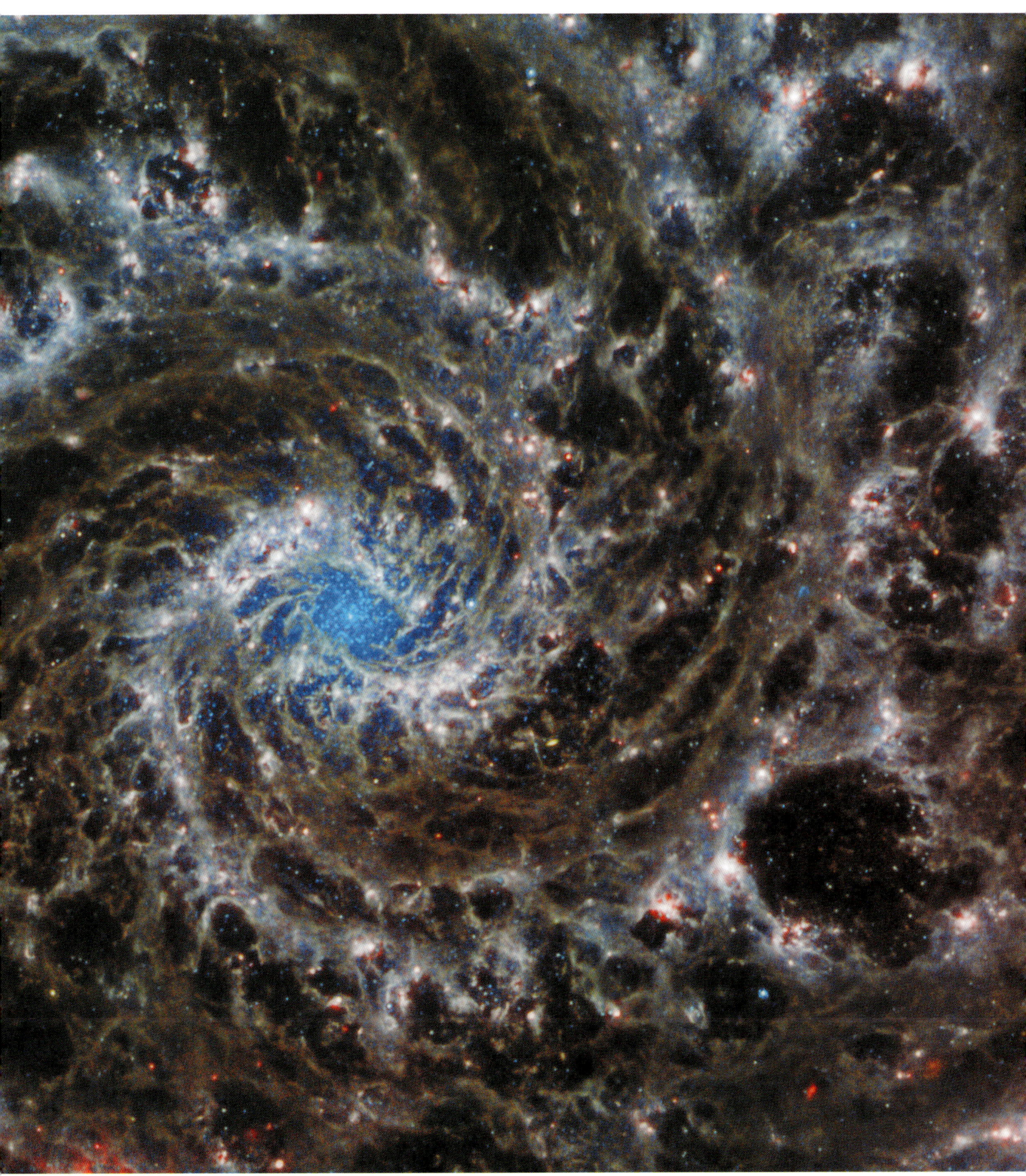

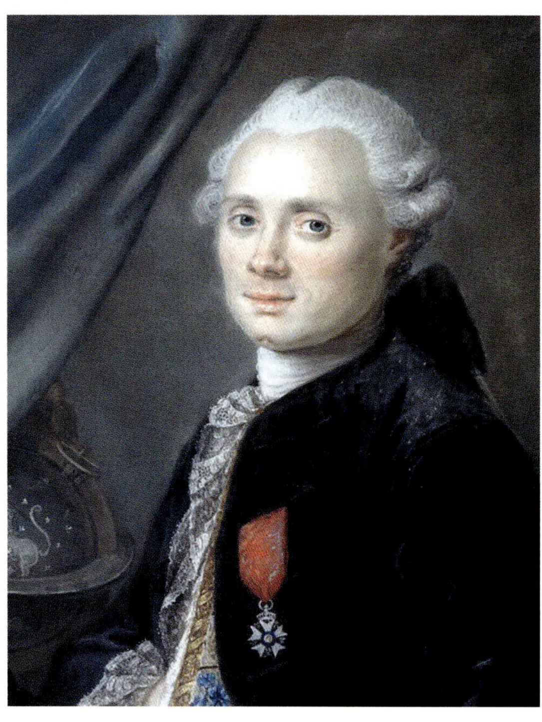

ABOVE: French astronomer Charles Messier published what became an eponymous astronomical catalogue.

RIGHT: Webb and Hubble's view of spiral galaxy NGC 628.

THIS GALAXY IS CALLED THE Phantom because its surface brightness is low. It is one of the astronomical objects listed in the Messier catalogue, an inventory of celestial objects compiled by the French astronomer Charles Messier in the 1700s.

Messier constructed the catalogue to help him and other astronomers distinguish between permanent and non-permanent celestial objects – and particularly things that could be mistaken for comets. Comets were of great interest to astronomers at this time because naming them resulted in a degree of immortalization, but they frequently caused confusion as they were easily misidentified as other forms of celestial entities.

The Messier catalogue lists 110 astronomical objects, including galaxies, nebulae and star clusters. These objects are identified by the letter 'M' and cover the entire celestial sphere. Examples of well-known Messier objects are the Orion Nebula (M42), the Andromeda Galaxy (M31) and the Whirlpool Galaxy (M51).

Messier objects are popular among amateur astronomers and astrophotographers because of their visibility and frequently spectacular appearance. The relative dimness of the Phantom Galaxy, or M74, makes it one of the most difficult to find. However, this dimness is not intrinsic to the galaxy itself but is a consequence of the angle at which we observe it. The Phantom Galaxy sits face-on to observers on Earth, which means its brightness is distributed over an extended area. This orientation makes it dim to observe with a small telescope, but with professional equipment or a space telescope we can study the structure of its spiral arms and galactic centre in fine detail. We can see that new stars are forming in the brilliant heart of this galaxy, and in the infrared we can see the dots of young stars (coloured blue) surrounding its core.

Thanks to its large mirror, Webb has a superior ability to distinguish features that would appear blurry to other telescopes. This ability is known as resolving power – it allows the telescope to image the intricacy of this quintessential spiral galaxy. In fact, the Phantom Galaxy is known as a 'grand-design spiral' because its arms are so obvious. You can distinctly see the swirls of the two spiral arms. But Webb shows the lace-like scaffolding of gas and dust that they are made of, and the features within this scaffold have scientists particularly excited. Webb looked at the Phantom Galaxy as part of an international collaboration, known as PHANGS (Physics at High Angular resolution in Nearby GalaxieS), in which astronomers are surveying galaxies' star-forming regions. In the image, nestled within the arms of the Phantom Galaxy, you can see these regions as scattered pink and blue dots.

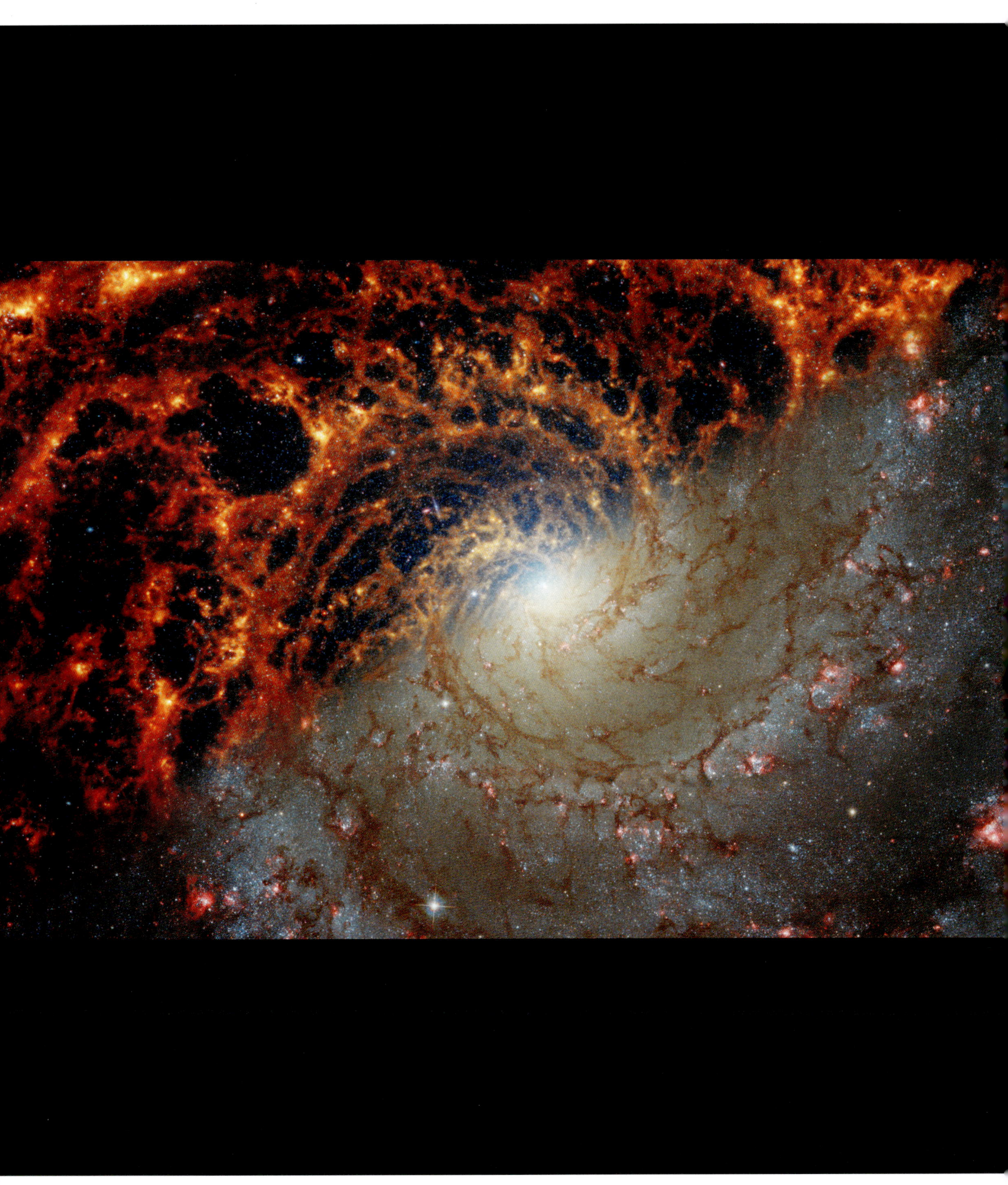

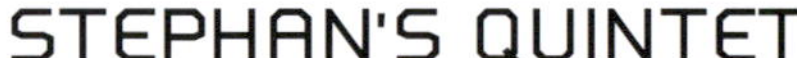

STEPHAN'S QUINTET

Constellation: Pegasus
Distance from Earth: NGC 7320 is about 40 million light-years, while the other four galaxies are about 290 million light-years

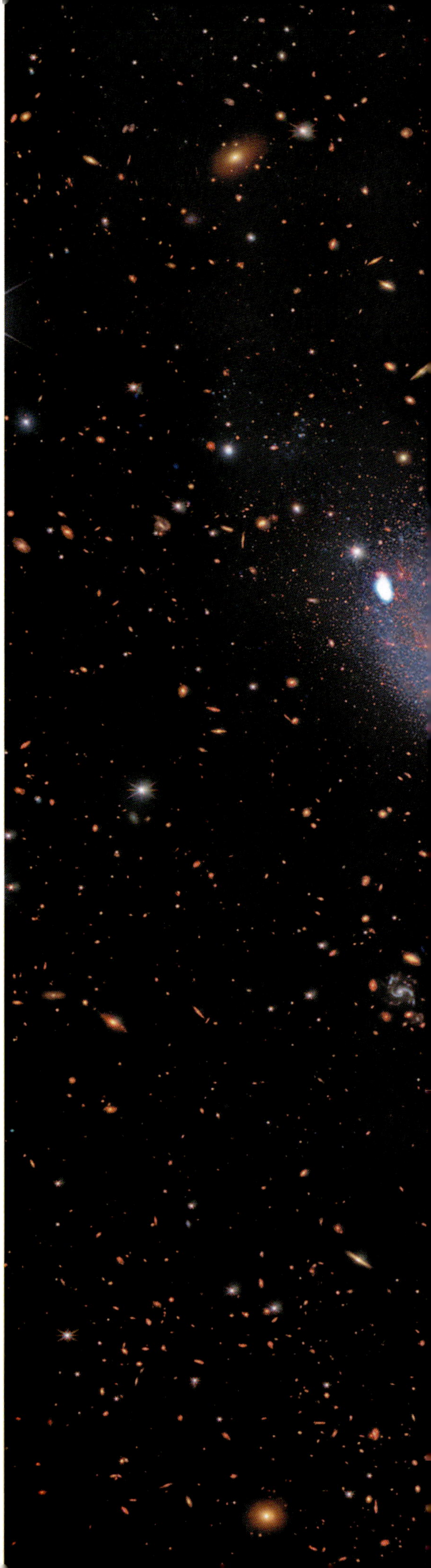

By combining Chandra's X-ray vision with Webb's infrared capabilities, this image has captured the majesty of the five galaxies in Stephan's Quintet. Only four of the galaxies are actually near each other. The fifth, which is on the left, is actually much closer.

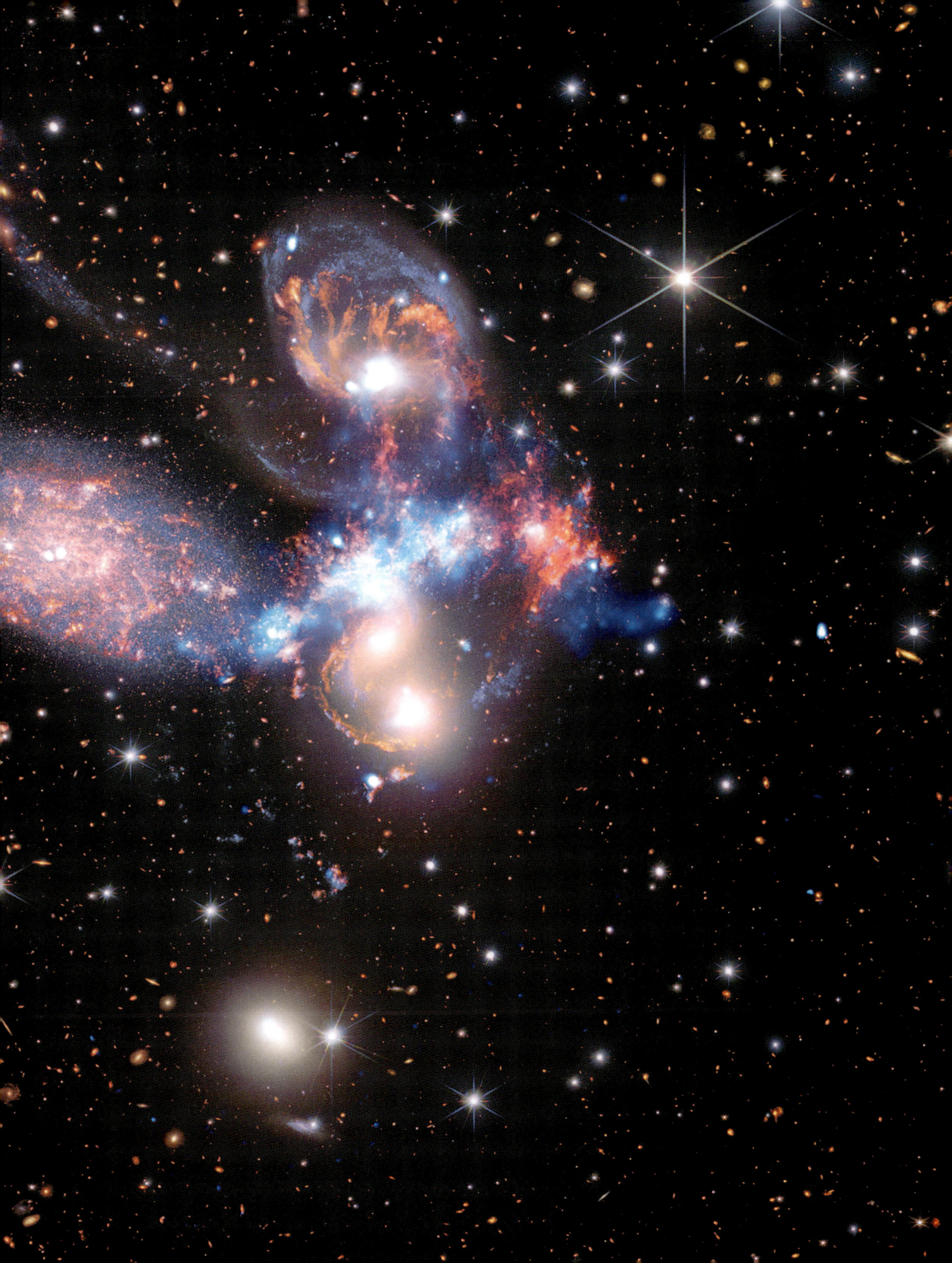

IN THIS IMAGE, WE CAN see the magic that happens when two observatories look at the same phenomena. The image on the previous page is a composite of Webb (which is in the infrared) and Chandra (which looks at the more energetic X-rays). The infrared wavelengths are red, orange, yellow, green and blue, while Chandra's X-rays are light blue.

From Earth, it looks like there are five galaxies in Stephan's Quintet, but one of the galaxies is actually much, much closer. NGC 7320 is the spiralling spray of the colour on the left. It contains extensive H II regions, where new stars are being born. The other four galaxies, known as the Hickson Compact Group 92, are what's called a compact galaxy – the first one ever discovered. They are the extremely white bright spots that bisect the image in a crooked line. The galaxy at the top, NGC 7319, harbours a supermassive black hole at its centre and is actively pulling matter towards it.

With the data from Webb, you can see their wispy tails of gas and the bright explosions of stars being born. Chandra's light-blue pops show a shockwave that happens when one of the galaxies passes by another, heating the gases to millions of degrees Celsius.

EXPLODING FIREWORKS

When looking at Stephan's Quintet just using Webb's MIRI, as in the image on the right, you get a very different perspective. It resembles a giant fireworks display in the night's sky. The red regions show large dust clouds, many of which contain stars being born. Stars without dust are blue, while the areas that look as though they've been spray-painted blue contain hydrocarbon molecules.

GALACTIC DANCE

Specialists constructed this composite of MIRI and NIRCam data from almost 1,000 image files – and it was totally worth it (see page 200). It shows us millions of young stars, and luminous swathes of dust and gas being manipulated as the four neighbouring galaxies dance around each other. You can see where the central galaxies directly interact in flames of red and gold. (MIRI wavelengths are coloured in warmer tones, while NIRCam's wavelengths are blue and white.) It is suspected that this cosmic dance will end with the galaxies colliding and ultimately merging.

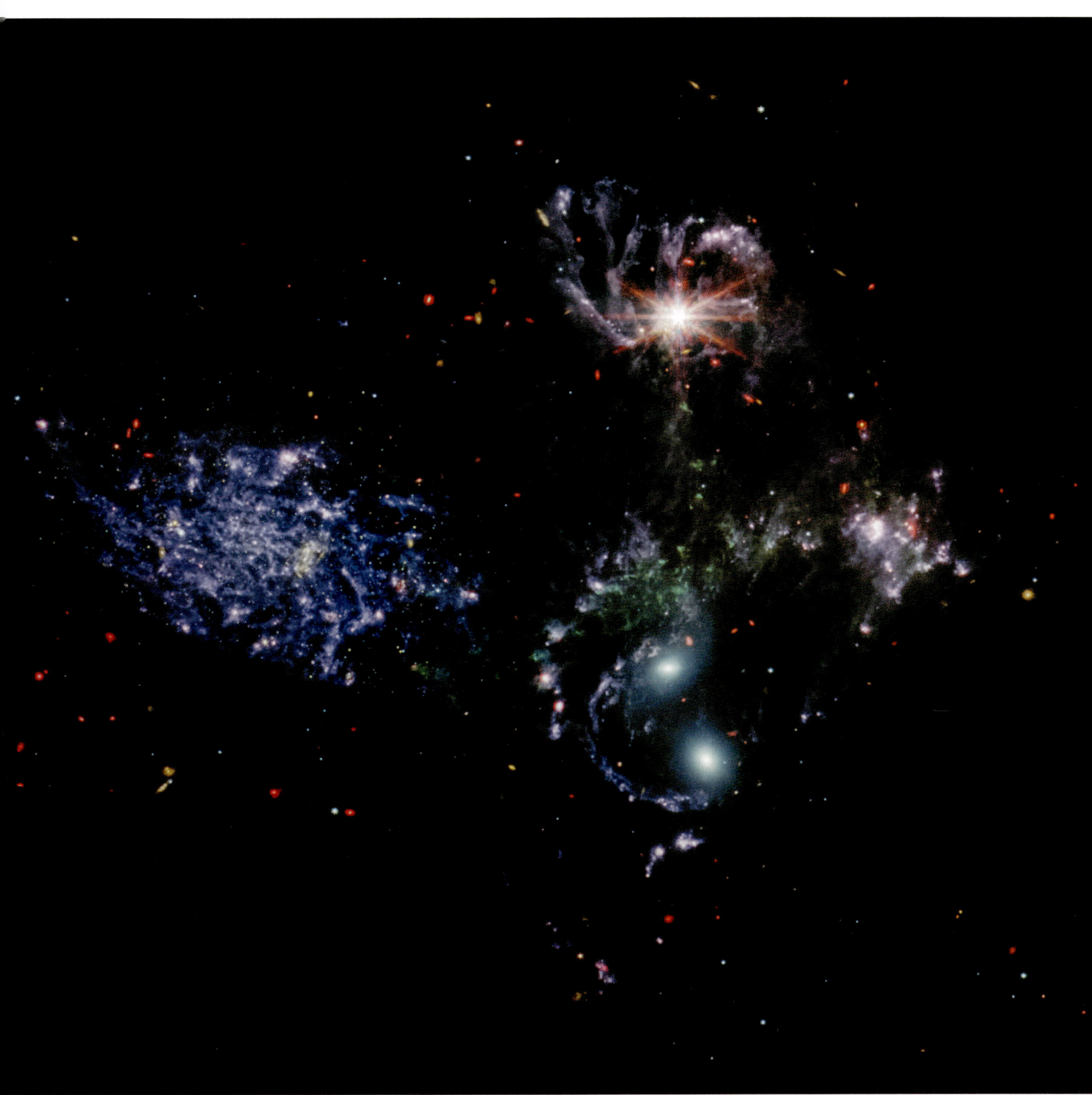

MIRI's perspective of Stephan's Quintet.

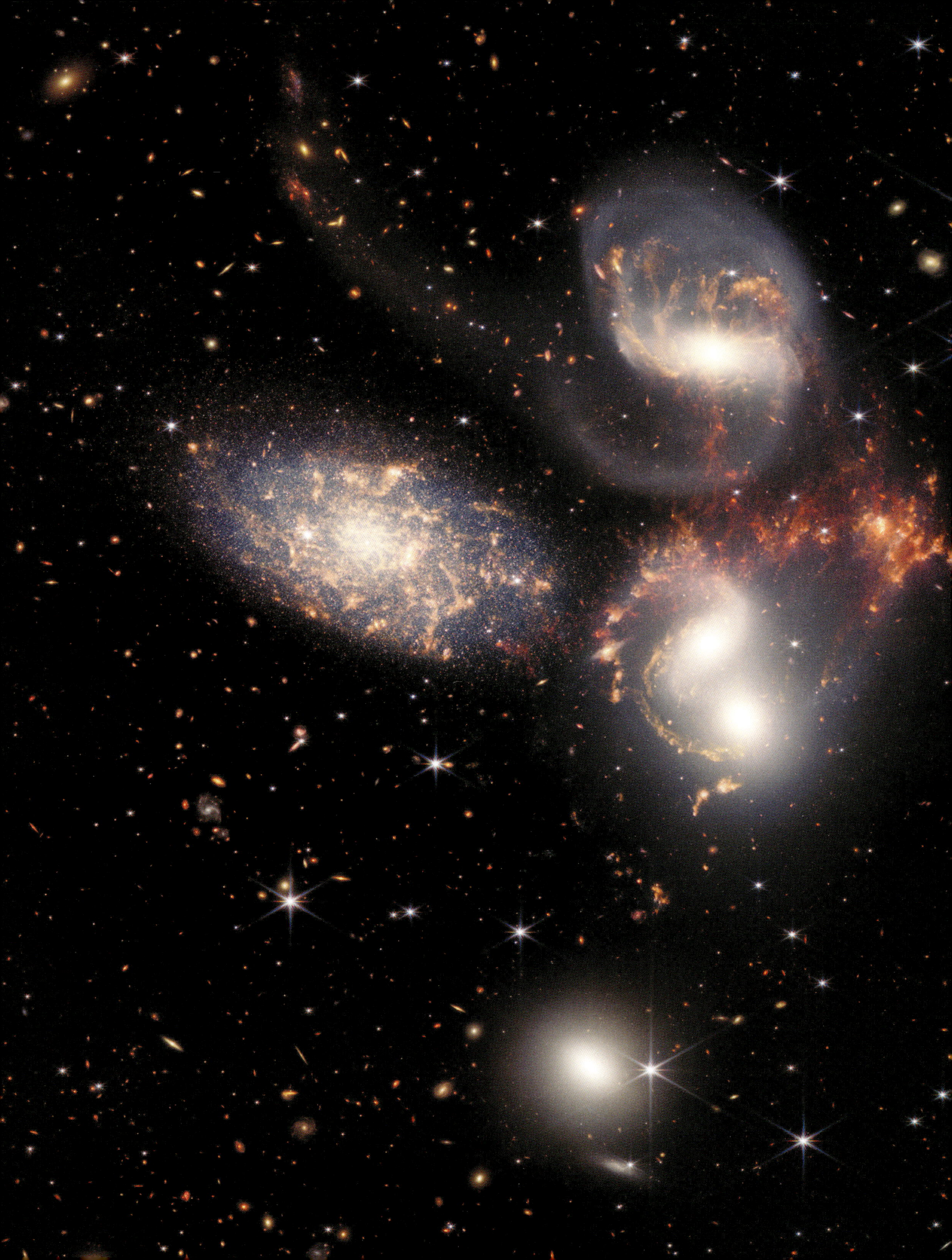

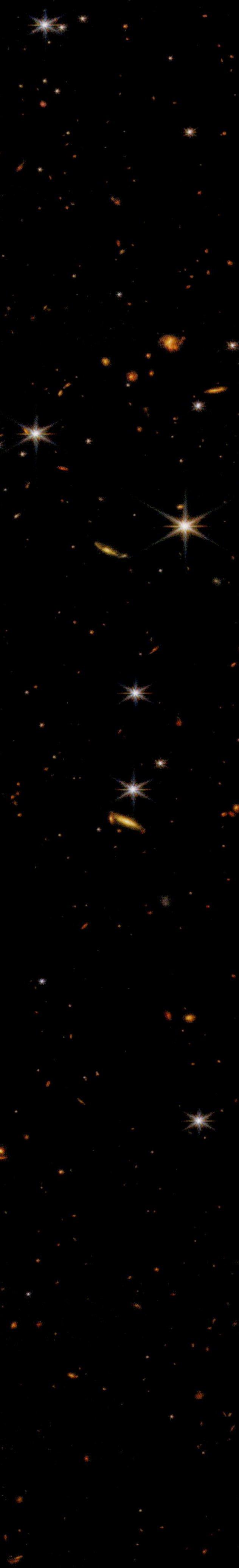

CUBES OF DATA

Webb's instruments are collecting a mind-boggling amount of information about the universe and the objects being born, exploding, merging and existing within it. A major challenge is making sense of this data and packaging it in a way that can be used to give a greater understanding of various phenomena taking place in the cosmos.

One instrument that was designed to do this from the outset is NIRSpec's integral field unit (IFU) (see page 58). This combines a camera with a spectrograph and integrates their information to form a data cube.

The multifaceted IFU 'slices' the view up and enables the spectra of each slice to be generated, enhancing the data that Webb is able to gather on an object. It's a bit like having a map of a country where every point on the surface is also associated with another variable, such as its temperature at that exact point. This combination of spectral and spatial data helps observers understand, for example, the structures within an object, their composition and how fast they are moving.

THE SPECTRA OF A BLACK HOLE

Webb scientists used NIRSpec's IFUs to create a data cube of the supermassive black hole at the centre of the topmost galaxy in Stephan's Quintet, NGC 7319. The cube combines images of the galactic core with its spectral features, effectively creating a 3D block. Each point inside the cube has both spatial and wavelength data, as well as spectral features. Doing this enabled scientists to glean greater knowledge than would have been possible from just looking at the images.

A composite image of Stephan's Quintet using MIRI and NIRCam.

NGC 6822

ALSO KNOWN AS BARNARD'S GALAXY

Constellation: Sagittarius
Distance from Earth: 1.5 million light-years

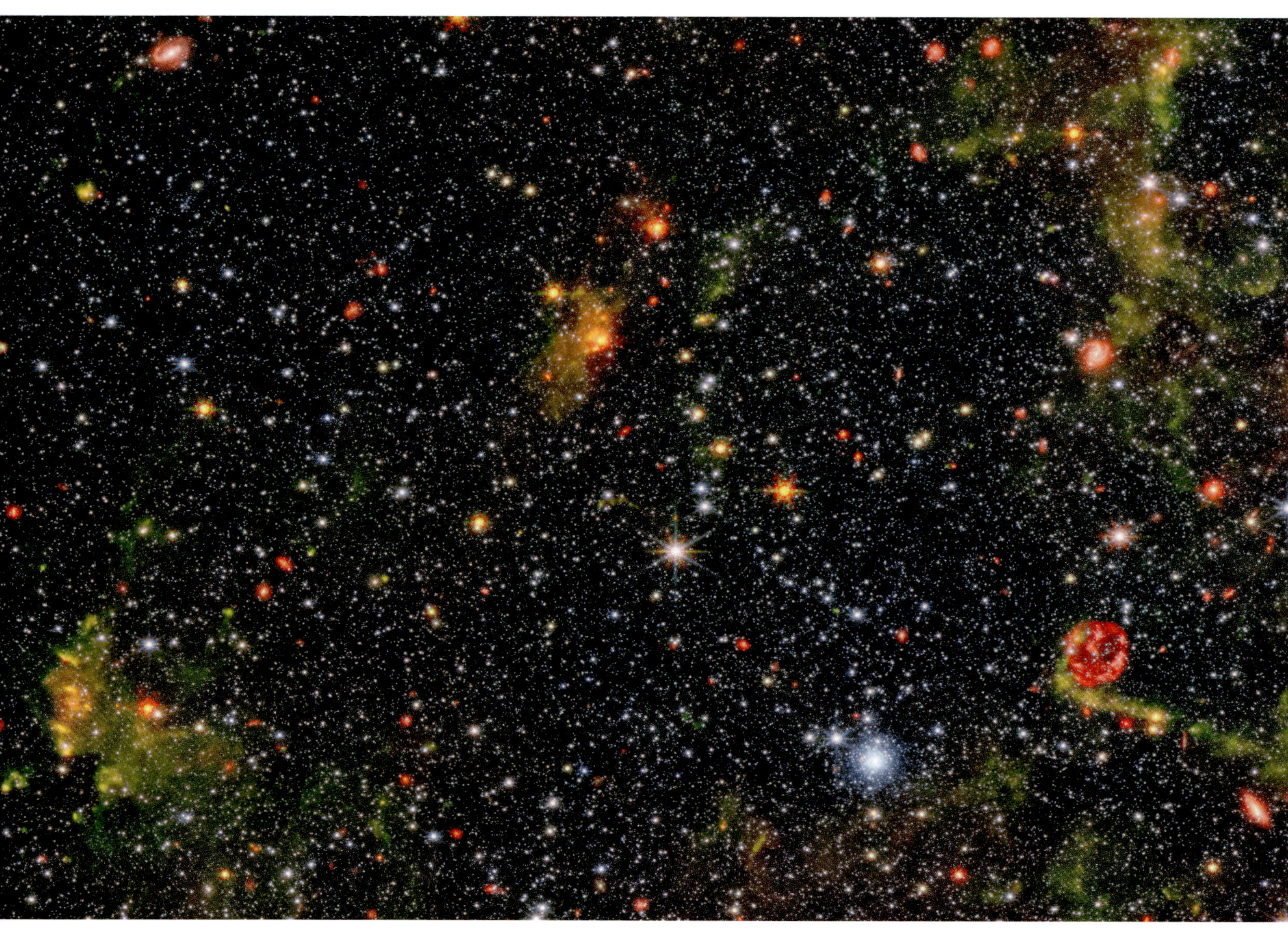

Barnard's Galaxy, also known as NGC 6822, is the Milky Way's closest galactic neighbour. This image is a composite of NIRCam and MIRI data.

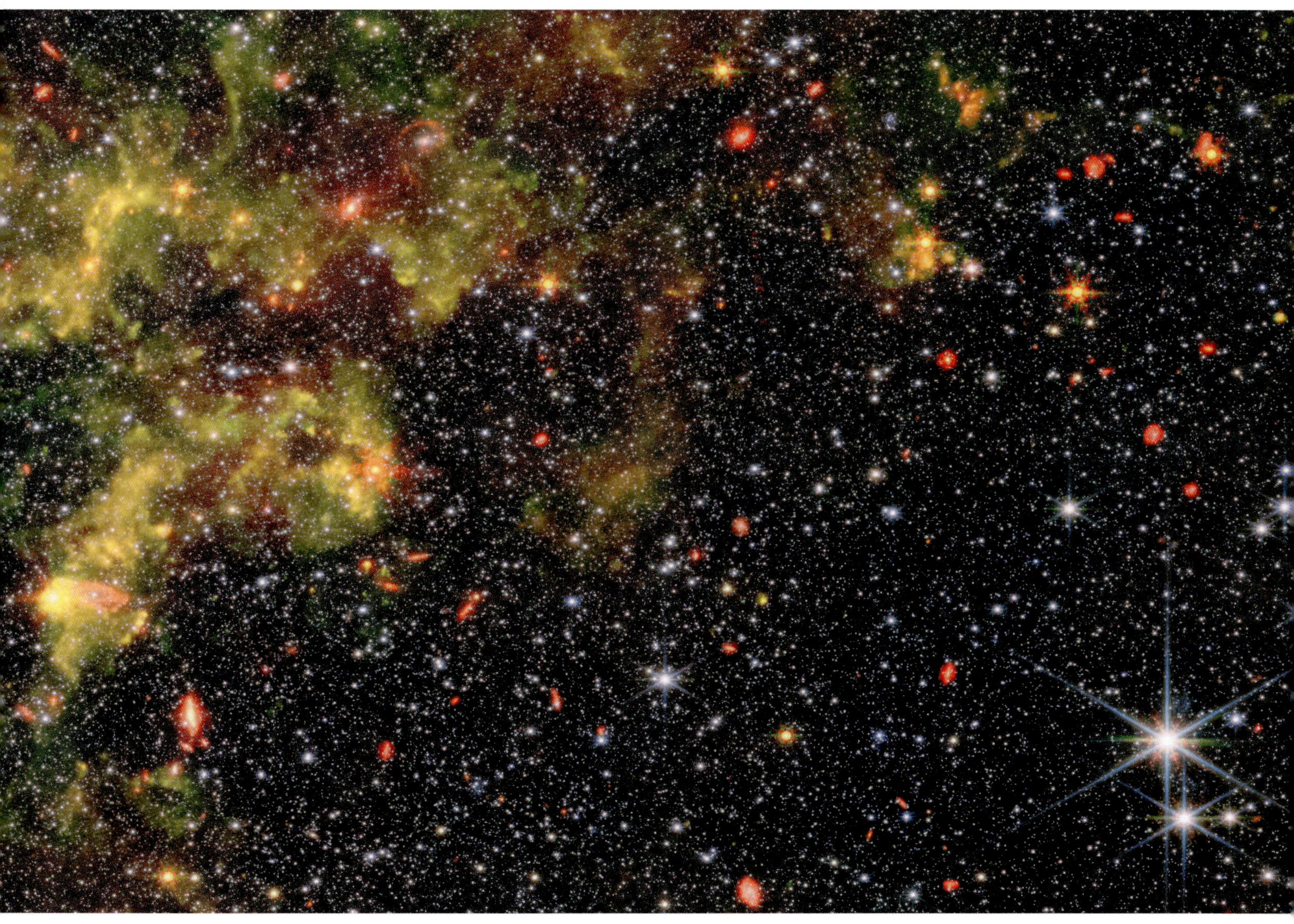

NGC STANDS FOR NEW GENERAL Catalogue of Nebulae and Clusters of Stars, which is another list of deep-sky objects. NGC 6822 was first discovered in 1884 by American astronomer E. E. Barnard and is sometimes called Barnard's Galaxy. At the time, there was much confusion about its size, brightness and other features. In fact, it was initially classified as a nebula. Back then, astronomers did not have the tools and knowledge to understand how the same object could look different through different telescopes.

NGC 6822 offered a milestone in our understanding of the cosmos. In 1925, Edwin Hubble described NGC 6822 as 'the first object definitely assigned to a region outside the galactic system'. Today, we know much more about this important object, which astronomers have been studying in detail for almost 140 years. NGC 6822 is about 7,000 light-years in diameter and classified as an 'irregular galaxy' because it has no distinct, regular shape. It is also the closest non-satellite galaxy to ours, meaning that it does not appear to orbit our Milky Way.

Containing approximately 1 million stars, NGC 6822 is a dwarf galaxy and is tiny compared to the Milky Way with its 100 billion stars. Nevertheless, it is counted in the Local Group, a collection of galaxies that includes our galaxy.

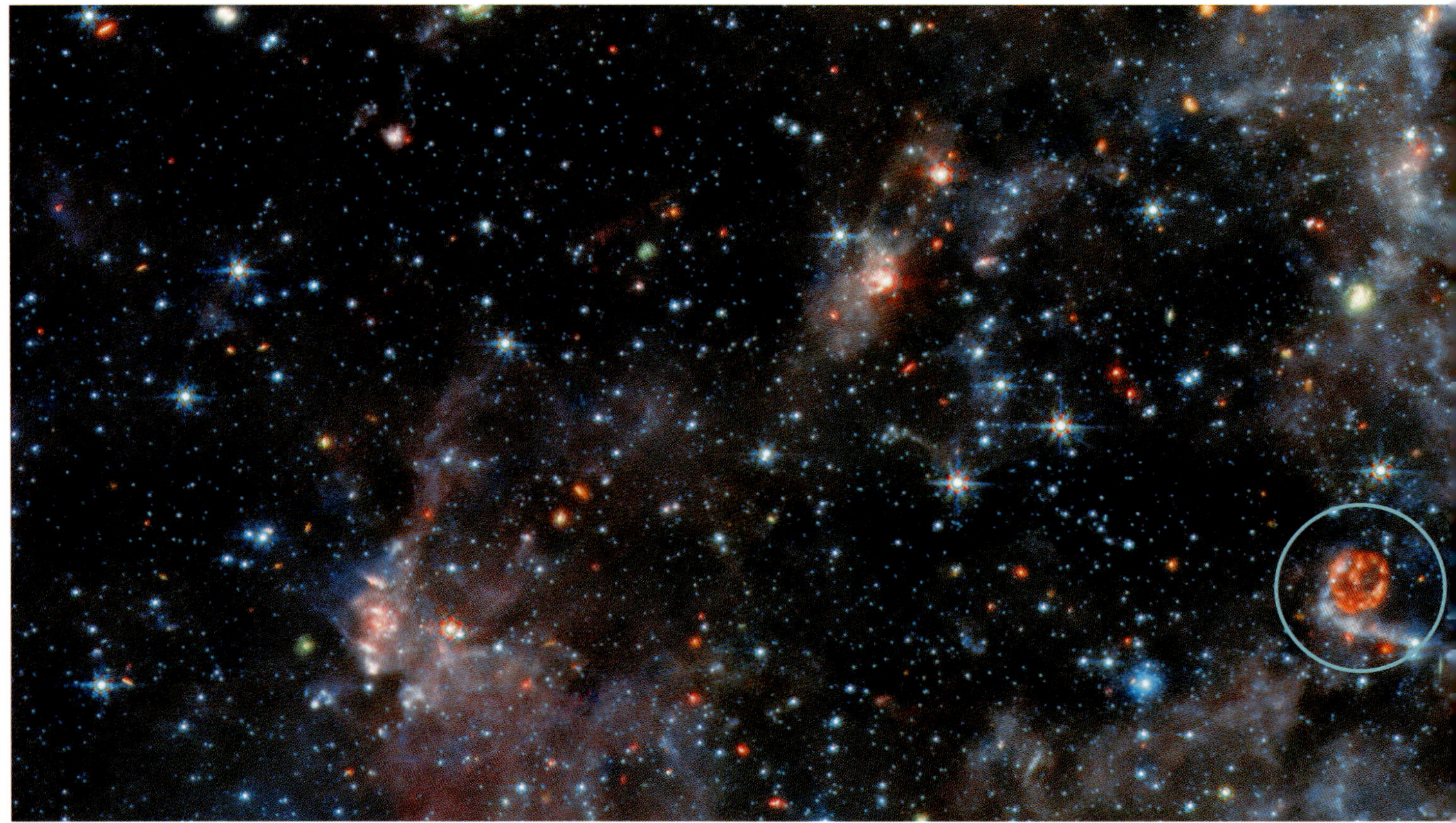

NGC 6822 as seen with MIRI's mid-infrared gaze. The circled area highlights the red remnants of a giant supernova.

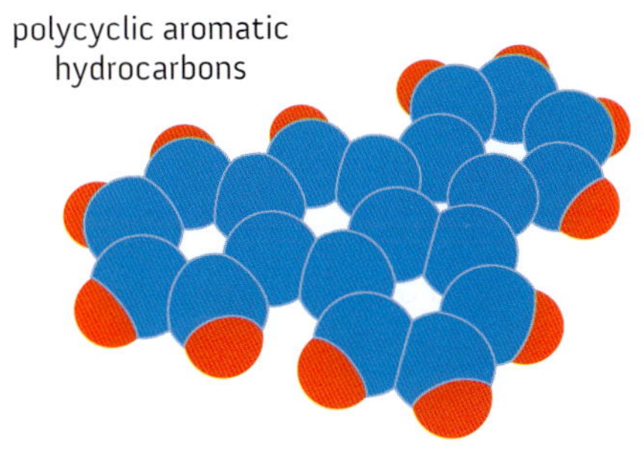

Polycyclic aromatic hydrocarbons are a type of organic molecule and are widespread in the universe.

The Webb images derived from MIRI and NIRCam offer us an unprecedented view of our cosmic neighbour. Importantly, they also illustrate what NGC 6822 looks like in different wavelengths – each of which foregrounds distinct astrophysical phenomena.

LOOKING INTO A DUST CLOUD

MIRI's mid-infrared detectors are particularly sensitive to gas emissions. Swirling white clouds of dust and gas eddy throughout the image, with explosions represented by the red and magenta flares that manage to penetrate the veil. These bright colours denote areas of active star formation. In the bottom centre of the image there is a red bull's eye, showing the remnants of a giant supernova.

The cyan indicates cooler dust, while more orangey colours point to warmer dust. Polycyclic aromatic hydrocarbons pop out of the image in bright blue. Distant galaxies glow in a warm orange, while closer galaxies contain signal-emitting dust that has been coloured green.

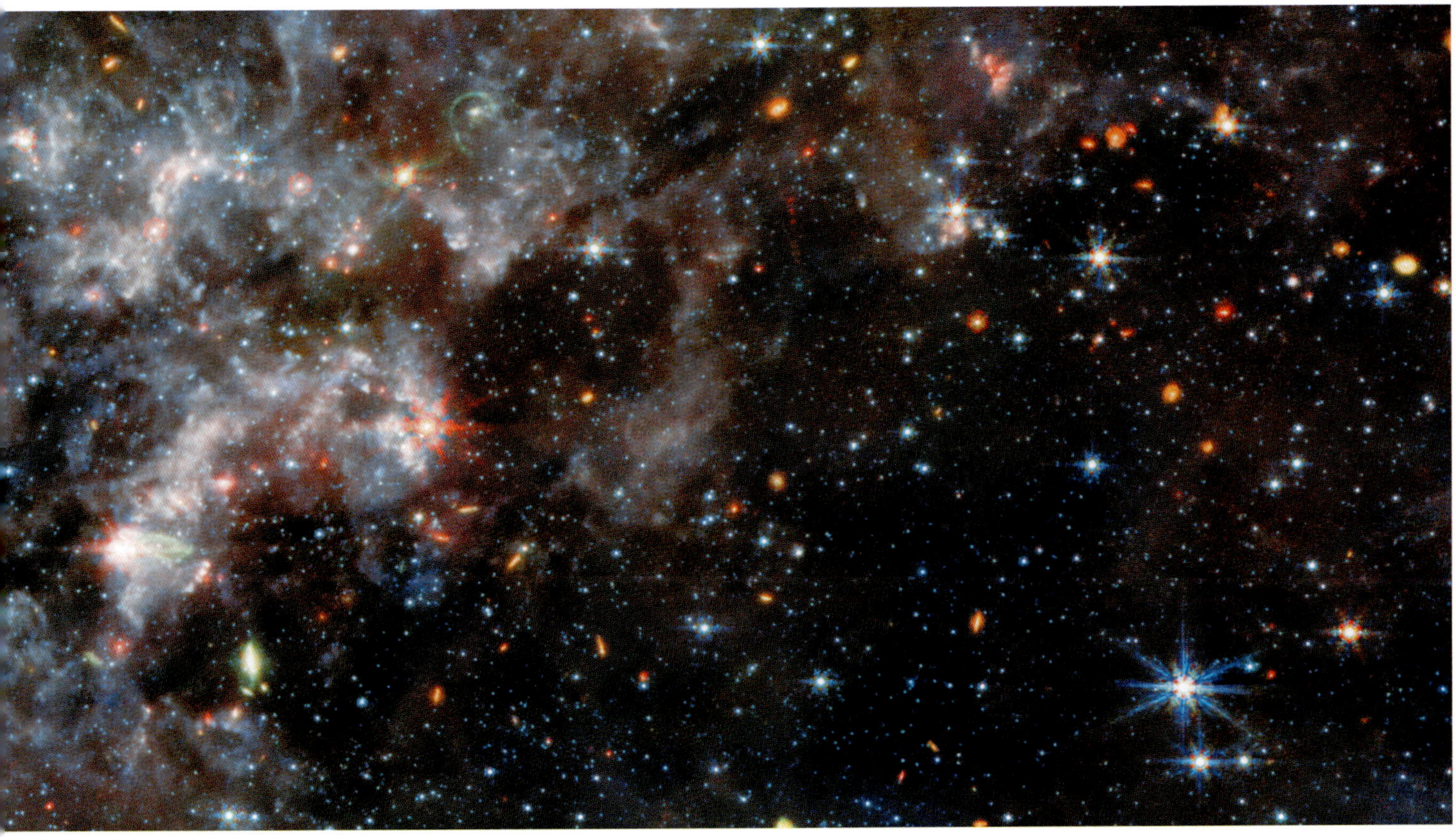

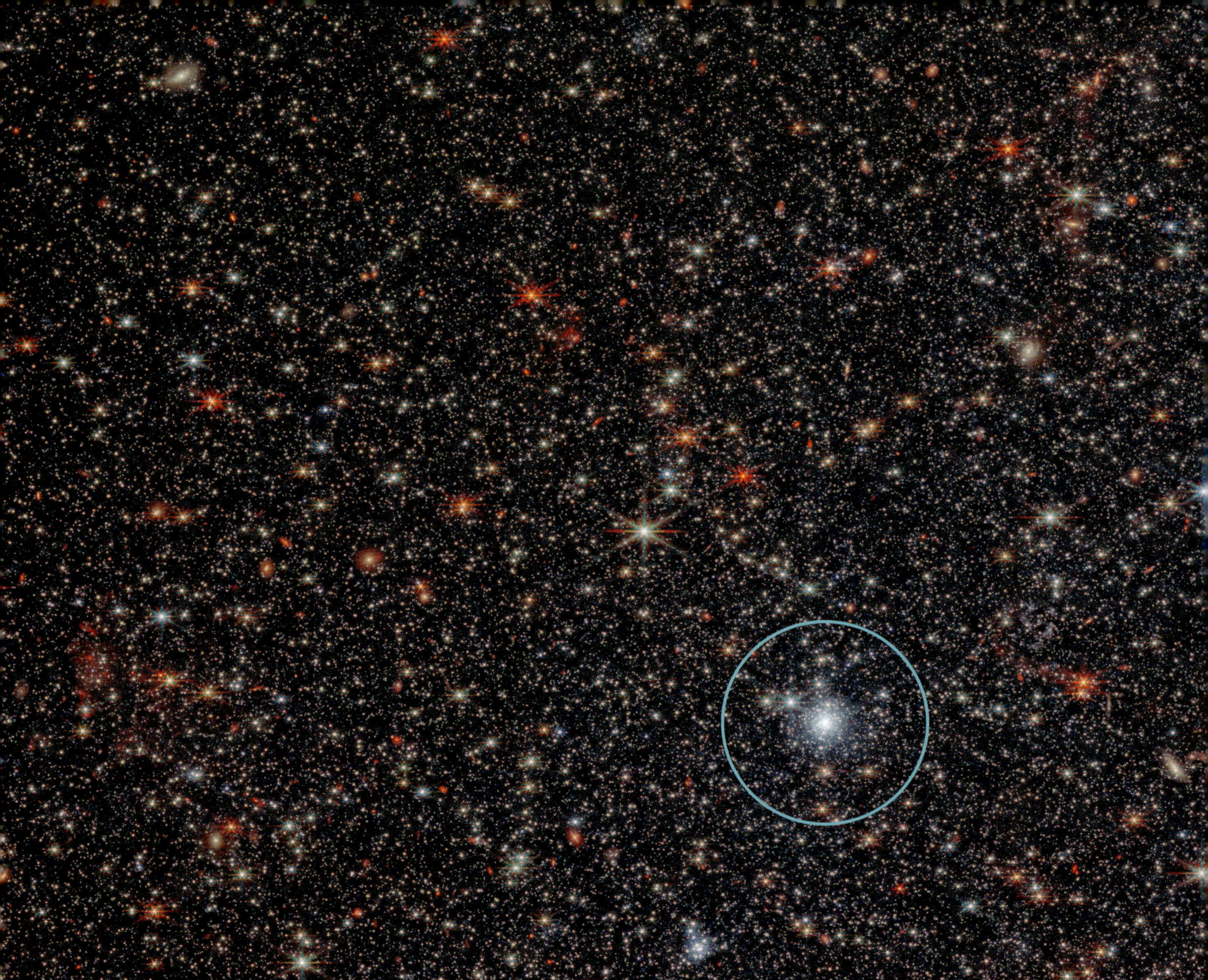

NIRCam's view of Barnard's Galaxy, showing its stars in amazing detail. Circled on the image is the globular cluster.

Notably, the MIRI image has brought an interesting and previously unknown characteristic of NGC 6822 to astronomers' attention. It shows us that the galaxy is mainly made up of hydrogen and helium, the lightest elements in the known universe, rather than heavier metallic elements that result from the fusion process happening within the heart of stars.

Many stars have these heavier elements because of the explosion of earlier stars that then go on to make new stars. This low 'metallicity' means that the stars in this area are probably first-generation stars, not made up of stars that have been before. The stars and dust of the early universe would also have had a low metallicity, making NGC 6822 an intriguing case study for scientists interested in the birth and evolution of stars and galaxies, as well as the lifecycle of interstellar dust.

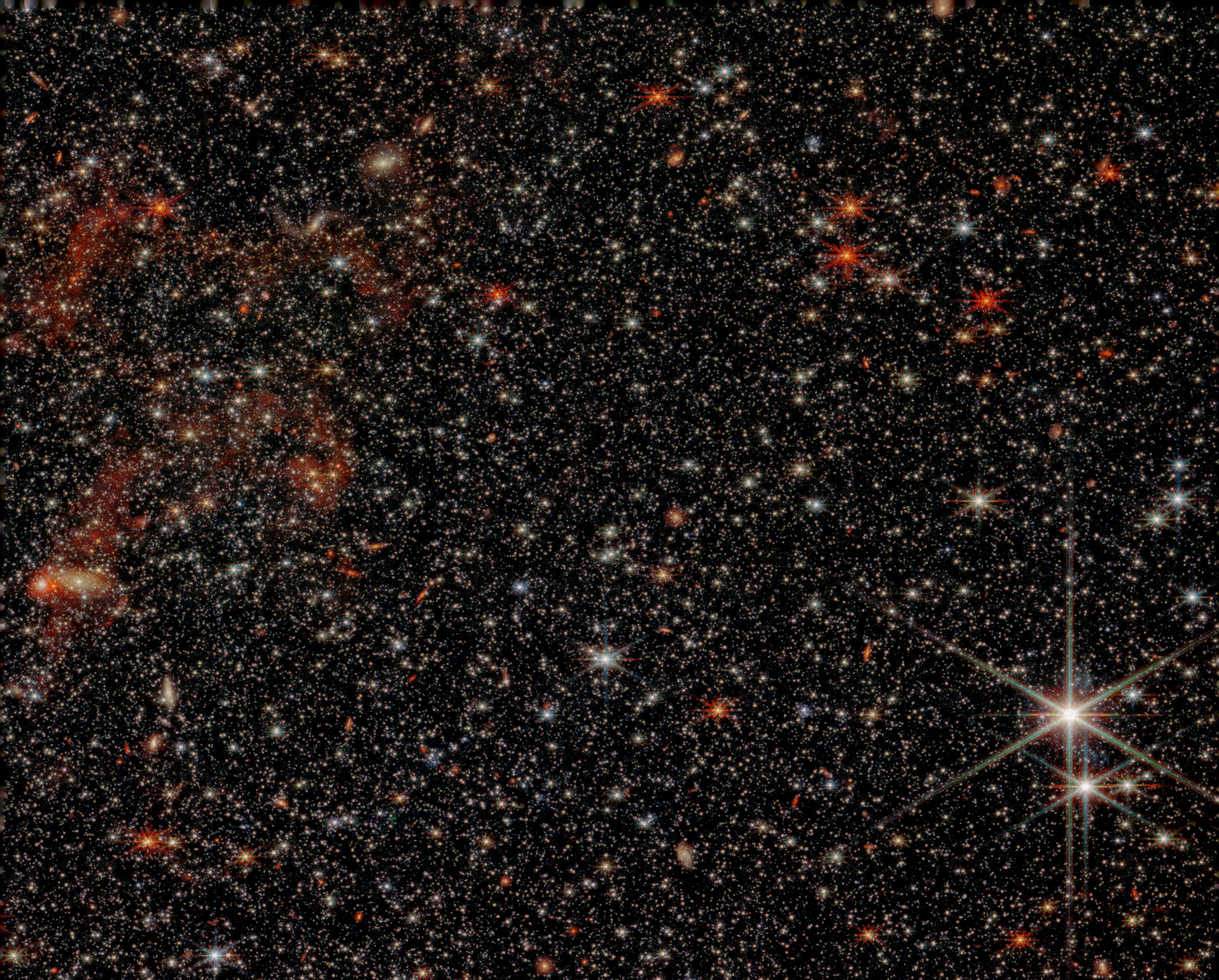

FIELDS OF STARS

In the image above, NIRCam's near-infrared eye reveals a galaxy of stars. Usually, it is very difficult to see stars in swirling clouds of cosmic gas and dust, but in this NIRCam image the dust and gas are reduced to faint red wisps. The shortest wavelengths – indicating the most energetic signals – are coloured in blue and cyan, and these are the brightest stars in the galaxy.

The bright blue-white sphere in the lower half of the image is a globular cluster, a collection of stars bound together by their own gravity. Fainter stars in the picture appear in orange and red.

NGC 7469

Constellation: Pegasus
Distance from Earth: about 220 million light-years

NGC 7469, which is about 220 million light-years away, has an active galactic nucleus at its centre. This image is a composite of MIRI and NIRCam data.

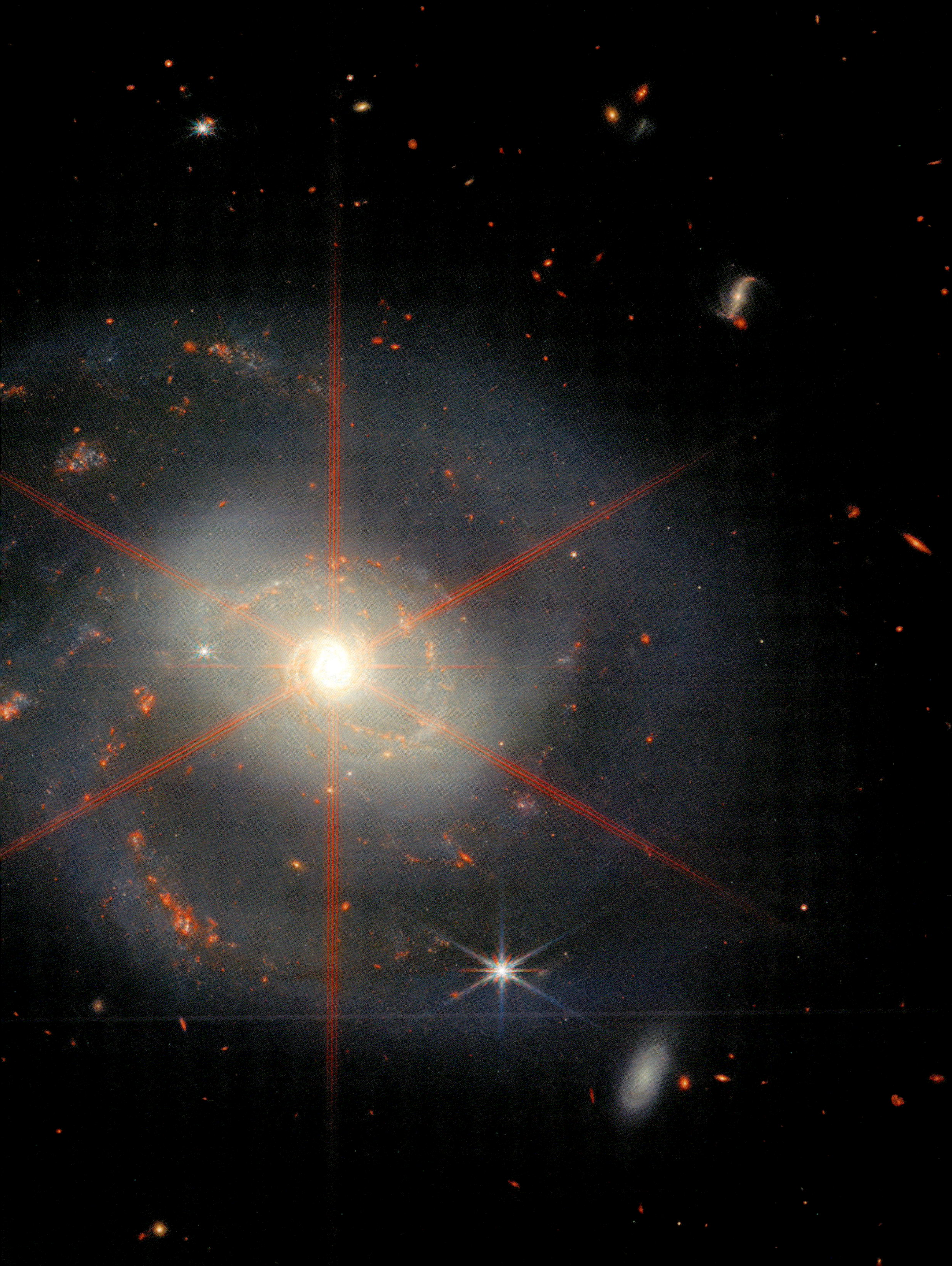

IT TAKES LIGHT 90,000 YEARS to travel from one edge of this vast and luminous spiral galaxy to the other. It is thought that there may be a supermassive black hole – more than 6 million times the mass of our sun – squatting in the centre of this charismatic galaxy. In this Webb image, it appears as an incredibly bright region in the centre. Webb's striking eight diffraction points dominate the picture, showing the sheer quantity of radiation bursting from NGC 7469's active galactic nucleus (AGN). AGNs are the dense region at the heart of galaxies that emit a broad range of signals across the spectrum – that variation of radiation is how astronomers know that an AGN is not merely an area of very energetic star formation.

THE LIGHT OF BLACK HOLES

Most AGNs are powered by supermassive black holes, which are among the most enigmatic and fascinating objects in the universe. While light cannot escape from these oubliettes, they are also,

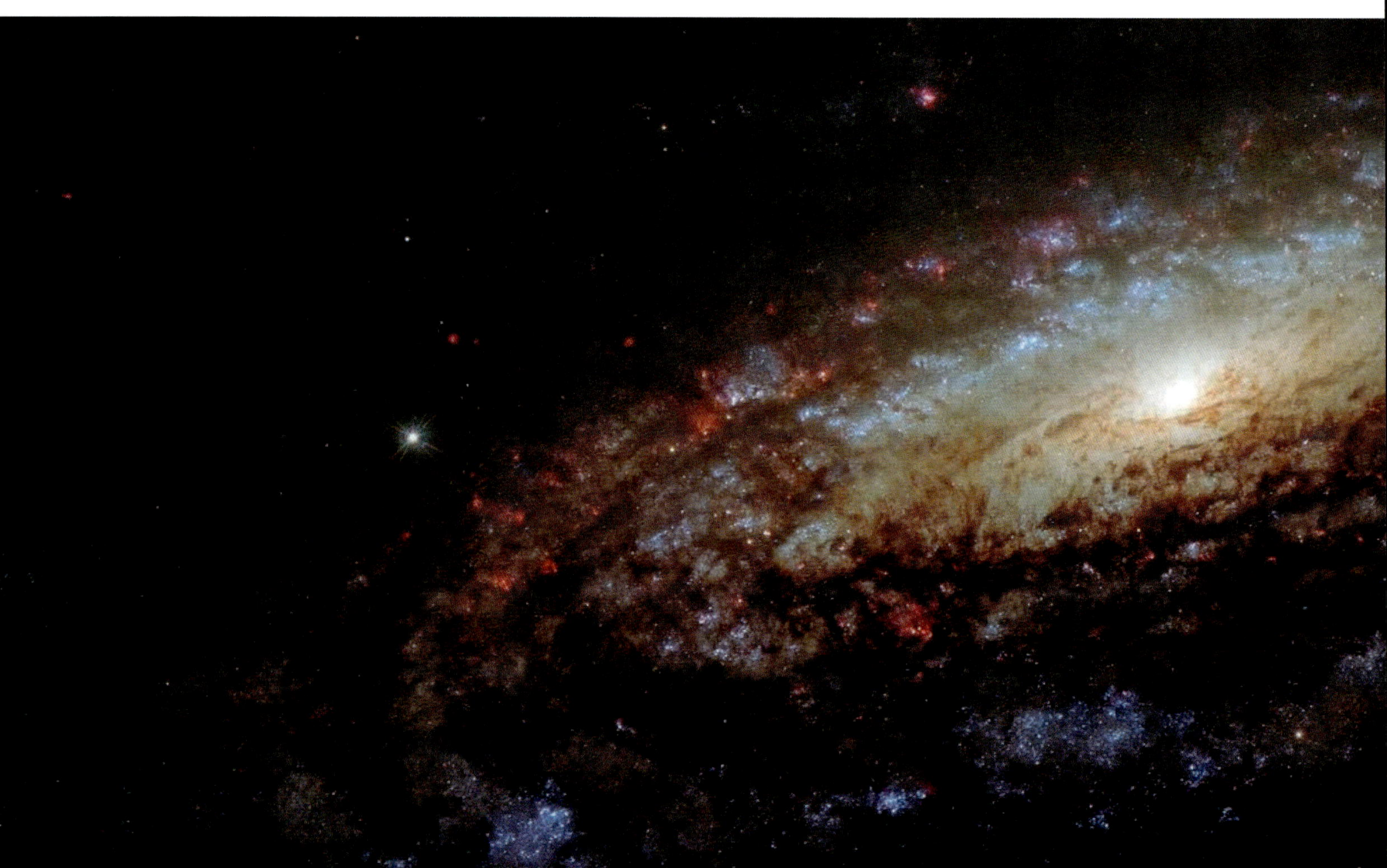

Hubble's view of spiral galaxy NGC 7331 in the Pegasus constellation.

ironically, some of the brightest objects in the universe – or rather, the matter around them is. Black holes have voracious appetites and grow larger by consuming nearby matter: stars, dust, gases and sometimes even other black holes. The matter attracted to the black hole's gravitational forces can form an accretion disk around it, just as with planetary formation around a protostar. The tremendous gravitational and magnetic forces at play at the edges of a black hole can heat the matter in the accretion disk, causing it to fall into the hole. Under these extreme conditions, this matter emits radiation that can be observed. This is why supermassive black holes can appear luminous despite emitting no light themselves.

A PECULIAR GALAXY

NGC 7469 has been of interest for decades, since instrumentation was able to show that there was something rather strange about it. In fact, this galaxy, along with its nearby companion IC 5283, is on a list known as the Atlas of Peculiar Galaxies, where its designation is Arp 298. The galaxy is peculiar for several reasons, one of which is highlighted in this image. Just around its compact glowing centre, it looks as though its bright core is surrounded by a necklace of glowing pearls. This is actually a ring of starbursts – a phase of intense and exceptionally rapid star formation within a galaxy. Several factors can trigger a starburst, including gravitational interactions with other galaxies, mergers between galaxies or the presence of an abundant supply of molecular gas. Gravitational interactions and collisions can compress gas clouds within a galaxy, causing them to collapse and form new stars.

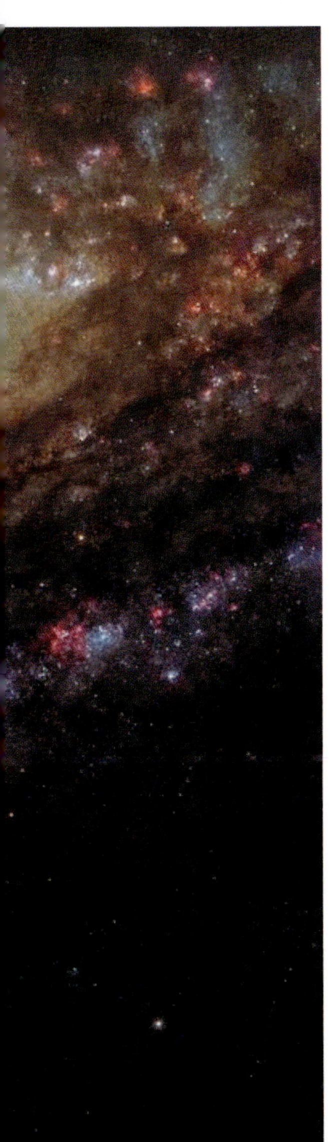

In this case, the rapid star formation is taking place remarkably close to the galactic nucleus. In cosmic terms, they are basically sitting on top of each other – just 1,500 light-years apart.

The primary reason that NGC 7469 found its way into the Atlas of Peculiar Galaxies is its smaller companion, IC 5283, which is mostly cut out of the frame at the bottom left of the image. Scientists think that the interaction between NGC 7469 and the other spiral galaxy may be a factor in NGC 7469's fecund star formation, although they will need more observations and data to really understand what is going on.

NGC 7496

Constellation: Grus
Distance from Earth: about 60 million light-years

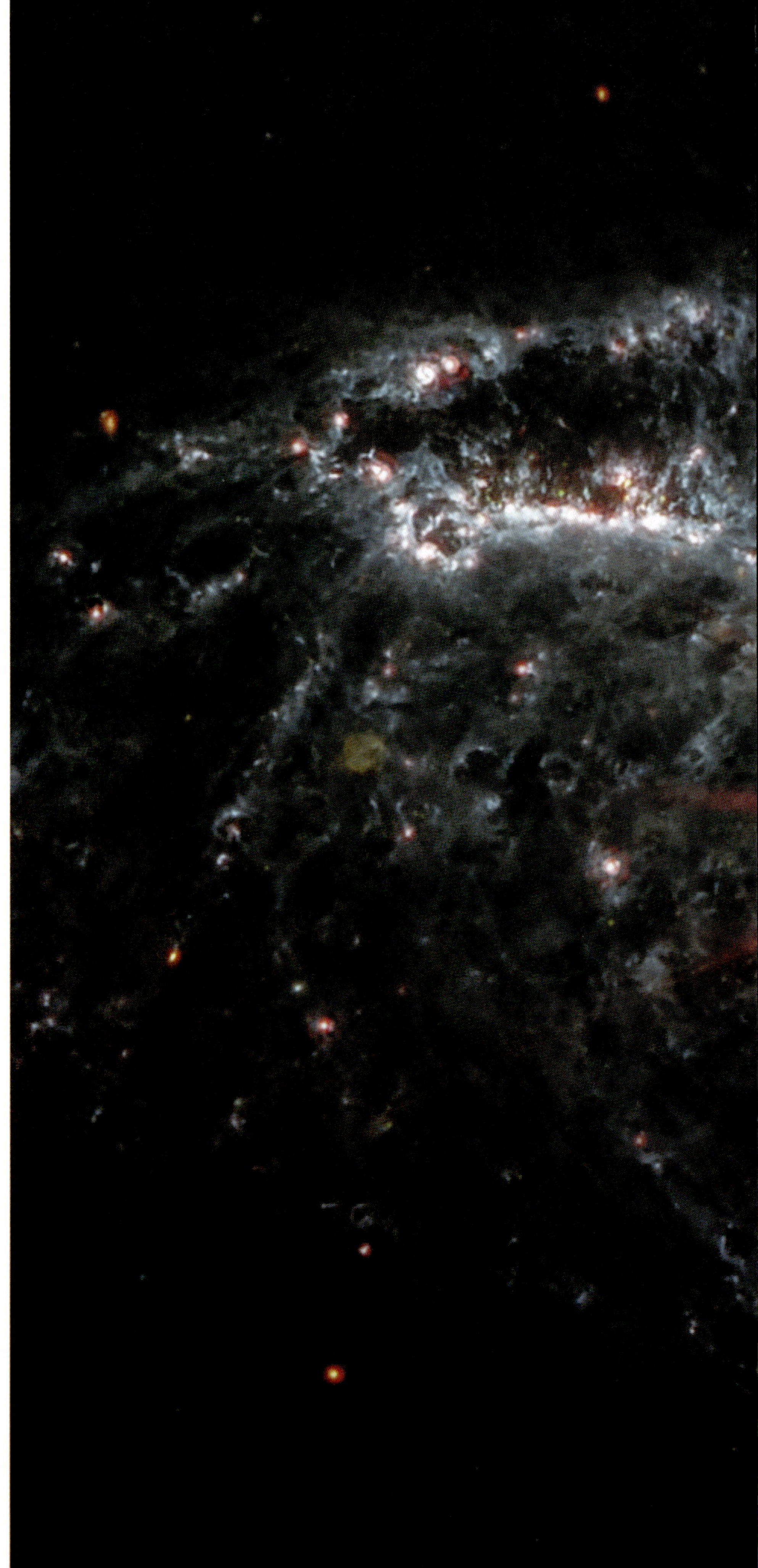

NGC 7496 was one of nineteen galaxies studied by the Physics at High Angular resolution in Nearby GalavieS (PHANGS) collaboration. This MIRI image foregrounds the black hole at its heart.

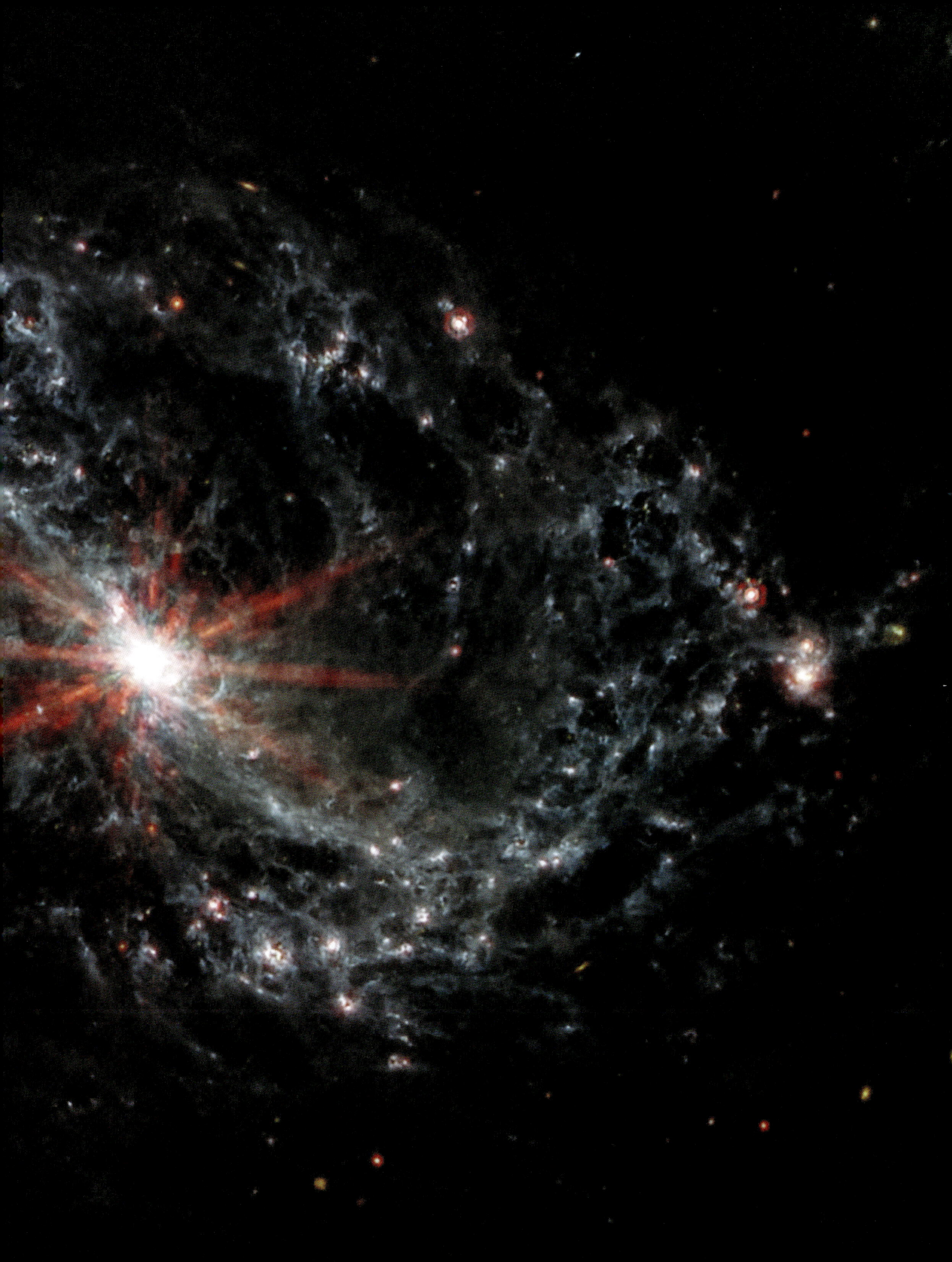

THIS SPIRAL GALAXY SPINS THROUGH space like the propeller of an old-fashioned aeroplane. It's called a 'barred' galaxy because it looks like a single curved bar turning against a black backdrop. It pivots around a black hole, which we can see shining in the centre of this Webb image. At least that is what we assume, since it is very difficult to snap a picture of a black hole.

In 2019, radio astronomers published the first direct image of a black hole. Although scientists and mathematicians hypothesized their existence more than a century before (thanks to Albert Einstein's general theory of relativity), all the evidence up until that point had been indirect. In other words, scientists assumed the presence of a black hole in the centre of galaxies based on how matter behaves around them.

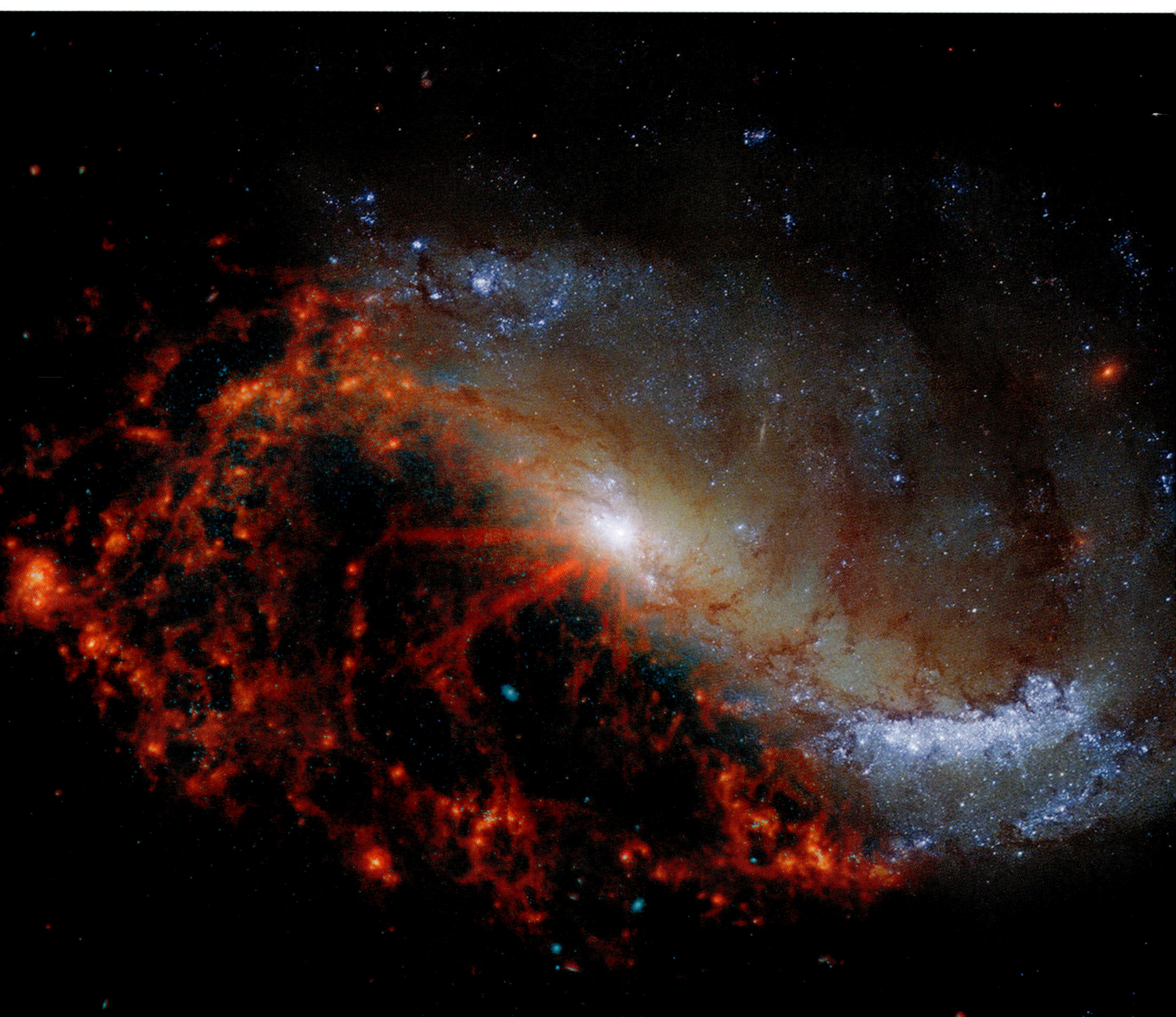

Half of this image of NGC 7496 – the red part – was imaged using Webb data, while the blue half was thanks to Hubble. The two gazes offer very different perspectives on the spiral galaxy.

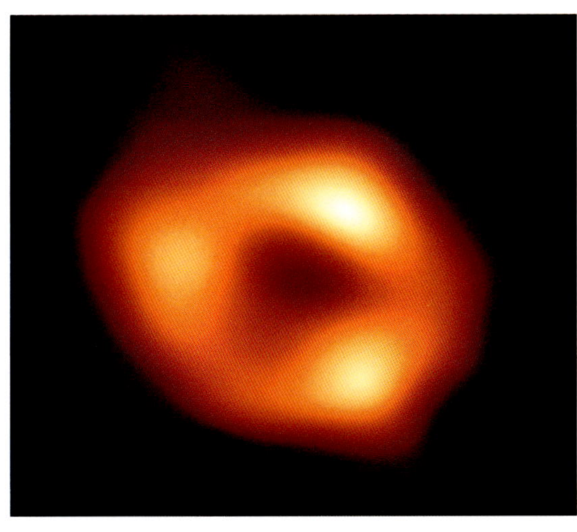

In 2019, the Event Horizon Telescope collaboration published the first image of a black hole, the one at the centre of galaxy Messier 87.

They used data from a network of synchronized radio telescopes that spans the globe, known as the Event Horizon Telescope. The miraculous photo shows the supermassive black hole at the centre of M87, a galaxy that is about 55 million light-years away from us. In their image, you can see the growing accretion disk – a bit reminiscent of a doughnut – and the event horizon, which is the black hole's edge, encircling the dark centre.

STARING INTO THE DUST CLOUD

The hearts of galaxies are chaotic places, where high pressures, strong magnetic fields and exorbitant temperatures kick up a lot of dust. This dust obscures much of the activity in these exciting areas, and it is here that Webb comes into its own. With its MIRI instrument, Webb has gleaned unprecedented insight into NGC 7496. In the image on the previous page, you can see the glowing beacon of the active galactic nucleus, whose emissions are so strong that it is surrounded by Webb's signature diffraction spikes, which have been coloured red. Its arms are filled with bubbles and caverns, many of them overlapping. These are caused by young stars that are manipulating their local environment, releasing large quantities of energy with their stellar winds howling through the dust around them. At the same time, they are accreting matter from the surrounding gas and dust. Wispy filaments of dust suffuse the galaxy like smoke.

STAR HUNTING

Webb's precision and resolution have enabled observations of very young stars. As part of the PHANGS collaboration, scientists are investigating how young stars warp their stellar neighbourhood and ultimately shape the evolution of the galaxies they are in. In these stellar nurseries, however, the very young stars often sit in clouds of dust and gases that are opaque to visible light. But it is now possible to image some of NGC 7496's baby stars for the first time, thanks to Webb's infrared capabilities. So far, sixty new potential star clusters have been identified. These stars could be some of the youngest in this barred spiral galaxy. Meanwhile, the MIRI instrument can also detect PAHs, which are noticeably abundant in the galaxy's spiral arms.

ABELL 2744

ALSO KNOWN AS PANDORA'S CLUSTER

Constellation: Sculptor
Distance from Earth: about 3.5 to 4 billion light-years

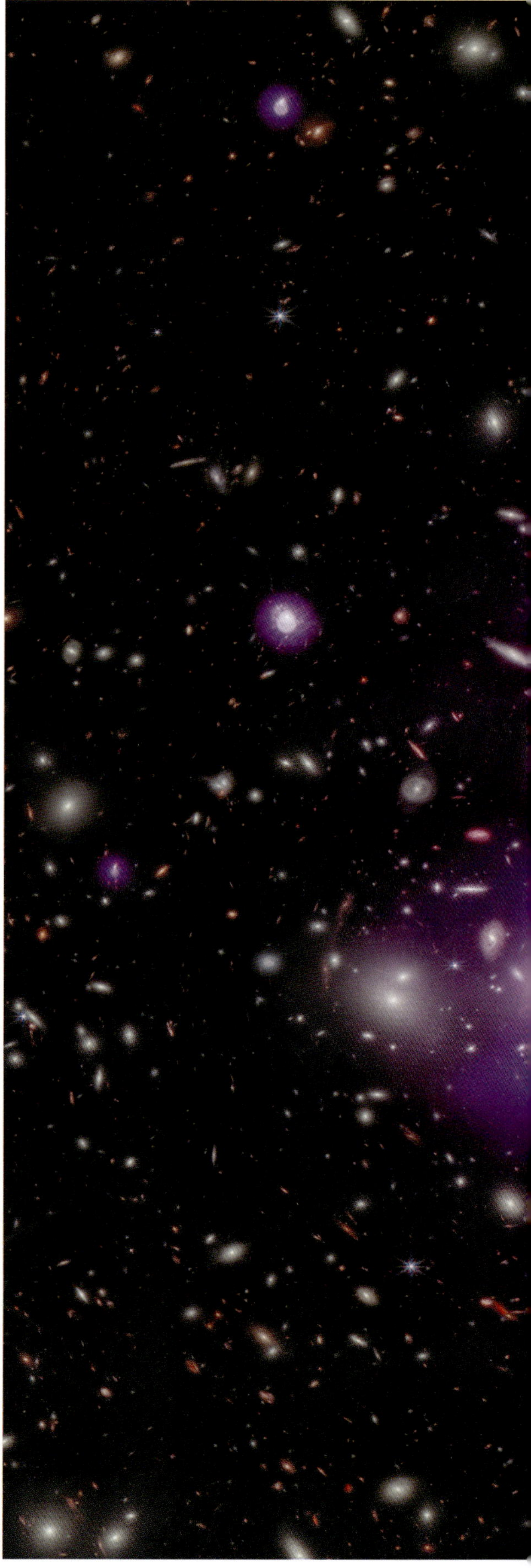

This image of Pandora's Cluster combines Chandra and Webb data and shows the most distant black hole in the galaxy UHZ1, which is much further away than Pandora's Cluster.

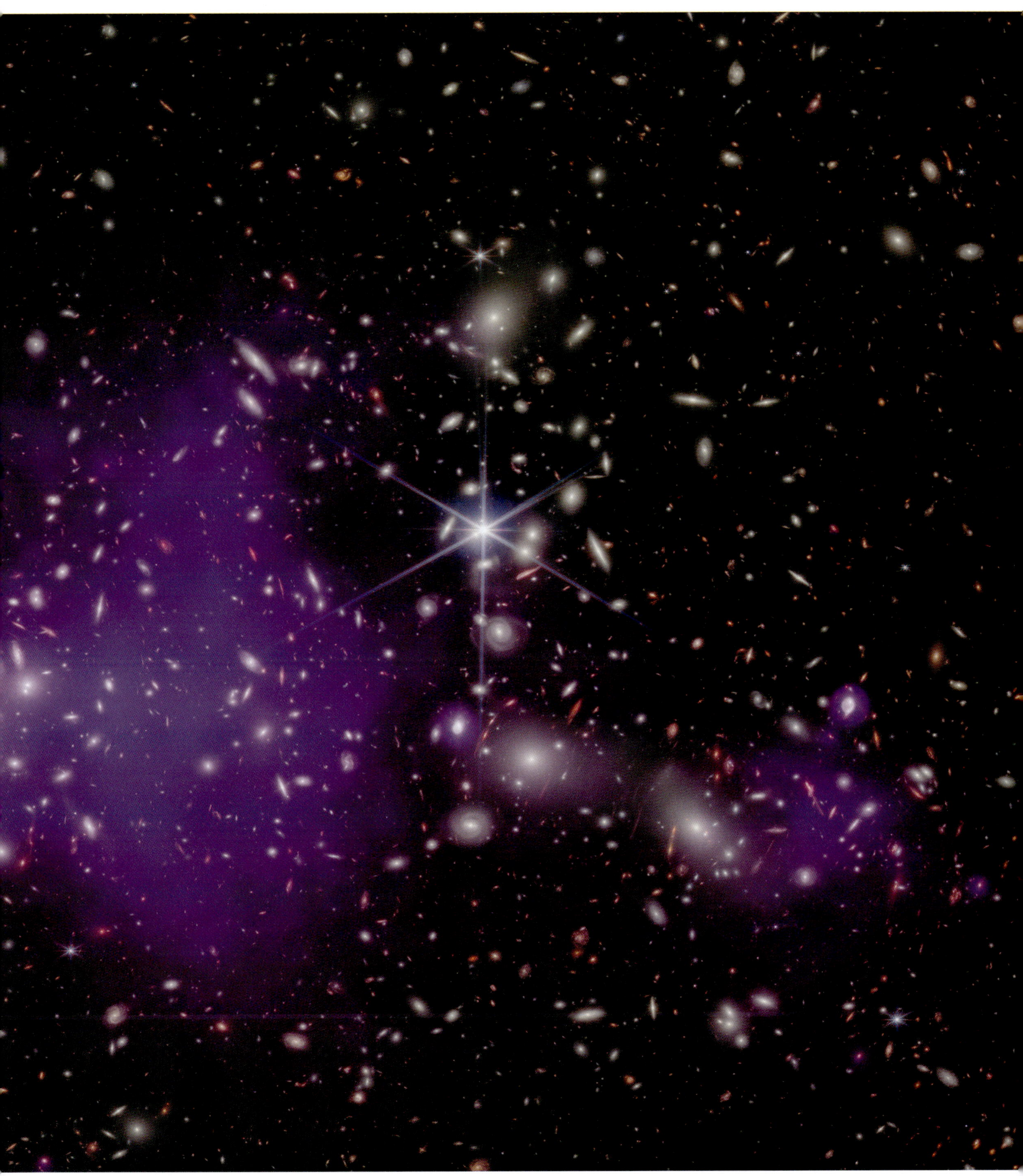

THERE ARE SO MANY STRANGE and different things happening in Abell 2744 that its discoverers decided to call it Pandora's Cluster. This giant galaxy cluster is the cosmic equivalent of a multi-car pile-up on a motorway: with at least four separate galaxies colliding with each other over a period of 350 million years. This image is a composite of data from the Chandra X-ray Observatory and Webb. The vivid purple infusing the image shows billowing clouds of hot gas, observed using Chandra, while Webb's data has been coloured red, green and blue.

One of the (many) interesting characteristics of Abell 2744 is that its visible galaxies make up only a small fraction (about 5 per cent) of its mass. More than three-quarters of its mass is dark matter. So far, we have been unable to observe dark matter directly; it's a theoretical type of matter that doesn't emit, absorb or reflect electromagnetic radiation. Similar to black holes, we know that it's there because of how visible matter and light behave around it.

DARK STARS

In 2023, a group of scientists reported that they had detected 'dark stars' using Webb. Dark stars are a hypothetical type of star that would have existed in the very early universe before traditional stars were able to form. These dark stars would still have contained a lot of normal matter, but also a large amount of dark matter. Instead of the fusion process that we see in stars today, they would have been powered by the interactions between normal and dark matter. Although called dark they are likely to be quite bright, enabling detection. This is one theory as to how the early stars

BELOW LEFT: Chandra's view of Abell 2744.

BELOW RIGHT: Webb's perspective on Pandora's Cluster.

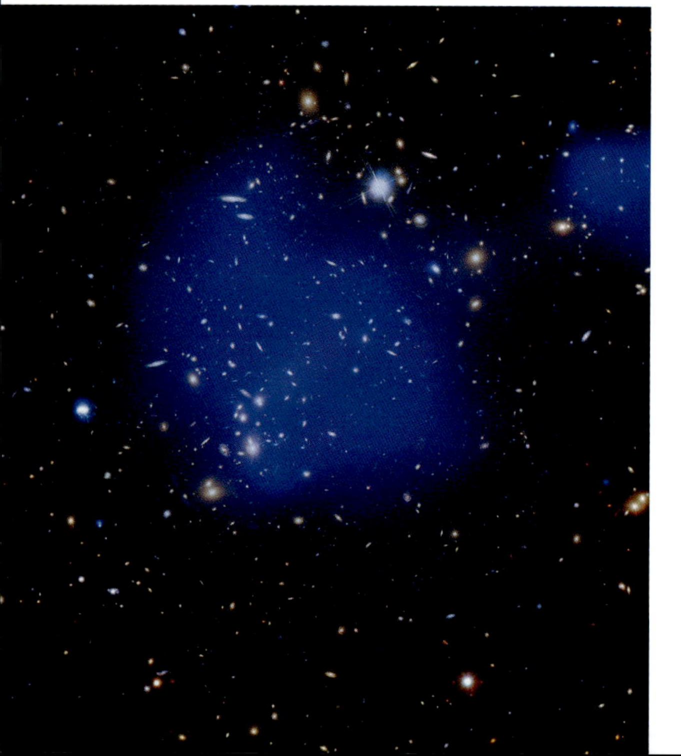

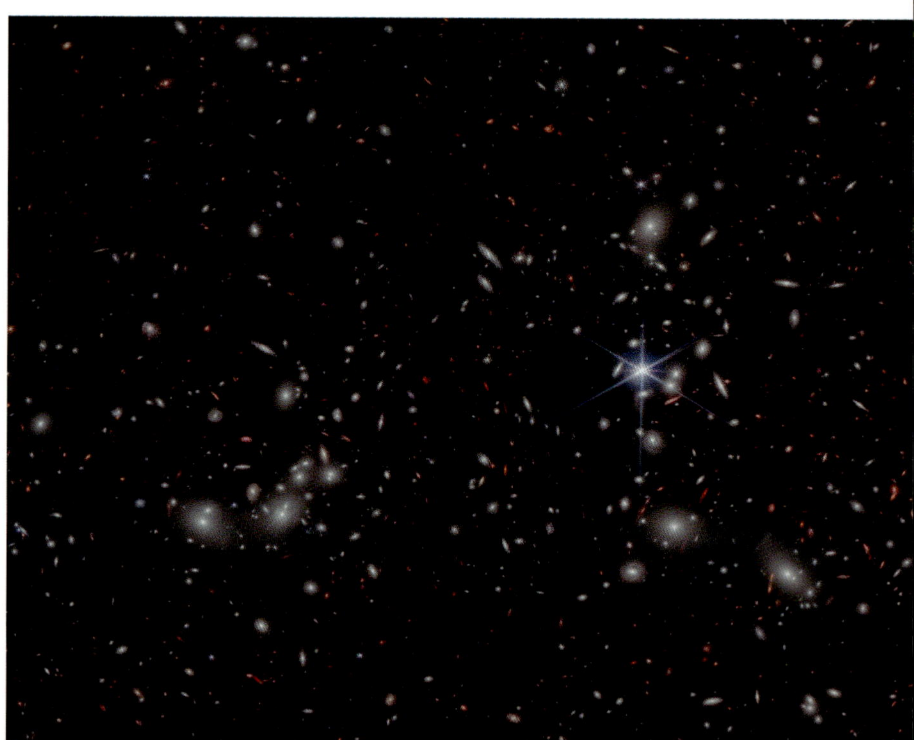

formed. Using Webb's ability to probe the early universe, it was argued that some of the oldest galaxies we see could actually be dark stars.

OLDEST BLACK HOLE

There are no reported dark stars in this image of Abell 2744, but there is something equally tantalizing. Due to the sheer quantity of matter – both normal and dark – in the galaxy cluster, there are massive gravitational forces at play. Consequently, there is a lot of gravitational lensing in this image. That lensing means that many of the colourful galaxies in the background are much further away than they appear in the image.

Chandra looked in Abell 2744's direction for more than two weeks, collecting data. One object in particular has caused a great deal of excitement. Even in this image, it looks tiny; you'd be forgiven for thinking it was a spec of lint on the page. But in the centre of this composite picture of the sky is the oldest black hole we've observed, at the heart of the UHZ1 galaxy. In the image, Chandra's purple X-ray emissions show the material close to UHZ1's supermassive black hole, while Webb shows infrared signals from the galaxy.

UHZ1 contains a quasar, which is an extremely luminous active galactic nucleus powered by a supermassive black hole. What makes it exciting is UHZ1's age. It's more than 13 billion light-years away. The universe was about 470 million years old when UHZ1 emitted the light that Chandra and Webb are observing, meaning that supermassive black holes already existed in the early universe. The genesis of black holes, especially the first ones, is an open question in astrophysics. Were they born out of collapsing gas clouds? Did they start out small when the first star imploded and grow larger as they consumed more material, including other black holes?

The evidence suggests UHZ1's supermassive black hole at least was born massive. Its mass – between 10 million and 100 million suns – is similar to that of the stars in the galaxy around it. This has led astronomers to suspect that it formed when a giant gas cloud collapsed in on itself (as opposed to a supernova). With Webb, we will likely be able to find more ancient black holes and expand our understanding of these enigmatic objects.

PICTURE CREDITS

p. 2: NASA, ESA, CSA, STScI, Matt Tiscareno, Matt Hedman, Maryame El Moutamid, Mark Showalter, Leigh Fletcher, Heidi Hammel
pp. 4–5: ESA/Webb, NASA and CSA, A. Martel.
pp. 6–7: X-ray: Chandra: NASA/CXC/SAO, XMM: ESA/XMM-Newton; IR: JWST: NASA/ESA/CSA/STScI, Spitzer: NASA/JPL/CalTech; Optical: Hubble: NASA/ESA/STScI, ESO
pp. 8–9: X-ray: NASA/CXC/Penn State Univ./L. Townsley et al.; IR: NASA/ESA/CSA/STScI/JWST ERO Production Team
p. 10: NASA/Chris Gunn
p. 11: NASA/Bill Ingalls
p. 12: NASA/Chris Gunn
pp. 14–15: X-ray: Chandra: NASA/CXC/SAO, XMM: ESA/XMM-Newton; IR: JWST: NASA/ESA/CSA/STScI, Spitzer: NASA/JPL/CalTech; Optical: Hubble: NASA/ESA/STScI, ESO
p. 16: NASA Earth Observatory/Lauren Dauphin
p. 17: Al Fenn/The LIFE Picture Collection/Shutterstock
p. 19: NASA, ESA
p. 20: NASA
p. 23: NASA, ESA, STScI, AURA, Hubble Heritage Project (STScI, AURA)
p. 24: NASA/CXC/SAO
p. 25: NASA/CXC/SAO
p. 26: NASA/JPL-Caltech
p. 27: NASA/JPL-Caltech
p. 29: Hubble: NASA, ESA, The Hubble Heritage Team (STScI/AURA); Webb: NASA, ESA, CSA, STScI
pp. 30–31: NASA
p. 32: Xinhua/Alamy Stock Photo
p. 33: top: NASA; bottom: Northrop Grumman Aerospace Systems
p. 34: NASA/Chris Gunn
p. 37: NASA, ESA, CSA, STScI
pp. 40–41: NASA, ESA, CSA, STScI, Samuel Crowe
pp. 42–3: John Fowler
p. 44: Bettmann/Getty Images
p. 45: ESA/Euclid/Euclid Consortium/NASA
p. 47: NASA, ESA, CSA, STScI, Nolan Habel (NASA-JPL).
pp. 48–9: ESA/Hubble & NASA, J. Greene
pp. 50–51: NASA/JPL-Caltech/Univ. of Ariz., NOAO/AURA/NSF
pp. 52–3: ESA/Webb, NASA & CSA, O. Nayak, M. Meixner
p. 54: Nasa/SwRI/MSSS/AndreaLuck
p. 56: top: NASA; bottom: NASA/JPL-Caltech/MSSS
p. 59: ESA/Webb, NASA & CSA, L. Armus, A. Evans
p. 61: all images: NASA, ESA, CSA, STScI
pp. 62–3: NASA, ESA, CSA, STScI, Janice Lee, Thomas Williams and the PHANGS team
pp. 64–5: NASA, ESA, CSA, STScI
pp. 66–7: NASA/Earth Science and Remote Sensing Unit/NASA Johnson Space Center
p. 68: ESA & NASA/Solar Orbiter/EUI Team
pp. 70–71: NASA, ESA, CSA, Jupiter ERS Team
p 72: NASA, ESA, CSA, STScI, R. Hueso, I. de Pater, T. Fouchet, L. Fletcher, M. Wong, J. DePasquale
p. 73: NASA, ESA, CSA, Gerónimo Villanueva, Samantha K Trumbo
pp. 74–5: NASA, ESA, CSA, STScI, Matt Tiscareno, Matt Hedman, Maryame El Moutamid, Mark Showalter, Leigh Fletcher, Heidi Hammel
p. 76: NASA/JPL-Caltech/Space Science Institute
p. 77: NASA, ESA, CSA, A. Pagan (STScI)
pp. 78–9: NASA, ESA, CSA, STScI
p. 81: top: NASA, ESA and M. Showalter L. Sromovsky, P. Fry, H. Hammel and K. Rages; bottom: NASA, ESA, CSA, STScI
pp. 82–3: NASA, ESA, CSA, STScI
p. 85: top: NASA/JPL/USGS; bottom: NASA/JPL-Caltech
pp. 86–7: NASA, ESA, CSA, STScI, Kevin Luhman, Catarina Alves de Oliveira
p. 88: NASA, ESA, CSA, L. Hustak
p. 89: NASA
p. 90: NASA, ESA, CSA, Alyssa Pagan
p. 92: NASA, ESA, CSA, STScI, Kevin Luhman, Catarina Alves de Oliveira
p. 93: X-ray: NASA/CXC/IASF Palermo/M.Del Santo et al.; Optical: NASA/STScI
p. 94: NASA, ESA and A. Simon; sun and low-mass star: NASA, SDO; brown dwarf: NASA, ESA and JPL-Caltech; Earth: NASA; infographic: NASA and E. Wheatley
pp. 96–7: ESA/Webb, NASA, CSA, M. Zamani (ESA/Webb), and the PDRs4All ERS Team
pp. 98–9: NASA, ESA, CSA, STScI
pp. 100–101: NASA, ESA, CSA, STScI
pp. 102–3: NASA, ESA, CSA, STScI, Webb ERO Production Team
p. 104: left: NASA, ESA, CSA, STScI, Webb ERO Production Team; right: X-ray: NASA/CXC/Penn State Univ./L. Townsley et al.; IR: NASA/ESA/CSA/STScI/JWST ERO Production Team
pp. 106–7: NASA, ESA, CSA, STScI, T. Temim
p. 108: Science History Images/Alamy Stock Photo
p. 109: NASA, ESA, CSA, STScI, T. Temim
pp. 110–11: NASA, ESA, CSA, STScI
p. 112: NASA, Jeff Hester, Paul Scowen
p. 113: left: Wellesley College; right: Public domain.
p. 114: top left: NASA, ESA, Hubble Heritage Team (STScI/AURA); top right: NASA, ESA, Hubble, Hubble Heritage Team; bottom left: NASA, ESA, CSA, STScI, Joseph DePasquale (STScI), Anton M. Koekemoer (STScI), Alyssa Pagan (STScI); bottom right: NASA, ESA, CSA, STScI
p. 115: NASA, ESA, CSA, STScI; Joseph DePasquale (STScI), Anton M. Koekemoer (STScI), Alyssa Pagan (STScI)
pp. 116–17: NASA, ESA, CSA, M. McCaughrean, S. Pearson
p. 118: NASA, ESA, CSA, M. McCaughrean, S. Pearson
p. 119: ESA/Webb, NASA, CSA, Mahdi Zamani (ESA/Webb), PDRs4ALL ERS Team
pp. 120–21: NASA, ESA, CSA, STScI, Klaus Pontoppidan (STScI)
p. 122: NASA, ESA, CSA, Kellen Lawson (NASA-GSFC), Joshua E. Schlieder (NASA-GSFC)
p. 123: NASA, ESA, CSA, STScI, Klaus Pontoppidan (STScI)
pp. 124–5: ESA, Webb, NASA, CSA, A. Hirschauer, M. Meixner et al.
pp. 126–7: NASA, ESA, CSA, and M. Zamani, M. K. McClure, F. Sun, Z. Smith, the Ice Age ERS Team
p. 129: NASA, ESA, K. Luhman, T. Esplin et al., ESO
pp. 130–31: NASA, ESA, CSA
p. 133: ESO, Bo Reipurth
pp. 134–5: ESA, Webb, NASA, CSA, T. Ray
p. 137: ESA, Webb, NASA, CSA, T. Ray
pp. 138–9: NASA, ESA, CSA, STScI
p. 140: NASA, ESA, CSA, STScI
p. 141: NASA, ESA, CSA, W. R. M. Rocha (LEI), Leah Hustak (STScI)
pp. 142–3: NASA, ESA, CSA, STScI, Webb ERO Production Team
p. 144: ESO, Joseph R. Callingham

pp. 146–7: NASA, ESA, CSA, STScI, JPL-Caltech
p. 148: NASA, ESA, Joseph Olmsted (STScI)
pp. 150–51: NASA, ESA, CSA, STScI
p. 153: top: NASA, ESA, CSA, O. De Marco; bottom left: NASA, ESA, CSA, O. De Marco; bottom right: NASA, ESA, CSA, STScI
pp. 154–5: NASA, ESA, CSA, STScI, D. Milisavljevic, T. Temim, I. De Looze
p. 157: NASA, ESA, CSA, Danny Milisavljevic, Tea Temim, Ilse De Looze
pp. 158–9: ESA, Webb, NASA, CSA, M. Barlow, N. Cox (ACRI-ST), R. Wesson
p. 160: ESA, Webb, NASA, CSA, M. Barlow, N. Cox (ACRI-ST), R. Wesson
p. 162: NASA, ESA, CSA, STScI, Janice Lee, Thomas Williams, PHANGS Team, Elizabeth Wheatley
pp. 164–5: NASA, ESA, CSA, STScI, S. Crowe
p. 167: NASA, ESA, SSC, CXC, STScI
p. 169: NASA, ESA, CSA, O. Jones, G. De Marchi, M. Meixner
p. 170: ESA, NASA, Hubble
p. 171: NASA, CXC, JPL-Caltech, STScI
pp. 172–3: NASA, ESA, CSA, STScI
p. 175: left: NASA, ESA, CSA, STScI; centre: NASA, ESA, CSA, STScI; right: NASA, ESA, S. Beckwith, the HUDF Team
p. 176: NASA, ESA, CSA, STScI
pp. 178–9: NASA, ESA, CSA, STScI, Webb ERO Production Team
p. 180: NASA, ESA, CSA, STScI, Webb ERO Production Team
p. 181: NASA, JPL, Caltech, P. Appleton et al.
pp. 182–3: ESA, Webb, NASA & CSA, L. Armus & A. Evans
p. 185: NASA, ESA, the Hubble Heritage Team, ESA/Hubble Collaboration, A. Evans
pp. 186–7: NASA, ESA, CSA, and A. Riess (STScI)
p. 188: Ian Dagnall Computing/Alamy Stock Photo
p. 189: NASA, ESA, the Hubble Heritage Team (STScI/AURA), Hubble/Europe Collaboration
p. 191: NASA/WMAP Science Team
pp. 192–3: ESA, Webb, NASA & CSA, J. Lee and the PHANGS-JWST Team
p. 194: Public domain
p. 195: NASA, ESA, CSA, STScI, Janice Lee, Thomas Williams, PHANGS team
pp. 196–7: X-ray: NASA/CXC/SAO; IR (Spitzer): NASA/JPL-Caltech; IR (Webb): NASA/ESA/CSA/STScI
p. 199: NASA, ESA, CSA, STScI
pp. 200–201: NASA, ESA, CSA, STScI
pp. 202–3: ESA/Webb, NASA & CSA, M. Meixner
pp. 204–5: ESA/Webb, NASA & CSA, M. Meixner
pp. 206–7: ESA/Webb, NASA & CSA, M. Meixner
pp. 208–9: ESA/Webb, NASA & CSA, L. Armus, A. S. Evans
p. 210: ESA/Hubble & NASA/D. Milisavljevic
pp. 212–13: NASA, ESA, CSA, Janice Lee (NSF's NOIRLab)
p. 214: NASA, ESA, CSA, STScI, Janice Lee, Thomas Williams, PHANGS Team
p. 215: Event Horizon Telescope collaboration
pp. 216–17: X-ray: NASA/CXC/SAO/Ákos Bogdán; Infrared: NASA/ESA/CSA/STScI
p. 218: left: X-ray: NASA/CXC; Optical: NASA/STScI; right: X-ray: NASA/CXC/SAO/Ákos Bogdán; Infrared: NASA/ESA/CSA/STScI

INDEX

A

Abell 2744 – Pandora's Cluster 216–19
absolute zero 34
absorption bands, chemical reaction 177
accretion disks 68, 140, 211, 215
active galactic nuclei (AGNs) 210, 215, 219
al-Ṣūfī, Abd al Rahmān 49
ammonia 76, 81, 84, 128
Andromeda Galaxy – M31 49, 194
Aperture Mask Interferometry (AMI) 60
Argo Navis constellation 100
argon 156, 157
Aristarchus of Samos 67
astrobiology 56
astronomical units (AU) 91
astronomy, historical development of 43–5, 49
Atlas of Peculiar Galaxies 211
atmosphere
  Earth's 16–17, 19, 25, 54
  Jupiter's 73
  Neptune's 84, 85
  Saturn's 76
  Titan's 77
  Uranus's 80
auroras 72, 73, 80

B

Baby Cas A 156
Barnard, E.E. 204
Bell Laboratories 43
Bevis, John 108
Big Bang 30, 45, 46, 190–91
black dwarfs 93, 160
black holes 21–2, 25, 50, 161, 166, 184, 201, 210–11, 214–15, 219
blueshift 46, 58, 177
Bouvard, Alexis 84
brown dwarfs 93–5

C

cameras, infrared *see* Mid-Infrared Instrument (MIRI); Near-Infrared Camera (NIRCam)
carbon 26, 56, 76, 88, 89, 95, 118, 119, 123
  see also polycyclic aromatic hydrocarbons (PAHs)
carbon-based life *see* life
carbon monoxide 114
carbonyl sulphide 128
Carina Nebula 96–101
Cartwheel Galaxy 178–81
Cartwheel Galaxy Group 181
Cassiopeia A 154–7
Cepheid Variables 188, 190, 191
Cepheus constellation 188
Chamaeleon complex 128
Chamaeleon I dark cloud 126–7, 128
Chandra X-Ray Observatory 25, 44, 104, 198, 218, 219
Chandrasekhar, Subrahmanyan 25
Class 0 protostars 141
comets 194
Copernicus 67
coronagraphs 55, 57, 91
Corrective Optics Space Telescope Axial Replacement (COSTAR) 21
Cosmic Cliffs – NGC 3324 100–101
cosmic microwave background (CMB) 190–91
Cosmic Noon 171
cosmogony 67
cryogenic cooling systems 26, 34, 38

D

dark current 34
dark energy 51
dark matter 22, 51, 175, 218
dark nebulae 97, 128
dark stars 218–19
Deep Space Network 65
Dione 77
Duncan, John Charles 113
dust shells 148, 149
dwarf galaxies 163, 170–71, 204
dwarf planets 84, 85

E

$E=mc^2$ 125
early
  black holes 219
  galaxies 174–7, 219
  stars 30, 46, 50, 60, 206, 218–19
  telescopes 43
  universe 30, 49, 57, 58, 60, 163, 169–70, 175, 219
Earth
  atmosphere 16–17, 19, 25, 54
  circumference of 145
  destruction by the Sun 160
  life 53–4
  magnetic field 72, 80
  orbit 67
  wind speeds 84
Einstein, Albert 214
electromagnetic spectrum
  wavelengths 18–19, 21, 22, 25, 44–5, 46, 51, 69, 87, 105, 113, 125, 177, 180, 210, 218
  see also gamma rays; infrared radiation; microwaves; radio radiation; ultraviolet radiation; visible radiation; X-rays
Enceladus 77
Europa 73
European Space Agency (ESA) 77
Event Horizon Telescope 215
event horizons 215
exoplanets 26, 28, 55–6, 57, 60, 87
  HIP 65426b – Najsakopajk 90–91
  IC 348 92, 94–5
  KELT-9b 87
  LHS 475b 88
  WASP-80b – Wadirum 89
expansion of the universe 22, 51, 58, 190–91
exploding stars 25

F

fusion 68–9, 125, 141, 160, 206

G

galaxies 37, 49–50, 163
  Abell 2744 – Pandora's cluster 216–19
  active galactic nuclei (AGNs) 210, 215, 219
  Andromeda – M31 49, 194
  barred 214, 215
  Cartwheel Galaxy – E50 350-40/PGC 2248 178–81
  chemical composition 50
  collisions 180–81, 184–5, 211, 218
  diffuse 176, 188
  dwarf 163, 170–71, 204
  early 46, 50, 57, 60, 170, 174–7
  elliptical 168
  First Deep Field Image 174–7
  flocculent spiral 188
  Hickson Compact Group 92 198
  IC 1623 – interacting galaxies VV114E & W 182–5
  IC 5283 211
  irregular 163, 170, 204
  lenticular ring 180–81
  the Local Group 204
  M87 215
  mergers 184–5, 211
  Milky Way 68–85, 163, 164–7, 171
  NGC 346 168–71
  NGC 5584 186–90, 191
  NGC 6822 – Barnard's Galaxy 202–7
  NGC 7469 208–11
  NGC 7496 212–14, 215
  number in the universe 18, 125, 163, 174
  Phantom – M74/NGC 628 192–4
  SMACS 0723 172–3, 175
  spiral 163, 180, 181, 188, 194, 210–11, 214, 215
  Stephan's Quintet 196–8, 201
  UHZ1 219
  Webb Telescope First Deep Field image 174–6
  Whirlpool 194
Galilei, Galileo 43, 118
gamma rays 18
Ganymede 73
gas giants, Milky Way 72, 76
Gemini ground-based telescope 73
general theory of relativity 214
Goldilock's zones 28, 54
Goodricke, John 188
gravitational forces 21, 36, 77, 91, 114, 125, 141, 160, 166, 175, 211, 219
  gravitational accretion 136
  gravitational collapse 53, 97, 108, 148, 160
gravitational lensing 175, 219
Great Red Spot, Jupiter 72
ground-based telescopes 73, 113, 132, 215

H

H II regions 97, 104–5, 198
halo, Neptune's 84
Haro, Guillermo 132
helium 26, 69, 76, 84, 97, 125, 141, 171, 206
Herbig, George 132
Herbig–Haro (HH) objects 130–37

Herbig–Haro 46/47 130–32
Herbig–Haro 211 136–7
Herbig–Haro 797 134
Herschel, William 80
Hickson Compact Group 92 198
'hot Jupiter' exoplanets 55
Hubble constant 190
Hubble, Edwin 21, 190, 204
Hubble–Lemaître Law 190
Hubble Space Telescope 13, 17, 18, 21–2, 29, 35, 49, 50, 58, 65, 73, 113, 180, 185, 188, 191
Deep Field (HDF) imaging 125, 163, 174
Huygens space probe, ESA's 77
hydrocarbons 95, 105, 118, 123, 180, 198
see also polycyclic aromatic hydrocarbons (PAHs)
hydrogen 46, 69, 72, 76, 84, 88, 94, 95, 97, 104, 114, 122, 125, 141, 161, 166, 171, 177, 206
hypergiants 125

I

IAU100 NameExoWorlds project 89
IC 1623 – interacting galaxies VV114E and W 182–5
IC 5283 211
ice giants, Milky Way 80, 84, 85
images/imaging specialists, production of public 60–61
infrared-dark clouds (IRDCS) 167
infrared radiation (IR) 13, 18–19, 29–30, 39, 49, 50, 53, 57, 58, 60–61, 76, 84, 91, 95, 97, 104, 105, 114, 118, 122, 128, 132, 140, 145, 156, 161, 180, 185, 194, 198, 215
infrared telescopes 13, 19, 26
see also Spitzer Space Telescope; Webb Space Telescope
International Astronomical Union (IAU) 73, 89
inverse-square law 188
iron 125

J

James Webb Space Telescope *see* Webb Space Telescope
Jansky, Karl 43–4
Jupiter and moons 55, 70, 72–3

K

Kelvins 34
Kepler, Johannes 28
Kepler Space Telescope 28, 55, 87

L

L2 Lagrange point 36
La Silla telescope 132
de Lacaille, Nicolas-Louis 100
Large Megellanic Cloud 163, 170
Leavitt, Henrietta Swan 188
Leavitt's Law 188
Lemaître, Georges 190
lenticular ring galaxies 180–81
life 53, 55–6, 87, 89, 95, 119, 128
light echoes 156
low Earth orbit 21

M

M87 215
magnetic fields 58, 72, 73, 80, 108, 156, 166
magnetospheres 72, 73, 80
Mars 56
Matza 91
Mayans 118
Mercury 160
Messier catalogue 194
Messier, Charles 194
methane 61, 77, 81, 84, 85, 88, 89, 128
methanol 128
methyl cation (CH_3^+) 119
microwaves 19
Mid-Infrared Instrument (MIRI) 34, 58, 60–61, 89, 91, 105, 109, 119, 145, 149, 151, 161, 180, 185, 198, 205–6, 215
Milky Way – our solar system 163, 171, 204
formation of the 68–9
infrared-dark clouds (IRDCS) 167
Sagittarius C 164–7
stellar evolution 167
supermassive black hole – Sagittarius 166
mirrors, telescope 21, 26
Webb Space Telescope 32–3, 36, 39, 60, 65, 194
moons
Jupiter's 73
Neptune's 85
Saturn's 77
Uranus's 81
Mount Wilson Telescope 113
multiwavelength technology, early 43–4

N

NameExoWorlds project 89, 91
NASA (National Aeronautics and Space Agency) 21, 28, 29, 36, 85
Near-Infrared Camera (NIRCam) 34, 57, 60–61, 73, 81, 84, 91, 95, 109, 119, 136, 145, 152, 156, 161, 174–5, 180, 185, 188, 198, 205
Near-Infrared Imager Slitless Spectrograph (NIRISS) 34, 58, 60
Near-Infrared Spectrograph (NIRSpec) 13, 34, 45, 57–8, 88, 95, 177, 185
Integral Field Unit (IFU) 58, 201
nebulae 53, 68, 94, 97
Carina Nebula 96–101
Cosmic Cliffs – NGC 3324 100–101
diffuse 161
M1-67 144, 145
Orion Nebula – Herbig-Haro (HH) objects 132
Orion Nebula – Messier 42/NGC 1976 116–19, 194
Pillars of Creation, Eagle Nebula 110–15
planetary 97, 152–3, 160–61
Rho Ophiuchi cloud complex 120–23
Ring Nebulae – M57/NGC 6720 158–61
Southern Ring Planetary Nebula 151–3
Tarantula Nebula 102–5
neon 156, 157
Neptune and moons 80, 82, 84, 91
neutron stars 108, 161
New General Catalogue of Nebulae and Clusters of Stars (NGC) 204
Newton, Sir Isaac 44
NGC 346 – Small Magellanic Cloud 168–71
NGC 5584 186–90, 191
NGC 6822 – Barnard's Galaxy 202–7
NGC 7319 198, 201
NGC 7320 198
NGC 7469 208–11
NGC 7496 212–14, 215
nitrogen 16
nuclear processes 141, 148
see also fusion

O

Optical Telescope Element (OTE) 36, 39
Orion constellation 118
Orion molecular cloud complex 118
Orion Nebula – M42 116–19, 132
oxygen 16, 156, 157, 177

P

Parsons, William 108
Penzias, Arno 190
Petra World Heritage Site 89
PHANGS (Physics at High Angular resolution in Nearby GalaxieS) collaboration 194, 215
Phantom Galaxy – M74/NGC 628 192–4
photometers 28
Pillars of Creation, Eagle Nebula 110–14
planet formation 55, 122, 123, 128
planetary nebulae 97, 152–3, 160–61
plasma 125, 141
'plerions' – pulsar wind nebula 108
Pluto 84
polycyclic aromatic hydrocarbons (PAHs) 118, 123, 149, 161, 184, 205, 215
prism experiment. Newton's 44–5
protoplanetary disks 119, 122
protostars 53, 68, 104, 123, 136–7, 167
Protostar L1527 138–9
pulsars 25, 108–9

Q

quasars 219

R

radiation 18, 30, 32, 33, 34, 45, 54, 160, 174, 177, 210, 211
see also gamma rays; infrared radiation; microwaves; radio radiation; ultraviolet radiation; visible radiation; X-rays
radio radiation 19, 43–4, 108

radio telescopes 214–15
Rayet, Georges 145
red dwarfs 119
red giants 169
red light 29, 39, 46, 84
redshift 46, 58, 177
reflecting telescopes, early 44
reflection nebula 94
Rho Ophiuchi cloud complex 120–23
Ring Nebula – M57/NGC 6720 158–61
rings, planetary 76, 81, 85
rogue planets 87, 91, 95

S
S1 122
Sagittarius A* 166
Sagittarius C 164–7
Saturn and moons 74, 76–7, 95
Science Instrument Module (ISIM) 36, 38, 39
Scott, James 44
silicate dust 180
Single-Object Slitless Spectroscopy 60
singularity point 46
SMACS 0723 36, 172–3, 175
Small Magellanic Cloud 163, 170–71
solar system formation, our
  see under Milky Way – our solar system
Southern Ring Planetary Nebula 151–2
Space Shuttle programme 21
Space Telescope Science Institute 65, 174
space telescopes overview 17–18
spaxels 58
spectra and spectral analysis 57–8, 177, 201
spectrographs 36, 57, 58, 89, 201
spectrometers 26
spectroscopy 44–5, 49, 50, 55, 57–8, 60, 128, 145
Spitzer, Lyman 17, 26
Spitzer Space Telescope 26
stars 125
  Cepheid Variables 188, 190, 191
  Chamaeleon complex 128
  cluster IC348 136–7
  dark 218–19
  death of stars (*see* supernovae)
  dust shells 148, 149
  early 30, 46, 50, 60, 206, 218–19
  eruptions/expulsion of materials 136, 137, 141, 144–5, 152
  formation of 50, 53, 97, 101, 104–5, 113, 118, 122, 123, 128, 132, 136–7, 140–41, 149, 152, 156, 163, 166, 167, 171, 176, 180, 184, 194, 198, 211
  Herbig–Haro 46/47 130–32
  Herbig–Haro 211 136–7
  Herbig–Haro 797 134–6
  hypergiants 125
  Proxima Centauri 28
  pulsars and neutron stars 108–9
  S1 122
  shockwaves 137, 141, 156
  starbursts 184, 211
  Trapezium Cluster 118–19
  Wolf–Rayet stars 144–9
  WR 124 144–5
  WR 140 146–9
  see also nebulae; protostars; supernovae
stellar nurseries
  see nebulae
stellar winds 94, 145, 215
Stephan's Quintet 196–8, 201
storms on Jupiter 72, 73
sulphur 109, 125, 156
the sun 18, 30, 38, 53, 54, 76, 80, 94, 108, 136–7, 140, 141, 160
'super-Jupiter exoplanet' 91
supermassive black holes 22, 25, 50, 166, 198, 201, 210–11, 215, 219
supernovae 144, 145, 152, 161, 180
  Cassiopeia A 154–7
  Crab Nebula 108
  destruction of the Pillars of Creation 114
  remnants 25, 97, 108, 156–7, 205
  Ring Nebula 160–61
  Type 1a 190
synchrotron radiation 109, 156

T
Tarantula Nebula 102–5
Tethys 77
Titan 77, 95
transit method, exoplanet detection 28, 55, 87, 89
Trapezium Cluster 118–19
Trappist-1 star system 26
Triton 85

U
UHZ1 219
ultraviolet (UV) wavelengths 18–19, 119
underground ocean, Ganymede 73
Uranus and moons 39, 78–9, 84

V
Venus 160
visible light 13, 18, 19, 46, 53, 58
  telescopes 13, 19, 29, 30, 43, 44, 53, 122, 128, 132, 136, 141, 184, 215
Volans constellation 37
Voyager 2, NASA's 85

W
WASP-12b planet 26
water, presence of 54, 73, 76, 77, 89, 128
wavelengths, electromagnetic spectrum
  see electromagnetic spectrum wavelengths
Webb Space Telescope
  attitude control subsystem 38
  camera filters 61
  combined information from other telescopes 25, 44, 73, 104, 188, 194, 198, 215, 218
  command and data handling system 38
  communication system 38
  dark stars 218–19
  data cubes 201
  deployment in space 35
  development 13, 30, 32–4
  diffraction spikes 85, 118, 136, 144, 152, 167, 184, 210, 215
  electrical power subsystem 38
  about exoplanet images 87, 88, 89, 91, 94, 95
  extent of data collection 65, 201
  Fine Guidance Sensor (FGS) 58
  First Deep Field image 174–7
  first image released 36
  fuel consumption 36
  about galaxy images 166, 167, 171, 174–7, 180, 181, 184, 185, 188, 190, 191, 194, 198, 205–7, 210, 214, 215, 218, 219
  halo orbit 36
  instrument overview 57–8, 60
  instrument temperature 34, 36, 38
  Jupiter and moons 73
  launch 11–12, 36
  location 36
  main components overview 36, 38–9
  microshutter array 58, 95, 177
  mirrors 32–3, 36, 39, 60, 194
  about nebulae images 97, 100–101, 104–5, 113–14, 118–19, 122–3
  Neptune 83, 84, 85
  power and propulsion 36, 38
  project goals 49–51, 53, 55, 56
  Saturn and moons 76–7
  spacecraft bus 36, 38
  about star images 128, 132, 136, 137, 140, 141, 144, 145, 148, 149, 152, 156, 157, 161
  sunshield system 33, 39
  telemetry 36
  temperature control 33–4, 36, 38, 39
  Uranus 80, 81
  weight at launch 36
  see also Mid-Infrared Instrument (MIRI); Near-Infrared Camera (NIRCam); Near-Infrared Imager Slitless Spectrograph (NIRISS); Near-Infrared Spectrograph (NIRSpec)
Whirlpool Galaxy – M51 194
white dwarfs 93, 152, 160
Wide-Field Slitless Spectroscopy 60
Williams, Robert 174
Wilson, Robert 190
Wolf, Charles 145
Wolf–Rayet stars 144–9
WR 124 142–5
WR 140 146–9

X
X-rays 18, 25, 44, 104, 198, 219